ENCYCLOPAEDIA OF ENGINEERING PHYSICS-4

X-RAYS AND WAVE MECHANICS

By

Dr. Shalender Singh

DISCOVERY PUBLISHING HOUSE PVT. LTD.

NEW DELHI-110 002

Reprinted - 2019

First Published - 2009

ISBN: 978-81-8356-417-5

X-Rays and Wave Mechanics

Published by:

DISCOVERY PUBLISHING HOUSE PVT. LTD.

4383/4B, Ansari Road, Darya Ganj

New Delhi-110 002 (India)

Phone: +91-11-23279245, 23253475; 43596065

E-mail: discoverybooksindia@gmail.com

discoverypublishinghouse@gmail.com

web: www.discoverypublishinggroup.com

Printed at:

Infinity Imaging Systems

Delhi

Preface

The present book "Encyclopaedia of Engineering Physics" meant for engineering students. The present book is an attempt to fulfil the need of all engineering students of U.P.T.U. as well as for engineering students of other states. It covers the complete syllabus of physics prescribed by Technical Universities. The treatment given in simple, lucid and comprehensive. The language used is simple.

Though nothing can be claimed as original but the subject matter has been arranged in my own style.

Suggestion for improvement of the book will be greatly acknowledged.

Author

Preface

CONTENTS

1

X-Rays

INTRODUCTION

In 1895 *Wilhelm Roentgen,* while investigating the fluorescence of various materials in a gas discharge tube in which cathode rays were passed at a high potential difference, accidentally observed that a very penetrating invisible radiation was being emitted by a discharge tube. These radiations produced fluorescence on screen several metres away even when the tube was completely wrapped in a dark paper. Roentgen called these invisible penetrating radiation as *X-rays.* They are also known as *Roentgen rays.*

X-rays are produced when fast moving electrons in cathode rays are suddenly stopped by some obstructions like the glass wall of a discharge tube or a solid metal piece of high atomic weight like *tungsten, platinum or molybdenum* X-rays are of the same nature as light rays, but it is invisible and of very short wavelength. X-rays are identical with light rays in all respects, except that they occupy narrow space of the spectrum beyond the ultraviolet region. The discovery of X-rays has opened a new field of study particularly in the fields of engineering, medicine, surgery, industry etc.

Soft X-rays

X-rays having longer wavelength greater that 4 Å smaller frequency and hence smaller energy are called soft X-rays, due to their low penetrating power. Soft X-rays are produced at comparatively low potential difference and high pressure.

Hard X-rays

X-rays having low wavelength of the order of 1 Å, having high frequency and hence high energy are called Hard X-rays, due to "their high penetrating power. The hard x-rays are usually produced at comparatively low pressure and high potential difference.

Continuous X-rays

Some of the fast moving electrons may go deep into the interior of the atoms of the target material of X-ray tube and pass close to their nuclei and experience strong force of attraction due to positive nucleus and is suddenly slowed down, suffering a deflection in its path as shown in Fig. 1. In this way, the electrons are decelerated and their velocities are reduced According to classical theory, an accelerated or decelerated charge radiates energy.

The energy lost during the deceleration is given off in the form of X-rays. As the electrons suffer collisions at all angles, right from the glancing collision to the direct hit, they suffer varying decelerations and hence the radiations or *X-rays of all possible continuous wavelengths within a certain range are emitted.* The continuous X-ray radiations are also called *Bremsstrahlung.*

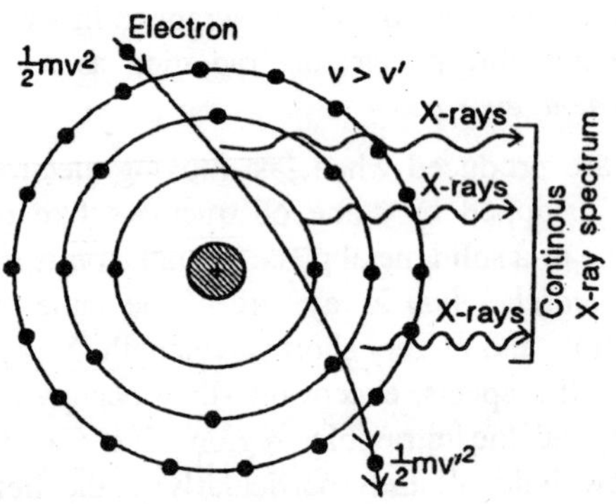

Fig. 1

When these X-rays of continuous wavelengths emitted from Coolidge tube are analysed by a Bragg's spectrometer, a continuous spectrum as shown in Fig. 2 is obtained. The continuous spectrum has a definite short wavelength limit λ_{min} or maximum frequency limit υ_{max}, below which there is no radiation.

This limit of wavelength does not depend on the target material but depends on the voltage applied to the cathode and anode of X-ray tube.

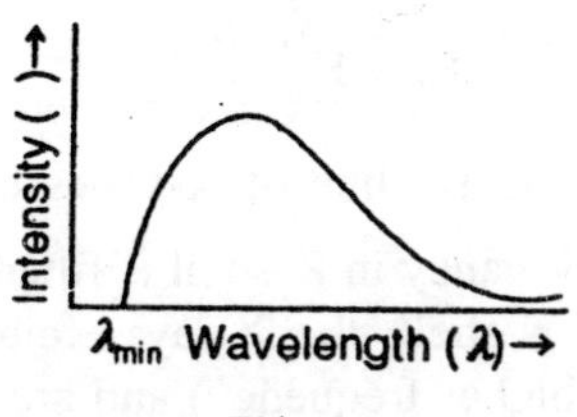

Fig. 2

Characteristic X-rays

When the speed of bombarding electrons are extremely high (about one-tenth of the velocity of light) some of these electrons may penetrate deep into the atoms of the target and collide with tightly bound electrons of the inner shells of the atoms. It may be possible that m collision the bombarding electron transfers all its energy to an electron bound in one of the inner shells, such as K or L, of the target atom, and the same bound electron is ejected from its orbit. The removal of an electron from the inner shell creates a vacancy (or hole) in the shell or in its low energy level. To fill this low energy vacancy, an electron from a higher energy level jumps in, and the excess energy is radiated in the form of X-ray photons. Thus, *the energy of X-ray photons is determined by the difference between the energies of two levels.* The X-rays emitted in this way are called the *characteristic X-rays* because their frequencies are the characteristic of the atoms of the material of the target from which they are ejected.

Suppose high speed bombarding electron knock off one electron from the K-shell, as shown in Fig. 3(a), so that a vacancy (or hole) is created in the shell. The electron from nearby L-shell jumps to occupy the hole in the K-shell [Fig. 3(b)] thereby emitting X-ray photon of energy $h\upsilon_{K_\alpha}$ and is given by

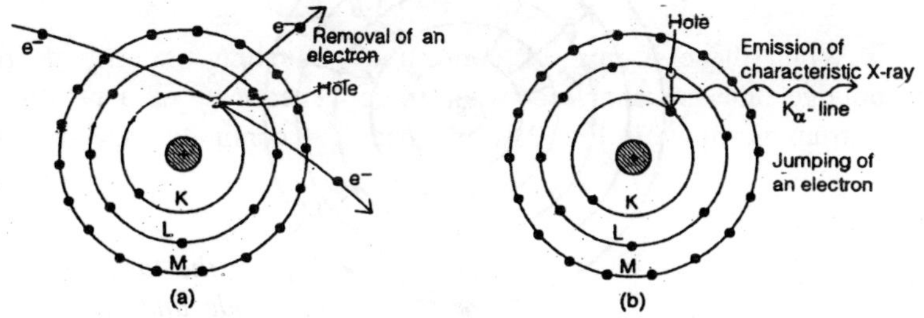

Fig. 3

$$h\upsilon_{K_\alpha} = E_K - K_L$$

This corresponds to K_α line of K-series or called *K_α X-ray.*

If, however, the vacancy in K-shell is filled up by an electron jump from the M-shell or N-shell, the X-rays emitted would be still more energetic (or still of higher frequency) and are called *K_β* and *K_γ X-rays* with energies

$$h\upsilon_{K_\beta} = E_K - K_M$$

$$h\upsilon_{K_\gamma} = E_K - K_N$$

and so on.

Such X-rays arising from millions of atoms produced K-lines as shown in Fig. 4. The holes so created in the L- and M-shells are filled by transitions of electrons from higher states, producing more X-rays. The process is continued until the vacancy reaches the outermost shell where it is filled by some free electrons. Therefore, *the emission of K-series is always accompanied by the emission of L-, M-, N-... series.*

Suppose the high speed bombarding electrons carry a lesser amount of energy, which is sufficient to knock off an electron from the L-shell. In this case an electron from outer-shells (e.g. M or N or...) jumps to the L-shell to fill the vacancy and excess energy is emitted as X-ray photons. These X-rays are less energetic than that of K-series and are called to form L-series of X-ray emission spectrum. *L-series of characteristic X-ray spectrum consists of L_α, L_β, L_γ... lines,* etc. lines caused by the jumping of electrons to fill the vacancy in L-shell from M, N, 0,... etc. shell respectively with energies

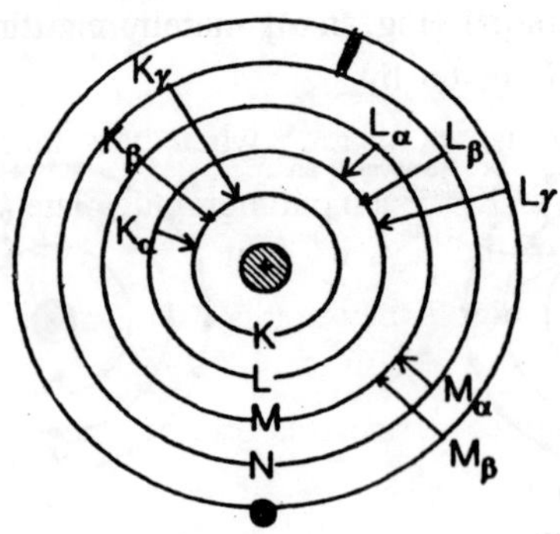

Fig. 4

$$h\upsilon_{L_\alpha} = E_L - E_M$$

$$h\upsilon_{L_\beta} = E_L - E_N$$

$$h\upsilon_{L_\gamma} = E_L - E_O$$

In a similar way the formation of M, N,... etc.-series of characteristic X-ray emission spectrum may be explained. Characteristic X-rays are produced in the same type of experimental set up as was used to produce continuous X-rays.

PROPERTIES OF X-RAYS

The important properties of X-rays may be summarised as follows:

1. Like visible light, X-rays are electromagnetic in nature. They show many properties which are resembled with the rays of light.
   (a) X-rays travel in straight lines with the speed of light.
   (b) They undergo reflection, refraction, interference, diffraction and polarisation.
2. The wavelength of X-rays are very short (100 Å – 1Å) as compared to the wavelength (4000 Å to 7500Å) of the rays of light.
3. X-rays are invisible.
4. X-rays do not deflect in electric and magnetic fields. It means that X-rays are chargeless or neutral.
5. X-rays produce fluorescence when fall on a certain chemicals like barium-platinocyanide, cadmium tungstate, zinc sulphide etc.
6. X-rays produce ionisation when pass through matter.
7. The affect of X-rays on photographic plate is much more intense than that of the rays of light.
8. X-rays are highly penetrating. They can pass through many solids, like wood, flesh etc., which are opaque to ordinary light.
9. X-rays produce destructive effects when fall on living tissues.
10. Similar to light rays, highly energetic X-rays liberate photo-electrons when fall on certain metals.

11. X-rays suffer Compton scattering.
12. When X-rays are incident on certain heavy metals, they produce secondary X-rays.
13. The penetrating power of X-rays depends on the voltage applied across the anode and cathode of the X-rays tube, and also on the atomic number of the material of the filamentary cathode. For high accelerating potential and cathode of high atomic number, the penetrating power is high.

DIFFERENCE BETWEEN X-RAY SPECTRA OF ELEMENTS AND OPTICAL SPECTRA

The striking features of the characteristic X-ray spectra which make it different from the optical spectra are as follows :

1. Characteristic X-ray spectra are obtained when the electrons from innermost shells, like K- or L-shell of the atom of the heavy elements are removed or excited to higher energy state, whereas optical spectra are produced when the valence electrons from the outermost shell are removed or are excited to higher energy states. Then after a short while when these electrons are de-excited or fall back to lower energy state, optical spectra are excited.
2. The frequencies in X-ray spectra are nearly a thousand times higher than optical frequencies in optical spectra. This clearly indicates that the emission of X-ray spectra are due to the transitions of very tightly bound electrons among the inner completed shells. This is in agreement with the fact that X-ray spectra of free elements and their chemical compound are same. The optical spectra are produced due to the transition of valence electrons of the outermost orbit.
3. The frequencies of the lines in the X-ray line spectra increase steadily from element to element. It is because of the fact that X-ray spectra of element depend on the binding energies of the inner most electrons. These energies increase smoothly and uniformly from element to element due to the increase in their nuclear charge. The optical spectra of the elements, on the other hand, show periodic character due to the periodic change in the electronic structure at the surface of the atom from element to element.

4. The X-ray emission spectra are different from X-ray absorption spectra. The X-ray emission spectra consist of discrete lines, whereas X-ray absorption spectra consist of continuous regions bounded by sharp edges. On the other hand optical emission and absorption spectra are identical. In the optical spectra, we get the absorption lines at the same frequencies as the corresponding emission lines.

DIFFRACTION OF X-RAYS

Just after the discovery of X-rays, the knowledge of physical nature of X-rays was in confusing stage. Some scientists were of the view that X-rays are high speed particles like cathode rays with high penetrating power while others regarded X-rays as electromagnetic waves of short wavelength. Many experiments were performed in this regards but non of them was able to determine the actual nature of the X-rays. To test the wave nature of X-rays attempts were made to observe interference and diffraction effects with them. On account of the short wavelengths of X-rays (of the order of 10^{-10} m) the man made transmission grating ruled with 6000 lines per inch Could not be able to produce any appreciable amount of diffraction with X-rays. To get considerable amount of diffraction. X-ray grating would require 40 million lines per cm. The construction of such a grating with 40 million lines per cm is quite tedius and almost impossible.

Max Von Laue, in 1913, suggested that atoms in a natural crystal are regularly arranged at equal distances and the separation between the successive layers is of the order of the wavelengths of X-rays. Therefore, natural crystal can be used as a closely spaced three dimensional natural diffraction grating for X-rays.

LAUE'S CRYSTAL DIFFRACTION OF X-RAYS

In 1913 Max Von Laue, in collaboration with *Friedrich and Knipping,* succeeded in obtaining diffraction pattern with X-rays by using a thin crystal as a three dimensional diffraction grating. A schematic diagram of Laue's experimental arrangement is shown in Fig. 5. The X-rays beam originated from a Coolidge tube is collimated into a fine pencil by passing through two pin-holed lead screens. Lead screens S_1 & S_2 absorb all rays except those which pass through small slits provided in them. This narrow beam when passed through a properly oriented crystal of zinc sulphide or sodium chloride, a symmetrical diffraction pattern on a photographic plate is obtained as shown in Fig. 5. The pattern consists of central dark spot where the direct beam struck the plate, and surrounded by a group of other symmetrical arranged fainter spots and is called *Laue pattern* and spots so obtained as Laue spots. Each spot in the Laue pattern corresponds to the constructive interference for a set of crystal planes satisfying the Bragg's equation, $2d \sin \theta = n\lambda$.

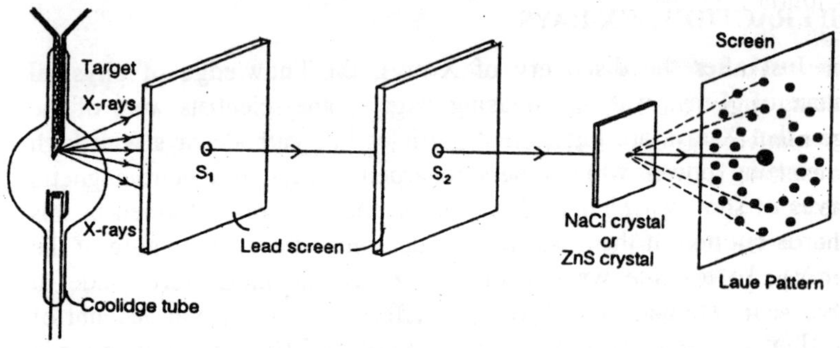

Fig. 5

The distribution of Laue's spots depends on the symmetry of the crystal and its orientation with respect to the incident X-ray beam. The occurrence of series of dark spots arranged in a symmetrical order on a photographic plate (diffraction pattern) shows that X-rays have a *wave nature.* The correct interpretation of X-rays diffraction pattern was proposed by *Dr. William Bragg.* According to him, *the diffraction pattern is the characteristic of crystal and Laue's spots are due to constructive interference between several reflected beams of X-rays from the various sets of parallel crystal planes which contained large number of atoms.* Laue experiment has established the following two important facts:

(i) X-rays are electromagnetic waves of short wavelength.

(ii) The atoms of a crystal are arranged in a regular three dimensional lattice.

Uses of Laue's Pattern

Laue's diffraction pattern are widely used for the following measurements:

1. It is used to determine the microscopic structure of matter in solid state.

2. It is used to determine angular relationship between different planes in a crystal lattice which are rich in atoms.

LAUE METHOD FOR THE STUDY OF CRYSTAL STRUCTURE

Laue's method is one of the ingenious method for the analysis of crystal structure. In this method a beam of polychromatic X-rays from an X-ray tube falls normally on the plane of a specimen crystal. During the journey in the crystal the rays are diffracted by various sets of Bregg's planes with different spacings d. These various families of the planes makes different angles θ with the direction of X-rays. For certain glancing angles the various sets of parallel crystal planes which contains large number of atoms satisfy the conditions of Bragg's reflection with the resulting increase in intensity of diffracted X-rays. As shown in Fig. 6 the rays which pass through the crystal undeviated forms black spot at the centre C of the photo- graphic plate XY. As the incident X-rays beam is polychromatic different wavelengths in the incident beam produce Laue's spot through some what less pronounced around the central spot. If (Fig. 6) P_1P_2 represents the position of one of the Bragg's plane, θ the corresponding angle and A the position of spot on the photographic plate, Then

$$AC = d \tan 2\theta$$

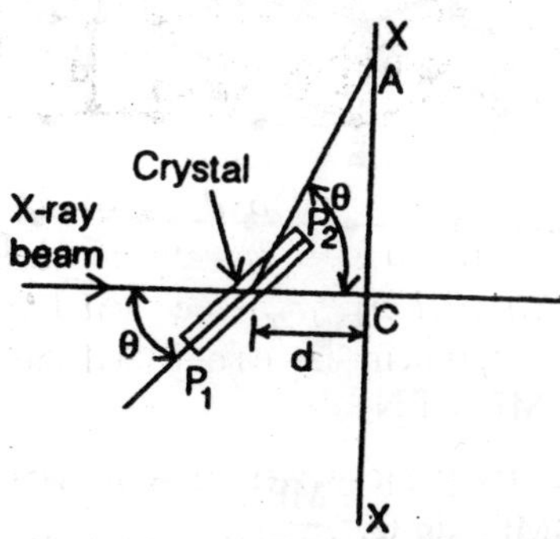

Fig. 6

By measuring AC and d the angle θ can be determined for the corresponding plane.

BRAGG'S LAW

Soon after Laue's experiment, *W.L. Bragg* devised another technique for the study of diffraction of X-rays by considering the scattering of

X-rays by the atoms in the crystal lattice. It should be remembered that Bragg's scattering is usually called Bragg's reflection.

Let us suppose that a narrow beam of monochromatic X-rays of wavelength λ be incident at a glancing angle θ on a set of parallel atomic planes of a crystal as shown in fig. 7. These rays are scattered by the atoms like B and E of two adjacent planes of separation 'd', in all directions. Constructive interference will occur only between those scattered waves which are reflected specularly (In specular reflection, the angle of incidence is equal to the scattering angle θ = θ' and the incident ray, scattered ray and normal to the lattice planes all lie in the same plane) from parallel planes of atoms, in a similar way as to the reflection of a light waves from a plane mirror, and have a path difference of nλ, where λ is the wavelength of X-rays and n is an integer. Let us consider that two incident X-rays, AB and DE, are reflected specularly along BC and EF respectively from crystal planes XX and YY as shown in Fig. 7. Draw normals BM on DE and BN on EF from B. The path difference between these two rays is

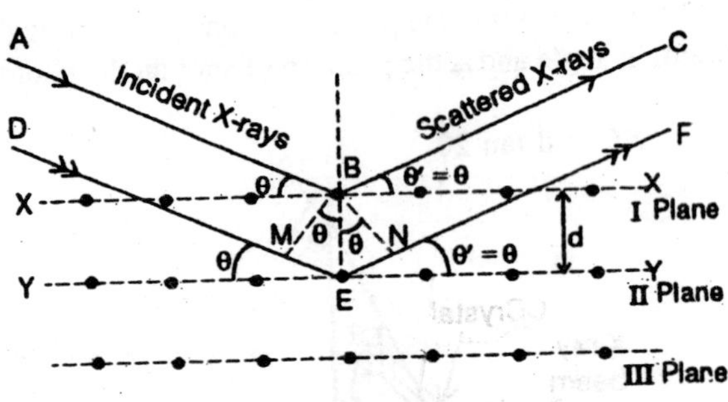

Fig. 7

$$\Delta = ME + EN$$

From Δ BME, $\sin\theta = \dfrac{ME}{BE}$

or $ME = BE \sin\theta = d \sin\theta$

Similarly, from Δ BNE, EN = BE sin q = θ sin θ

∴ Path difference, $\Delta = d\sin\theta + d\sin\theta = 2d\sin\theta$

For constructive interference, the path difference $\Delta = \pm 2n\lambda/2$

Hence, 2d sin θ = where n = 0, 1, 2, 3,...

This is well known *Bragg's equation for reflection or diffraction of X-ray by crystal planes,* and is called mathematical expression of *Bragg's law. n = 1,2,3,... etc. corresponds to first, second, third order spectrum and so on.*

From the Bragg's equation it is clear that if the wavelength of X-rays correspond to maximum intensity at a glancing angle θ is known, then the lattice spacing or spacing between successive lattice planes d may be found. Conversely, if the lattice spacing, d, is known, the wavelength λ of X-rays may be determined..

BRAGG'S SPECTROMETER AND EXPERIMENTAL DETERMINATION OF WAVELENGTH OF X-RAYS

For the detailed study of X-ray diffraction by crystals, *W.H. Bragg* and his son *Lawrence Bragg* devised an X-ray spectrometer in which a crystal is used as a reflection grating instead of transmission grating. The experimental arrangement of Bragg's X-ray spectrometer is shown in Fig. 8. A beam of monochromatic X-rays from a Coolidge tube is collimated into a sharp pencil by passing it through two lead sheets, S_1 and S_2 that absorbs all rays except those which pass through small slits provided in them. This collimated fine pencil of X-rays is made to fall upon a tiny crystal C mounted on the turn table T of the spectrometer (similar to that of an ordinary prism table) which is capable of rotation about a vertical axis passing through the centre of the turn table. The rotation of the crystal can be read from the circular scale provided with a vernier V_1. The X-rays beam after suffering a reflection from the crystal face, C enters an ionisation chamber *I* through a slit S_3. The ionisation chamber I which is filled with ethyl bromide is mounted on an arm R which is capable of rotation about the same axis as the crystal or turn table. The position of ionisation chamber can be recorded by a second vernier V_2. The ionisation chamber (or arm R) and circular table (or turn table) are so joined together that when the circular table (or crystal) rotates through an angle 6, the ionisation chamber (arm R) automatically turns through an angle 2θ (that is, double the angle). By this coupling, for every glancing angle (θ) at the crystal surface, the X-ray beam will always be reflected into the ionisation chamber. The X-rays entering the chamber ionise the gas causing a current in the associated circuit connected with the chamber. The resulting ionisation current, which is a measure of the intensity of X-rays reflected by the crystal, is measured by an electrometer E.

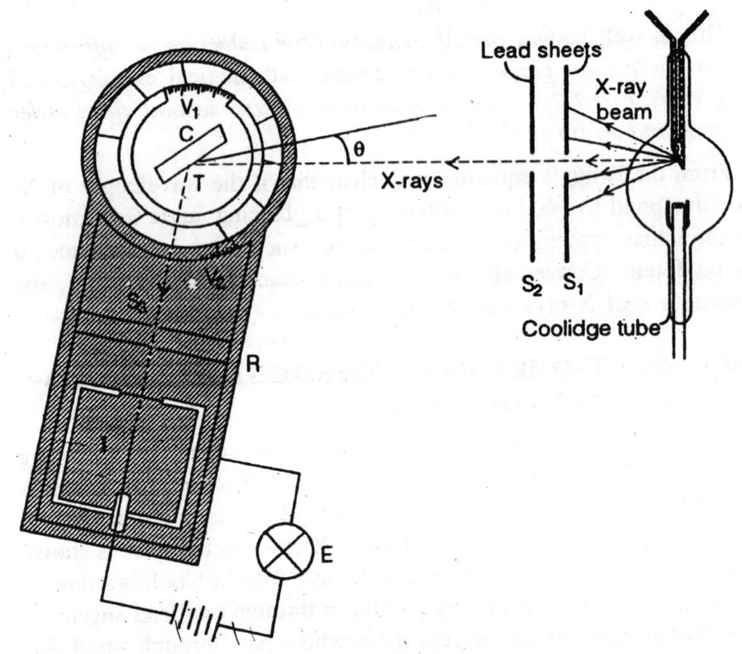

Fig. 8

For taking observations with the spectrometer, the glancing angle θ for the incident beam is increased in small steps, (starting from zero) by rotating the crystal and the corresponding ionisation current is recorded. Then a graph is plotted between the ionisation current I and glancing angle θ. A curve so obtained for sodium chloride crystal is shown in Fig. 9. It is clear from the curve that for certain values of θ, the ionisation current, I increases abruptly. The number of peaks (A_1, A_2, A_3, etc.) in the curve correspond to those glancing angles which satisfy the Bragg's equation

$$2d \sin \theta = n\lambda$$

The different peaks correspond to different orders (n = 1, 2, 3). Therefore, for first j, second and third order spectrum

$$2d \sin \theta_1 = \lambda,$$

$$2d \sin \theta_2 = 2\lambda$$

$$\text{and } 2d \sin \theta_3 = 3\lambda$$

or

$$\sin \theta_1 : \sin \theta_2 : \sin \theta_3 = 1 : 2 : 3$$

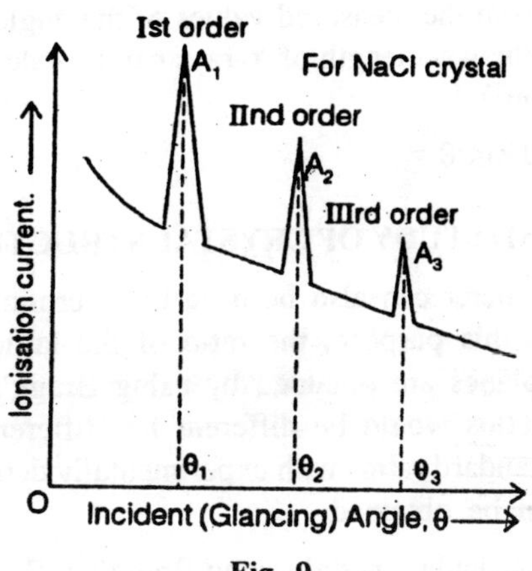

Fig. 9

Hence, the glancing angles θ_1, θ_2, θ_3, for the first, second and third order reflections are measured directly from the curve. If the X-ray beam consists of two nearby wavelengths λ_1 and λ_2, then in addition to the peaks A_1, A_2 and A_3 corresponding to wavelength λ_1, the other peaks B_1, B_2 and B_3 corresponding to wavelength λ_2 also appear near the peaks corresponding to wavelength λ_1 as shown in fig. 10. It is observed that for wavelength λ_1, Bragg's equation

$$\sin\theta_1' : \sin\theta_2' : \sin\theta_3' = 1 : 2 : 3$$

is satisfied. Similarly, if θ_1', θ_2', and θ_3 are the glancing angles for λ_2, then Bragg's equation

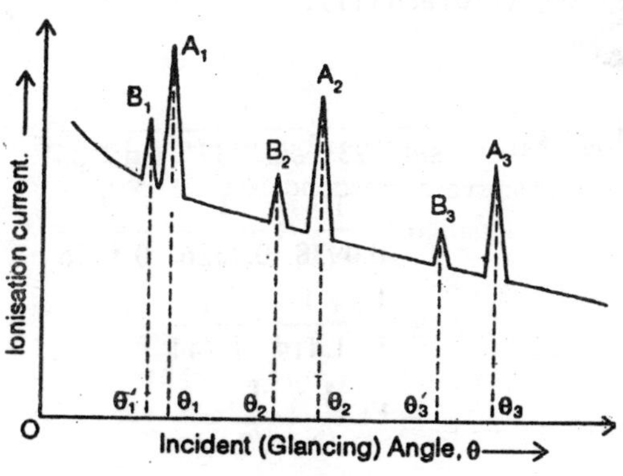

Fig. 10

$$\sin\theta_1' : \sin\theta_2' : \sin\theta_3' = 1 : 2 : 3$$

is again satisfied. From the measured values of the angle θ αvδ known value of spacing d, the wavelength of X-rays can be calculated by using the Bragg's equation

$$2d \sin \theta = \lambda$$

BRAGG'S LAW AND STUDY OF CRYSTAL STRUCTURE

Bragg's spectrometer can also be used to determine the structure of the crystals. For this purpose, the ratio of the lattice spacing for various groups of planes are obtained by using Bragg's equation, $2d \sin \theta = \lambda$. These ratios would be different for different crystals. By comparing known standard ratios with experimentally determined ratios, crystal structure can be obtained.

Suppose for a particular crystal strong Bragg's reflections from the three sets of planes with interplanar spacings d_1, d_2 and d_3 are obtained for glancing angles θ_1, θ_2 and θ_3 respectively in the first order (n = 1). Then from Bragg's equation, we have

$$2d_1 \sin \theta_1 = \lambda,$$

$$2d_2 \sin \theta_2 = \lambda$$

and $$2d_3 \sin \theta_3 = \lambda$$

$$\therefore \quad d_1 : d_2 : d_3 = 1/\sin \theta_1 : 1/\sin \theta_2 : 1/\sin \theta_3$$

For KCl crystal, Bragg obtained strong reflections of X-rays at glancing angles 5.23°, 7.37° and 9.25° respectively using three different reflecting planes (100), (110) and (111).

Hence, for KCl

$$d_{100} : d_{110} : d_{111} = \frac{1}{\sin 5°23'} : \frac{1}{\sin 7°37'} : \frac{1}{\sin 9°25'}$$

$$= \frac{1}{0.0938} : \frac{1}{0.1326} : \frac{1}{0.1636}$$

$$= \frac{1}{1} : \frac{1}{1.414} : \frac{1}{1.744}$$

$$= 1 : \frac{1}{\sqrt{2}} : \frac{1}{\sqrt{2}}$$

This corresponds to the theoretical result for a simple cubic lattice (SC) structure. Therefore, it is concluded that *the KCl crystal has a simple cubic structure.*

Example 1:

An X-ray tube with copper target is operated at 25 kV. The glancing angle for a NaCl crystal for the Cu K_α line is 15.8°. Find the wavelength of this line. Also find the glancing angle for photons at the shortest wavelengths limit (d for NaCl –2.82Å, h = 6.63 × 10^{-34} J-sec, c = 3 × 10^{-8} m/sec and e = 1.6 × 16^{-19} coulomb).

Solution:

If d is the crystal spacing and θ the glancing angle for NaCl crystal, then according to Bragg's equation for specular reflection,

$$2d \sin \theta = n\lambda$$

For the first order, n = 1

$$\therefore \quad \lambda = 2d \sin\theta = 2 \times 2.82 \times \sin 15.8°$$

$$= 2 \times 2.82 \times 0.2723 = 1.5358Å$$

The wavelength of X-ray photon at the shortest limit is given by

$$\lambda_{min} = \frac{hc}{eV} = \frac{6.63 \times 10^{-34} \times 3 \times 10^8}{1.6 \times 10^{-19} \times 25 \times 10^3}$$

Let θ_1 be the glancing angle for photons at the shortest wavelength limit, λ_{min}. Then

$$\lambda_{min}\ 2d \sin \theta_1$$

$$\text{or} \quad \sin \theta_1 = \frac{\lambda_{min}}{2d}$$

$$\text{or} \quad \sin \theta_1 = \frac{0.4972}{2 \times 2.82}$$

$$\text{or} \quad \theta_1 = \sin^{-1}(0.088) = 5.05°$$

Example 2:

In Bragg's reflection of X-ray, a reflection was found at 30^0 glancing angle with lattice planes of spacing 1.87Å. If this is a second order reflection. Calculate the wavelength of X-rays.

Solution:

According to Bragg's equation

$$2d \sin \theta = n\lambda$$

Here $\theta = 30^o, n = 2$

and d = 1.87Å

$$\lambda = \frac{2d\sin\theta}{n} = \frac{2\times1.87\times\sin30^0}{2}$$

$$= 1.87 \times 0.5 = 0.935Å$$

Example 3:

X-rays having wavelength of 2.6Å are diffracted at an angle of 20° for the second order in Bragg's spectrometer. Find the spacing constant of the crystal.

Solution:

If *n*th order maximum in Bragg's spectrometer is given in a direction θ, then according to Bragg's equation

$$2d \sin \theta = n\lambda$$

or $$d = \frac{n\lambda}{2\sin\theta}$$

Here $\lambda = 2.6Å = 2.6 \times 10^{-10}m$,

$n = 2, \theta = 20^\circ$ and $\sin 20^\circ = 0.3420$

$$d = \frac{2\times2.6\times10^{-10}}{2\times0.3420}$$

$$= 7.602 \times 10^{-10}m = 7.60Å$$

Example 4:

Calculate the longest wavelength that can be analysed by a rock salt crystal of spacing d = 2.82Å in the second order.

Solution:

According to Bragg's equation for specular reflection

$$2d \sin\theta = n\lambda$$

or $$\lambda = \frac{2d\sin\theta}{n}$$

For the longest wavelength, $(\sin\theta)_{max} = 1$

$$\lambda_{max} = \frac{2d}{n}$$

Here, $d = 2.82Å = 2.82 \times 10^{-10}m$ and $n = 2$

$$\therefore \qquad \lambda_{max} = \frac{2.82 \times 10^{-10} \times 2}{2} = 2.82 \times 10^{-10}m = 2.82Å$$

Example 5:

A beam of X-rays, λ = 0.842Å, is incident on a crystal at a grazing angle of 8°35' when the first order Bragg's-reflection occurs. Calculate the glancing angle for 3rd order reflection.

Solution:

According to Bragg's equation for specular reflection,

$$2d \sin\theta = n\lambda$$

For first order, n = 1 $\quad \therefore \quad 2d \sin\theta_1 = \lambda$ ...(1)

and for third order, n = 3 $\quad \therefore \quad 2d \sin\theta_3 = 3\lambda$ ...(2)

Dividing equation (2) by (1), we get

$$\frac{\sin\theta_3}{\sin\theta_1} = 3$$

or $\quad \sin\theta_3 = 3 \times \sin 8°35'$

$\therefore \quad \sin\theta_3 = 3 \times 0.15 = 0.45$

or $\quad \theta_3 = \sin^{-1}(0.45) = 26.7°$

Example 6:

An X-ray beam of wavelength 0.97Å is obtained in the third order after reflection at 60° from the crystal plane. Another beam is obtained in the first order after reflection at 30° from the same crystal plane. Find the wavelength of the second X-ray beam.

Solution:

According to Bragg's law of diffraction

$$2d \sin\theta = n\lambda \qquad ...(1)$$

For the first X-ray beam, θ = 60°,

n = 3 and λ = 0.97Å

$\therefore$ 2d sin 60°,

n = 3 and λ = 0.97Å

$$2d \sin 60° = 3 \times 0.97 \qquad ...(2)$$

Let λ' be the wavelength of the second X-ray beam. For the same crystal plane d remains the same, therefore

$$2d \sin \theta' = n' \lambda'$$

Here θ' = 30° and n = 1

$$\therefore \quad 2d \sin 30^\circ = 1 \times \lambda' \qquad ...(3)$$

Dividing equation (3) by (2), we get

$$\frac{\lambda'}{3 \times 0.97} = \frac{\sin 30^0}{\sin 60^0}$$

or $$\lambda' = 3 \times 0.97 \times \frac{(1/2)}{\left(\sqrt{3/2}\right)}$$

or $$\lambda' = \frac{3 \times 0.97 \times 1}{\sqrt{3}} = 1.68\text{Å}$$

Example 7:

X-rays of λ = 0.3Å are incident on a crystal with a lattice spacing 0.5Å. Find the angles at which second and third Bragg's diffraction maxima are observed.

Solution:

According to Bragg's law of diffraction

$$2d \sin \theta = n\lambda$$

or $$\sin \theta = n\lambda/2d$$

Here $\lambda = 0.3\text{Å}$

and d = 0.5Å,

For second order maxima, n = 2

$$\therefore \quad \sin\theta = \frac{2 \times 0.3}{2 \times 0.5}$$

and $\theta = \sin^{-1}(0.6) = 36.86^\circ$

For third order maxima, n = 3

$$\therefore \quad \sin \theta = \frac{3 \times 0.3}{2 \times 0.5} = 0.9$$

or $$\theta = \sin^{-1}(0.9) = 64.15^\circ$$

Example 8:

An X-ray spectrometer has a crystal of rock salt ($d = 2.82 \times 10^{-8}$cm) set at an angle of 14° to the beam coming from a tube operated at a constantly increasing voltage. An intense first line appears when the voltage across the tube is 9045 volt. Calculate the value of h.

Solution:

According to Bragg's relation for diffraction of X-rays of wavelength λ in a direction θ,

$$2d \sin \theta = n\lambda$$

For first order maximum, n = 1, θ = 14° and $d = 2.82 \times 10^{-8}$cm

$$\therefore \quad \lambda = 2 \times 2.82 \times 10^{-8} \times \sin 14°$$

$$= 2 \times 2.82 \times 10^{-8} \times 0.2419 = 1.3643 \times 10^{-8}\text{cm}$$

Since the first line appeared when the voltage across the tube is *V* (= 9045 volt), the energy of X-rays quanta must be equal to the energy gained by electrons when accelerated through these operating voltage V. Thus,

$$\frac{hc}{\lambda} = eV = 1.6 \times 10^{-12} \times 9045 \text{ erg}$$

$$\text{or} \quad h = \frac{1.6 \times 10^{-12} \times 9045 \times \lambda}{c} = \frac{1.6 \times 10^{-12} \times 9045 \times 1.3643 \times 10^{-8}}{3 \times 10^{10}}$$

$$\text{or} \quad h = 6.58 \times 10^{-27} \text{ erg-sec}$$

Example 9:

The first order reflection from the plane of NaCl is obtained at an angle 2θ = 20° with the incident beam. If d_o = 2.82Å, calculate the wavelength of X-ray used.

Solution:

According to Bragg's equation for the diffraction of X-rays,

$$2d_o \sin \theta = n\lambda$$

For first order n = 1,

$$\therefore \quad 2d_0 \sin\theta = \lambda$$

Here $2\theta = 20^o$ or $\theta = 10^o$

and $d_o = 2.82Å = 2.82 \times 10^{-10}$ m

$\therefore \quad \lambda = 2 \times 2.82 \times 10^{-10} \sin 10^o = 2 \times 2.82 \times 10^{-10} \times 0.1736$

$= 9.791 \times 10^{-10}$ m $= 0.9791Å$

Example 10:

The first order Bragg maximum of electron diffraction in a nickel crystal (d = 0.4086Å) occurred at a glancing angle of 65°. Calculate the de-Broglie wavelength of the electrons and their velocity.

Solution:

According to Bragg's equation for the diffraction of X-rays,

$$2d \sin\theta = n\lambda$$

For first order, n = 1 $\quad \therefore 2d \sin \theta = \lambda$

or $\lambda = 2 \times 0.4486 \times 10^{-10} \times \sin 65^o = 2 \times 0.4086 \times 10^{-10} \times 0.9063$

$= 0.7406$ Å $= 0.7406 \times 10^{-10}$ m

The velocity of electrons, $v = \dfrac{h}{m\lambda} = \dfrac{6.6\times10^{-34}}{9.1\times10^{-31}\times0.7406\times10^{-10}}$

$= 9.793 \times 10^6$ m/sec

Example 11:

Electrons are accelerated through 344 volts and are reflected from a crystal. The first reflection maximum occurs when glancing angle is 60°. Determine the spacing of the crystal. Given; h = 6.62 × 10^{-34} J-sec, e = 1.6 × 10^{-19} Coulomb and m = 9 × 10^{-31} kg.

Solution:

The wavelength λ associated with an electron of kinetic energy, E is given as,

$$\lambda = \frac{h}{mv} = \frac{h}{\sqrt{2mE}}$$

or $$\lambda = \frac{h}{\sqrt{2meV}}$$

$$= \frac{6.2\times10^{-34}}{\sqrt{2\times9\times10^{-31}\times1.6\times10^{-19}\times344}}$$

or $\lambda = 6.62 \times 10^{-9}$ m = 0.662Å

According to Bragg's law,

$$2d \sin\theta = n\lambda$$

or $$d = \frac{n\lambda}{2\sin\theta}$$

Here $\theta = 60^\circ$, n = 1

and $\lambda = 6.62 \times 10^{-9}$ m

or 0.38 Å

Example 12:

A set of lattice planes reflects X-rays of wavelength 1.32Å at a glancing angle of 9°30°. Deduce the possible spacing of this set of planes

Solution:

Bragg's law for the diffraction of X-rays is

$$2d \sin\theta = n\lambda$$

For first order reflection, n = 1

$$2d \sin\theta_1 = \lambda \qquad ...(1)$$

For second order reflection, n = 2

$$2d \sin\theta_2 = 2\lambda \qquad ...(2)$$

For third order reflection, n = 3

$$2d \sin\theta_3 = 3\lambda \qquad ...(3)$$

Similarly, for fourth order reflection, n = 4

$$2d \sin\theta_4 = 4\lambda \qquad ...(4)$$

Dividing equation (2) by equation (1), we get

$$\frac{\sin\theta_2}{2\sin\theta_1} = 2$$

or $$\sin\theta_2 = 2 \sin\theta_1 \qquad ...(5)$$

Similarly, from equations (3) and (1), and from equations (4) and (1), we get

$$\sin\theta_3 = 3 \sin\theta_1$$

and $\sin\theta_4 = 4\sin\theta_1$

For the first order reflection the spacing d is

$$\frac{d}{n} = \frac{\lambda}{2\sin\theta_1} = \frac{1.32\times10^{-10}}{2\times\sin\theta 9^\circ 30'}$$

or $$d = \frac{1.32\times10^{-10}}{2\times0.1650} = 4.0\times10^{-10}\,m$$

If $n = 2$, $$\frac{d}{n} = \frac{\lambda}{2\sin\theta_2} = \frac{\lambda}{2\times2\sin\theta_1} = \frac{1}{2}\left(\frac{\lambda}{2\sin\theta_1}\right)$$

$$\therefore \quad \frac{d}{n} = \frac{4.0\times10^{-10}}{2}m = 2.0\times10^{-10}\,m$$

Similarly, if n = 3

$$\frac{d}{n} = \frac{\lambda}{2\sin\theta_3} = \frac{\lambda}{2\times3\sin\theta_1} = \frac{1}{3}\times4.0\times10^{-10}m$$

or $$\frac{d}{n} = 1.33\times10^{-10}\,m$$

Similarly, if n = 4

$$\frac{d}{n} = \frac{\lambda}{2\sin\theta_4} = \frac{\lambda}{2\times4\sin\theta_1}$$

$$= \frac{1}{4}\times4.0\times10^{-10}\times = 1.0\times10^{-10}\,m \text{ and so on.}$$

Example 13:

Bragg found that for a potassium chloride crystal, strong reflection from the sets of planes (1,0,0); (1,1,0) and (1,1,1) are obtained for angles 5°23', 7°37' and 9°22'. Show that the potassium chloride crystal has a simple cubic crystal structure.

Solution:

According to Bragg's equation for X-rays diffraction

$$2d\sin\theta = n\lambda$$

For first order n = 1 $\therefore$ $2d\sin\theta = \lambda$

The strong reflection from set of plane (1, 0, 0) is obtained at an angle $\theta_1 = 5^\circ\,23'$

$$\therefore \quad 2d_{100} \sin\theta_1 = \lambda \qquad ...(1)$$

Similarly, for reflection at the plane (1, 1, 0), we have

$$2d_{110} \sin\theta_2 = \lambda \qquad ...(2)$$

And for reflection at the plane (1, 1, 1), we have

$$2d_{111} \sin\theta_3 = \lambda \qquad ...(3)$$

Comparing equations (1), (2) and (3), we get

$$d_{100} \sin\theta_1 = d_{110} \sin\theta_2 = d_{111} \sin\theta_3$$

or
$$\frac{1}{d_{100}}:\frac{1}{d_{110}}:\frac{1}{d_{111}} = \sin\theta_1 : \sin\theta_2 : \sin\theta_3$$

Here
$$\theta_1 = 5°23', \theta_2 = 7°37' \text{ and } \theta_3 = 9°22'$$

$$\therefore \quad \frac{1}{d_{100}}:\frac{1}{d_{110}}:\frac{1}{d_{111}} = \sin 5°23' : \sin 7°37' : \sin 9°22'$$

$$= 0.0939 : 0.1326 : 0.1628$$

$$= 1 : 1.142 : 1.733$$

or
$$\frac{1}{d_{100}}:\frac{1}{d_{110}}:\frac{1}{d_{111}} = 1 : \sqrt{2} : \sqrt{3} \qquad ...(4)$$

The condition represented by equation (4) is same as the condition for a simple cubic crystal. Hence, the crystal of potassium chloride is a simple cubic.

PRACTICAL APPLICATIONS OF X-RAYS

On account of their remarkable properties, X-rays have many valuable applications in industry, engineering, medicine and in other fields of our daily life. They are also widely used as a powerful tools in scientific research.

Industrial and Engineering Applications

1. X-rays are frequently used to identify manufacturing defects in tennis balls, rubber tyres etc. They are also used to check the defect in diamonds and other precious stones.
2. In the custom department, X-rays are used as a powerful detective for the detection of contraband goods in baggages, gold hidden in sealed parcel, pearls in oysters, human body, for investigating the distinction between artificial and real gems, etc.

3. X-rays are also used to detect cracks and blow holes in the body of the aeroplanes, motor cars, wood, porcelain and other insulator. They are also used to examine the flaws in metal castings, welding joints, insulating material etc. and for investigating the changes occurring in old oil painting.
4. The X-rays are also used to detect the flaws and air bubbles in iron-girders fitted in buildings and bridges. The accidents may be avoided by replacing defective girders.
5. X-rays may be used to analyse the structure of alloys, like cobalt-nickel, steel, bronze, duraluminium and other composite bodies.
6. X-rays diffraction pattern from the material like rubber, cellulose and plastic help in elucidating their basic structure.

Medical Applications

In the field of medicine, X-rays are widely used for the diagnosis of many diseases that can not be identified by any pathological test. It is also effectively used as a curative in many chronic diseases that cannot be cured by any other conventional treatments.

As a diagnostics in surgery

X-rays are commonly used to locate and detect fractured bones, tumours, diseased organs, the presence of foreign matter like bullets etc. Newly developed X-ray devices for radiological diagnosis is used to locate brain tumors and blood clots which are difficult to see with conventional techniques. Recently developed three dimensional scanners present sharp, detailed pictures of the heart, lungs, kidneys and other sophisticated organs. On account of differential absorption of X-rays between bones, tissues and mitals, radiographs are used for precise location of peptic ulcers and ruptures etc. in the internal organs of the human body. Similarly, radiographs are routinely used for the diagnosis of tuberculosis (T.B.), stones in kidneys and gall-bladders etc.

X-rays therapy or radiotherapy or treatment of-diseases by irradiation with X-ray

Is its ingenious utility. The controlled X-rays exposure of appropriate quality cure many types of skin diseases, malignant sores, internal cancer and tumours. The curative power of X-rays is due to the fortunate fact that diseased tissue is more susceptible to destruction than the surrounding healthy tissue. When an internal cancer is treated by communicating a sharp pencil of X-rays directly through the body, the diseased and the healthy tissue are simultaneously killed off. The normal tissue, however, grows in again while the cancerous tissue is destroyed.

Applications in Pure Scientific Researches

In the field of research, X-rays are used to study the structure of the crystalline solids, atoms and alloys.

X-rays are effectively used for the determination of atomic numbers of the element and for the identification of chemical elements.

X-rays diffraction pattern of complex organic molecules are used for analysing their structure.

COMPTON EFFECT

In 1921 *A.H. Compton,* while studying monochromatic X-rays scattered by carbon atoms, found that when a monochromatic beam of high energy photons (energy > 0.51 MeV) (such as X-rays or γ-rays) is scattered by a target rich in electrons, the scattered beam contain photons not only of the same wavelength is that of incident photons, but also the photons of longer wavelength (or shorter frequency). *The scattered radiation unchanged in frequency are called unmodified scattered radiation and such type of scattering is called coherent scattering, white the scattered radiation with changed frequency are called modified scattered radiation and such type of scattering is called incoherent scattering.* Compton further noticed that the difference in frequencies (or wavelengths) of scattered radiation increases with the increase of angle of scattering and it is independent of the frequency of the incident radiation and also of the nature of the scatterer. Thus, *the phenomenon of scattering with change in frequency in called the Compton effect.* Hence, in Compton effect there is still a photon after the collision but its energy is less than that of the incident photon, only a part of the energy of incident photon is given to the *Compton electron or recoil electron.* In fact *Compton effect is the outcome collision between the high energy photon and free electron.*

The classical theory (which predict that when electromagnetic radiation is scattered from a charged particles, the scattered radiation will have the same frequency as the incident radiation in all directions) failed to explain this change of frequency. Compton in 1923, however, gave a satisfactory explanation for this change of frequency on the basis of quantum theory of radiation, according to which incident radiation consists of photons or quanta of energy $h\upsilon$, where h is the Planck's constant and υ the frequency of radiation. These photons behave like a particle, move with the velocity of light, possess momentum $h\upsilon/c$ and obey all the laws of conservation of energy and momentum. When these photons strike with the free electrons of the target get scattered. According to Compton, *the phenomenon of scattering is due to an elastic collision between a photon of incident radiation and free electron of the scattering material which is initially at rest. During the collision photon of energy $h\upsilon$ transfers some of its energy to the target electrons which gains kinetic energy and recoil with some velocity in a direction other than the direction of incident of photons. Hence, the photons scattered in a direction other than the direction of incident will have lower energy (or lower frequency or longer wavelength) than incident photons.* The Compton effect offers one of the strongest evidence in support of the Planck's theory of radiation.

Theory

Compton derived an expression for the change in wavelength of scattered photon by considering the elastic collision between the incident photon and free electron of the scattering material and by applying the law of conservation of energy and momentum to the problem.

For simplicity he assumed that the electron of the atom of target material is free and is at rest before collision (Even if the electron is bound to the nucleus, a negligibie amount of energy is needed to free it). After collision the relativistic mass of the electron is considered.

Let a photon of energy $h\upsilon$ and momentum $h\upsilon/c$ collides with a free electron at rest and give up a fraction of its energy to the free electron and scattered. Let the scattered photon is emitted at an angle θ with the energy $h\upsilon'$ and momentum $h\upsilon'/c$, and the electron recoils at an angle ϕ with momentum $m\upsilon$ as shown in Fig. 11. According to the theory of relativity the rest mass energy of the electron is $m^{o}c^{2}$ and its momentum is zero before collision.

After collision, if υ is the velocity of the recoil electron, then its relativistic mass is

$$m = \frac{m_o}{\sqrt{1 - v^2 / c^2}}$$

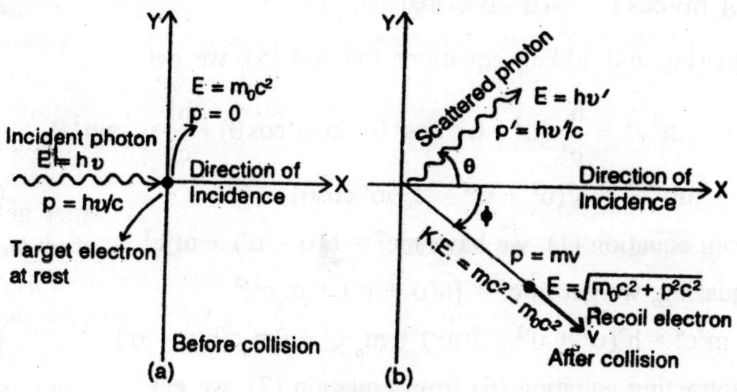

Fig. 11

The energy of the recoiled electron

$$= mc^2 = \frac{m_o c^2}{\sqrt{1 - v^2/c^2}}$$

and its momentum

$$= mv = \frac{m_o v}{\sqrt{1 - v^2/c^2}}$$

Now according to the principle of conservation of energy

Energy of the system before collision = Energy of the system after collision

$$h\upsilon + m_o c^2 = h\upsilon' + mc^2 \qquad ...(1)$$

Now, applying principle of conservation of momentum along and perpendicular to the direction of incident photon, that is,

Momentum before collision = Momentum after collision.

In a direction along the direction of incidence

$$\frac{h\upsilon}{c} + 0 = \frac{h\upsilon'}{c}\cos\theta + m\nu\cos\phi \qquad ...(2)$$

In a direction perpendicular to the direction of incidence

$$0 + 0 = \frac{h\upsilon'}{c}\sin\theta + m\nu\sin\phi \qquad ...(3)$$

Rearranging equations (2) and (3), we have

$$m\nu\cos\phi = \frac{h}{c}(\upsilon - \upsilon'\cos\theta) \qquad ...(4)$$

and $mv\cos\phi = \frac{h}{c}(\upsilon - \upsilon'\cos\theta)$ ...(5)

Squaring and adding equations (4) and (5), we get

$$m^2v^2 = \frac{h^2}{c^2}(\upsilon^2 + \upsilon'^2\cos^2\theta - 2\upsilon\upsilon'\cos\theta) + \frac{h^2}{c^2}\upsilon'^2\sin^2\theta$$

or $m^2v^2c^2 = h^2(\upsilon^2 + \upsilon'^2 - 2\,\upsilon\upsilon'\cos\theta)$ ...(6)

From equation (1), we have $mc^2 = h(\upsilon - \upsilon') + m_oc^2$

Squaring, we get $m^2c^4 = [h(\upsilon + \upsilon') + m_oc^2]^2$

or $m^2c^4 = h^2(\upsilon^2 + \upsilon'^2 - 2\upsilon\upsilon') + m_o^2c^4 + 2m_oc^2h(\upsilon - \upsilon')$ ...(7)

Subtracting equation (6) from equation (7), we get

$m^2c^4 - m^2v^2c^2 = h^22\upsilon\upsilon'(\cos\theta - 1) + 2h(\upsilon - \upsilon')m_oc^2 + m_o^2c^4.$

$m^2c^2(c^2 - v^2) = -2h^2\upsilon\upsilon'(1 - \cos\theta) + 2h(\upsilon - \upsilon')m_oc^2 + m_o^2c^4$

Substituting the value of *m* in relativistic terms as

$$m = \frac{m_o}{\sqrt{(1 - v^2/c^2)}}$$

$$\frac{m_o^2c^2}{1 - v^2/c^2}(c^2 - v^2) = -2h^2\upsilon\upsilon'(1 - \cos\theta) + 2h(\upsilon - \upsilon')m_oc^2 + m_o^2c^4$$

$\therefore\quad 2h(\upsilon - \upsilon')m_oc^2 = 2h^2\upsilon\upsilon'(1 - \cos\theta)$

or
$$\frac{\upsilon - \upsilon'}{\upsilon\upsilon'} = \frac{h}{m_oc^2}(1 - \cos\theta)$$

$$\frac{1}{\upsilon'} - \frac{1}{\upsilon} = \frac{h}{m_oc^2}(1 - \cos\theta) \quad ...(8)$$

As the values of h, m_o and c^2 are positive and the maximum value of $\cos\theta$ is 1, the equation (8) show that $\upsilon > \upsilon'$ that is, *the frequency of incident photon (or radiation) is always greater than the frequency of the scattered photon.*

Equation (8) can also be rewritten as

$$\frac{c}{\upsilon'} - \frac{c}{\upsilon} = \frac{h}{m_oc}(1 - \cos\theta)$$

or
$$\Delta\lambda = \lambda' - \lambda = \frac{h}{m_oc}(1 - \cos\theta) = \frac{2h}{m_oc}\sin^2\frac{\theta}{2} \quad ...(9)$$

This is the required expression for the Compton shift $\Delta\lambda$. It is clear from eqn. (9) that, *the Compton shift ($\Delta\lambda$) depends only on the angle of scattering and is independent of the wavelength of incident photon.* It is also clear from eqn. (9) that,

1. The wavelength, λ' of scattered photon is greater than the wavelength λ of the incident photon;
2. When $\theta = 0$, $\lambda' - \lambda = \Delta\lambda = 0$, that is, Compton shift is zero. It means that, *no scattering occurs along the direction of incident radiat}"ion.*
3. When $\theta = \dfrac{\pi}{2}$,

$$\lambda' - \lambda = \frac{h}{m_o c} = \frac{6.62\times10^{-34}}{9.1\times10^{-31}\times3\times10^{8}}$$

$$= 2.42 \times 10^{-12} = 0.0242 \text{ Å}$$

The quantity h/m_oc (= 0.0242Å) is called Compton wavelength and is denoted by λc. It is a constant quantity equal to h/m_oc.

4. When $\theta = \pi$, $\lambda' - \lambda = \dfrac{2h}{m_o c} = 2 \times 0.0242 = 0.04852$ Å

This shift is less than 0.001% of the initial wavelength for visible light which is undetectable. Hence, *as angle of scattering θ varies from 0 to 180°, the wavelength of scattered photon varies from λ to $\lambda + 2h/(m_oc)$.*

Direction of Recoiled Compton Electron

To find the direction of recoiled Compton electron, divide equation (5) by equation (4)

$$\tan\phi = \frac{\upsilon\sin\theta}{\upsilon - \upsilon'\cos\theta} \qquad ...(10)$$

$$\tan\phi = \frac{\dfrac{c}{\lambda'}\sin\theta}{\dfrac{c}{\lambda} - \dfrac{c}{\lambda'\cos\theta}}$$

or $$\tan\phi = \frac{\lambda\sin\theta}{\lambda' - \lambda\cos\theta}$$

For finding the value of υ', rearranging equation (8) as

$$\frac{1}{\upsilon'} = \frac{1}{\upsilon} + \frac{h}{m_o c^2}(1-\cos\theta) = \frac{1}{\upsilon} + \frac{2h}{m_o c^2}\sin^2\theta/2$$

or
$$\frac{1}{\upsilon'} = \frac{1+(2h\upsilon/m_o c^2)\sin^2\theta/2}{\upsilon}$$

or
$$\frac{1}{\upsilon'} = \frac{\upsilon}{1+(h\upsilon/m_o c^2)2\sin^2\theta/2}$$

or
$$\upsilon' = \frac{\upsilon}{1+2\alpha\sin^2\theta/2}, \text{ where } \alpha = \frac{h\upsilon}{m_o c^2} \qquad ...(11)$$

Substituting this value of υ', in equation (10), we get

$$\tan\phi = \frac{\upsilon\sin\theta/[1+2\alpha\sin^2\theta/2]}{\upsilon - \left(\dfrac{u}{1+2\alpha\sin^2\theta/2}\right)\cos\theta}$$

or
$$\tan\phi = \frac{\upsilon\sin\theta}{\upsilon(1+2\alpha\sin^2\theta/2)-\upsilon\cos\theta} = \frac{\sin\theta}{(1-\cos\theta)+2\alpha\sin^2(\theta/2)}$$

or
$$\tan\phi = \frac{2\sin\theta/2\cos\theta/2}{2\sin^2\theta/2+2\alpha\sin^2\theta/2} = \frac{\cos\theta/2}{\sin\theta/2(1+\alpha)} = \frac{\cot\theta/2}{1+\alpha}$$

or
$$\tan\phi = \frac{\cot\theta/2}{[1+(h\upsilon/m_o c^2)]} \qquad ...(12)$$

Equation (12) reveals that the angle of the recoil electron ϕ depends on the scattering angle θ. If $\theta = 0$, $\phi = 90°$ and if $\theta = 180°$, $\phi = 0$. It shows that the electron can recoil only in the onward direction at angles less than 90°, while a photon can scattered in any direction.

Kinetic Energy (K.E.) of Recoiled Electron

Since the incident photon gives up a part of their kinetic energy to the striking electron and then gets scattered. The kinetic energy of the recoiled electron is the difference between the energies of incident and scattered photon, that is,

$$\textit{K.E. of the recoiled electron} = h\upsilon - h\upsilon' \qquad ...(13)$$

Substituting the value of υ' from equation (11) in equation (13), we get

$$\text{K.E. of the recoiled electron} = h\upsilon - h\frac{\upsilon}{[1+2\alpha\sin^2\theta/2]}$$

$$\text{or K.E.} = h\upsilon\left(\frac{2\alpha\sin^2\theta/2}{1+2\alpha\sin^2\theta/2}\right) \quad ...(14)$$

This shows that *the kinetic energy of the recoiled electron depends on the scattering angle θ.* The kinetic energy of the recoil electron should be maximum when

$$\sin\theta/2 = 1$$

$$\text{or} \quad \theta/2 = \pi/2$$

$$\text{or} \quad \theta = \pi$$

Therefore, from eqn. (14) *the maximum kinetic energy of the recoil electron is,*

$$\text{Maximum K.E.} = h\upsilon\frac{2\alpha}{1+2\alpha}, \text{ where } \alpha = \frac{h\upsilon}{m_o c^2}$$

$$\therefore \quad (\text{K.E.})_{max} = \frac{h\upsilon(2h\upsilon/m_o c^2)}{(1+2h\upsilon/m_o c^2)}$$

$$\text{or Maximum K.E.} = \frac{2h^2\upsilon^2}{m_o c^2\left(1+\frac{\lambda h\upsilon}{m_o c^2}\right)} \quad ...(15)$$

This is the required expression for maximum kinetic energy of the recoil electron.

PRESENCE OF UNMODIFIED SCATTERED RADIATION ALONGWITH MODIFIED RADIATIONS FOR NON-ZERO SCATTERING ANGLE

The presence of unmodified scattered radiation could not be explained on the basis of Compton theory in which the scattering between incident radiation and free electron is considered. In fact, the target material contained free as well as bound electrons. When an incident photon instead of striking a free electron, strikes a bound electron, the electron does not detached by the collision of photon but remains bound so that the atom as a whole recoils. Therefore, to find the shift in wavelength of photon, we have to replace the rest mass of an electron m_o by the mass of an atom in the relation, $\Delta\lambda = h(1 - \cos\theta)/m_o c$.

For an aluminium target, the mass M_o of an atom is

$$M_o = 27\frac{m_H}{m_o}m_o = 27m_H \; 27\times1840\,m_o$$

where m_H is the mass of hydrogen atom. Then

$$\frac{h}{M_oc^2} = \frac{h}{27\times1840\,m_oc^2} = \frac{1}{27\times1840}\left(\frac{h}{m_oc^2}\right)$$

$$= \frac{1}{27\times1840}\times2\cdot426\times10^{-12}\,\text{mt}$$

$$\text{esea} = \frac{1}{27\times1840}\times2\cdot426\times10^{-12}\,\text{mt}$$

$$= 5.0\times10^{-17}\text{m}\; 5.0\times10^{-7}\text{Å}$$

Thus, for cases where electron is tightly bound to the atom, for an incident wavelength of the order of a few Angstroms the change in wavelength, $\Delta\lambda$ is negligible for all values of θ.

Since the mass of an atom is several thousand times greater than the rest mass of an electron, the Compton shift ($\Delta\lambda$) becomes too small to be detected. Therefore, when a photon collides with a bound electron, its wavelength does not change and thus the unmodified radiations are obtained. Hence, *both modified and unmodified radiations are simultaneously obtained at every non-zero scattering angle.*

For zero scattering angle only unmodified radiations are present;

Experimental Verification of Compton Effect

The schematic experimental arrangement for observing Compton scattering is shown in Fig. 12. A well collimated beam of monochromatic X-rays obtained from the X-ray tube is incident on a graphite target T.

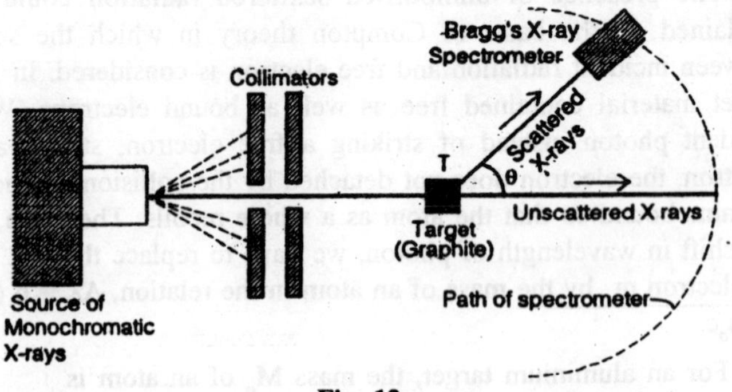

Fig, 12

The distribution of intensity with wavelength in the X-rays scattered at various angles θ ισ measured by means of a Bragg's X-ray speedometer. Fig. 13 (on the next page) shows the experimental results obtained for four scattering angles θ = 0°, 45⁰, 90⁰ and 135° For each value of θ, except θ = 0, there are two peaks the first peak appears at the same wavelength as the incident radiation (unmodified radiation), white the second peak corresponds to longer wavelength (modified radiation) or Compton shift (Δλ),

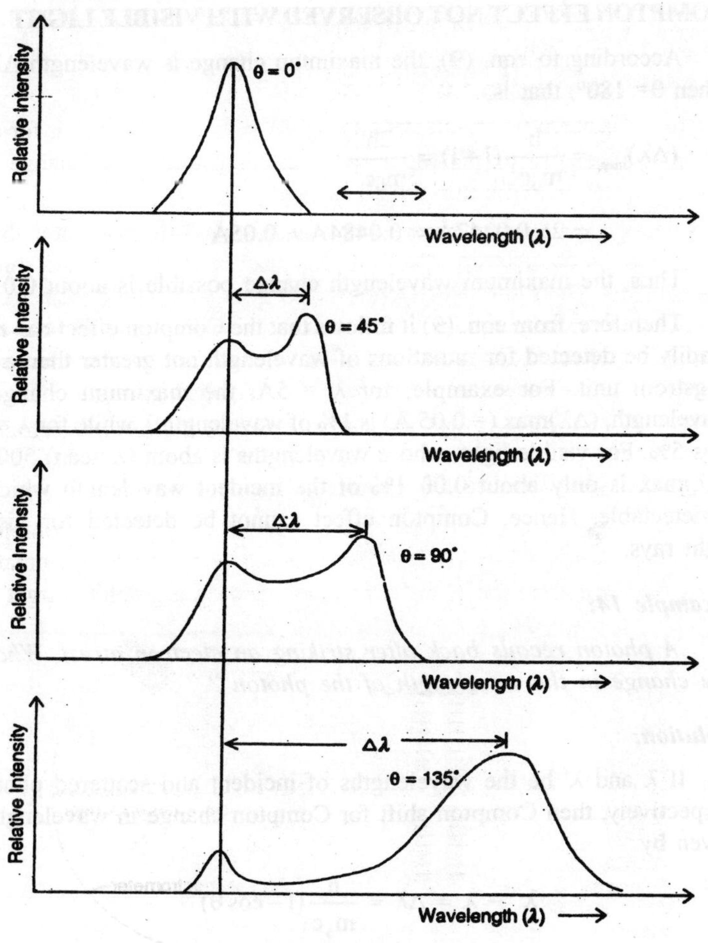

Fig. 13

From the experimental results it is clear that the wavelengths of the modified radiation peaks are found to depend upon the angle of scattering. The Compton shift for different angles of scattering calculated from theoretical equation agree well with those obtained experimentally. The appearance of first peak is due to the fact that the photon not only scattered by free electrons of the target material but also by very tightly bound electrons. In this type of collision, electron does not detached but the entire atom recoils instead of a single electron. In this case, the resulting Compton shift is so small as to be undetectable.

COMPTON EFFECT NOT OBSERVED WITH VISIBLE LIGHT

According to eqn. (9), the maximum change is wavelength $\Delta\lambda$ is, when $\theta = 180^\circ$, that is,

$$(\Delta\lambda)_{max} = \frac{h}{m_0c^2}(1+1) = \frac{2h}{m_0c^2}$$

$$= 2\times 0.0242\text{Å} = 0.0484\text{Å} \approx 0.05\text{Å}$$

Thus, the maximum wavelength change possible is about 0.05Å

Therefore, from eqn. (9) it follows that the Compton effect can most readily be detected for radiations of wavelength not greater than a few angstrom unit. For example, for λ = 5Å, the maximum change in wavelength, $(\Delta\lambda)$max (= 0.05 Å) is 1% of wavelength λ while for λ = 1Å it is 5%. For visible light, whose wavelengths is about (λmean) 5000Å. $(\Delta\lambda)$max is only about 0.00 1% of the incident wavelength which is undetectable. Hence, Compton effect cannot be detected for visible light rays.

Example 14:

.A photon recoils back after striking an electron atrest. What is the change in the wavelength of the photon ?.

Solution:

If λ and λ' be the wavelengths of incident and scattered photons respectively, then Compton shift for Compton change in wavelength is given by

$$\lambda' - \lambda = \Delta\lambda = \frac{h}{m_oc}(1-\cos\theta)$$

Since the photon recoils back, that is, the scattering angle θ = 180°, h = 6.63 × 10^{-34} joule-sec,

m_0 = 9.1 × 10^{-31} kg and c = 3.0 × 10m/sec

$$\text{Thus,}\quad \frac{h}{m_o c} = \frac{6.63\times10^{-34}}{9.1\times10^{-31}\times3.0\times10^{8}} = 2.43\times10^{-12}\,\text{m} = 0.0243\text{Å}$$

Since photon return, back, θ = 180° or cos 180° = –1.

Therefore, Δλ = 0.0243 (1 – cos 180°) = 0.0243 × 2 = 0.0486Å

Example 15:

Calculate Compton shift if X-rays of λ = 1.00Å are scattered from a carbon block. The scattered radiation is viewed at 99° to the incident beam.

Solution:

If λ and λ' be the wavelengths of incident and scattered X-ray photons, then Compton shift (Δλ) is given by

$$\Delta\lambda = \lambda' - \lambda = \frac{h}{m_o c}(1-\cos\theta)$$

Here λ = 1.00Å and θ = 90°, h = 6.63 × 10^{-34} J-sec,

c = 3.0 × 10^8/sec and m_o = 9.1 × 10^{-31} kg

$$\therefore\qquad \Delta\lambda = \frac{6.63\times10^{-34}}{9.1\times10^{-31}\times3.0\times10^{8}}(1-\cos 90^\circ)$$

$$\text{or}\qquad \Delta\lambda = 0.0243\times 1 = 0.0243\text{Å}$$

Example 16:

X-rays of wavelength 1.0Å are scattered from a carbon block. Find (i) the wavelength of the scattered beam in a direction making 90° with the incident beam, (ii) kinetic energy imparted to the recoil electron, (iii) direction of the recoil electron. (h = 6.63 × 10^{-34} Joule-see, c = 3.0 × 10^8 m/sec and 1 eV = 1.6 × 10^{-19} joule and m_o of electron = 9.1 × 10^{-31} kg).

Solution:

If λ and λ' be the wavelengths of incident and scattered photons respectively and θ the scattering angle, then Compton change in wavelength or Compton shift (Δλ) is given by

$$\Delta\lambda = \lambda' - \lambda = \frac{h}{m_o c}(1-\cos\theta)$$

where m_o is the rest mass of the electron

(i) The Compton shift, $\Delta\lambda = \lambda' - \lambda = \frac{6.63\times10^{-34}}{9.1\times10^{-31}\times3.0\times10^{8}}(1-\cos\theta)$

or $\Delta\lambda = 0.0243 \times 1 = 0.0243$ Å

Hence, the wavelength λ' of the scattered radiation

$\lambda' = \lambda + \Delta\lambda = 1.0 + 0.0243 = 1.0243$ Å

(ii) The kinetic energy imparted to the recoil electron = K.E. of incident X-ray photon – K.E. of scattered photon

or K.E. to the recoil electron $= h\upsilon - h\upsilon' = \frac{hc}{\lambda} - \frac{hc}{\lambda'}$

or $\text{K.E.} = \frac{hc(\lambda'-\lambda)}{\lambda\lambda'} = \frac{6.63\times10^{-34}\times3.0\times10^{8}\times0.0243\times10^{-10}}{1.0\times10^{-10}\times1.0243\times10^{-10}}$

$$= 4.72\times10^{-17}\text{ joule} = \frac{4.72\times10^{-17}}{1.6\times10^{-19}}\text{eV} = 295.0\text{eV}$$

(iii) We know that the direction of recoil electron ϕ is given by

$$\tan\phi = \frac{\lambda\sin\theta}{\lambda' - \lambda\cos\theta}, \text{Here } \theta = 90°$$

$$\therefore \quad \tan\phi = \frac{\lambda}{\lambda'} = \frac{1.0\times10^{-10}}{1.0243\times10^{-10}} = 0.98$$

or $\quad \phi = \tan^{-1}(0.98) = 44.4°$

Example 17:

γ-rays of energy 0.88 MeV are made to fall on a sheet of aluminium. Calculate the maximum energy of Compton recoil electrons (h = 6.62 × 10^{-34} joule-sec, c = 3.0 × 10^{8} m/sec and 1 eV = 1.6 × 10^{-19} joule).

Solution:

The wavelength of a γ-ray corresponding to the energy 0.88 MeV is

$$E = h\upsilon = \frac{hc}{\lambda}$$

or $\lambda = \frac{hc}{E} = \frac{6.62\times10^{-34}\times3.0\times10^{8}}{0.88\times10^{6}\times1.6\times10^{-19}} = 1.14\times10^{-12}\text{m} = 0.0141\text{Å}$

If θ is the scattering angle, then the Compton shift is given by

$$\Delta\lambda = \lambda' - \lambda = \frac{h}{m_o c}(1-\cos\theta) = \frac{6.62\times10^{-34}}{9.1\times10^{-31}\times3.0\times10^{8}}(1-\cos\theta)$$

The energy of the recoil electron will be maximum when $\theta = 180^o$

$\therefore (\Delta\lambda)_{max} = 0.0243\ (1 - \text{ccs}\ 180^o) = 0.0243 \times 2 = 0.0485\ \text{Å}$

The corresponding wavelength of scattered ray,

$\lambda' = \lambda + (\Delta\lambda)_{max} = 0.01'1 + 0.0486 = 0.0627\text{Å}$

The energy of the recoil electron is maximum for maximum Compton shift and is given by

Maximum K.E. of the recoil electron $= h\upsilon - h\upsilon' = \frac{hc}{\lambda} - \frac{hc}{\lambda'}$

$$(\text{K.E.})_{max} = \frac{hc(\Delta\lambda)_{max}}{\lambda\lambda'}$$

$$= \frac{6.62\times10^{-34}\times3.0\times10^{8}\times0.0486\times10^{-10}}{0.0141\times10^{-10}\times0.0627\times10^{-10}}$$

$$= 1.092\times10^{-13}\ \text{joule} = \frac{1.092}{1.6\times10^{-19}}$$

$\text{eV} = 0.682 \times 10^{6}\ \text{eV} = 0.682\,\text{MeV}$

Example 18:

An X-ray photon is found to have its wavelength doubled on being scattered through 90°. Find the wavelength and energy of the incident photon (m_o of electron = 9.0×10^{-31} Kg)

Solution:

If λ and λ' be the wavelengths of incident X-ray photon and scattered photon respectively and θ the angle of scattering, then

Compton shift, $\Delta = \lambda' - \lambda = \frac{h}{m_o c}(1-\cos\theta)$

Here $\theta = 90$ and $\lambda' = 2\lambda$ $\quad\therefore\ \Delta\lambda = 2\lambda - \lambda = \lambda$

Thus $$\lambda = \frac{h}{m_o c} = \frac{6.62\times10^{-34}}{9.0\times10^{-31}\times3.0\times10^{8}} = 0.0245Å$$

The energy of the incident photon,

$$E = h\upsilon = \frac{hc}{\lambda} = \frac{6.62\times10^{-34}\times3.0\times10^{8}}{0.0245\times10^{-10}} = 8.106\times10^{-14} \text{ joule}$$

Example 19:

A photon collides with a free electron and is scattered at right angles to its initial direction. Find the percentage change in its energy as a result of this Compton collision, when the photon is a :

(i) microwave photon of λ = 3.0 cm,

(ii) visible light photon of λ = 5000 Å,

(iii) X-ray photon of λ = 1.0 Å and

(iv) y-ray photon of λ = 0.0124 Å.

In view of your results discuss the relative importance of the Compton effect in the different regions of electromagnetic spectrum.

Solution:

The change in energy of the incident photon after suffering Compton scattering is given by

$$\Delta E = h\upsilon' = \frac{hc}{\lambda} - \frac{hc}{\lambda'}$$

or $$\Delta E = \frac{hc(\lambda'-\lambda)}{\lambda\lambda'} = \frac{hc\Delta\lambda}{\lambda\lambda'}$$

where $\Delta\lambda$ is the change in wavelength.

Thus, $$\Delta E = \frac{hc\Delta\lambda}{\lambda(\lambda+\Delta\lambda)}$$

The initial energy of the photon is

$$E = h\upsilon = \frac{hc}{\lambda},$$

Therefore, the fractional change in energy

$$\frac{\Delta E}{E} = \frac{\Delta\lambda}{\lambda+\Delta\lambda}$$

Now, the Compton shift $\Delta\lambda$ is given by

$$\Delta\lambda = \frac{h}{m_o c}(1-\cos\theta)$$

$$= \frac{6.63\times10^{-34}}{9.1\times10^{-31}\times3.0\times10^{8}}(1-\cos\theta)$$

$$= 0.0243\ (1 - \cos\theta)$$

or $$\Delta\lambda = 0.0243\ \text{Å for } \theta = 90^o$$

$$\therefore \quad \frac{\Delta E}{E} = \frac{0.0243}{\lambda+0.0243}$$

Let us now calculate the fractional change or percentage charge of energy for different type of photons.

(i) For microwave photon (λ = 3.0 cm = 3.0 × 10^8Å), we have

$$\frac{\Delta E}{E} = \frac{0.0243}{3.0\times10^{8}+0.0243}$$

$$= 8.1 \times 10^{-11} = 8.1 \times 10^{-9}\ \%.$$

$$\frac{\Delta E}{E} = \frac{0.0243}{5000+0.0243}$$

$$= 4.86 \times 10^{-6} = 4.86 \times 10^{-4}\ \%.$$

(iii) For the X-ray photon (λ = 1.0 Å), we have

$$\frac{\Delta E}{E} = \frac{0.0243}{1.0+0.0243} = 0.024 = 2.4\ \%.$$

(iv) For the γ-ray photon (λ = 0.0124 Å), we have

$$\frac{\Delta E}{E} = \frac{0.0243}{0.0124+0.0243}$$

$$= 0.66 = 66\ \%$$

Hence *the Compton effect is dominant only in the γ-ray region and shorter γ-ray region. It is not observable in the visible region and microwave region.*

Example 20:

X-rays of wavelength λ_o = 2.00Å are scattered from a block of carbon. The scattered X-rays are observed at an angle of 45° *to the incident beam. Find the fraction of energy lost by the photon in this collision.*

Solution:

If λ_0 and λ' be the wavelengths of incident and scattered X-rays respectively, then

$$\Delta\lambda = \lambda' - \lambda_0 = \frac{h}{m_0 c}(1-\cos\theta)$$

X-rays are scattered at an angle 45° to the incident beam. Therefore, the wavelength of scattered X-rays is,

$$\lambda = \lambda' - \lambda_0 = \frac{h}{m_0 c}(1-\cos 45^\circ)$$

Here,
$$\frac{h}{m_0 c} = \frac{6.62\times10^{-34}}{9.1\times10^{-31}\times3.0\times10^{8}}$$
$$= 0.0243 Å$$

$\therefore$
$$\lambda' = 2.00 + 0.243\ (1 - 1/\sqrt{2})$$
$$= 2.00 + \frac{0.0243\times0.4142}{1.4142}$$
$$= 2.00 + 0.007 = 2.007 Å = \lambda_0 + \Delta\lambda$$

The energy lost by the photon in the collision or the change in the energy of the incident photon is,

$$\Delta E = h\upsilon_0 - h\upsilon' = \frac{hc}{\lambda_0} - \frac{hc}{\lambda'}$$

or
$$\Delta E = \frac{hc\Delta\lambda}{\lambda_0\lambda'}$$

The initial energy of the photon,

$$E = h\upsilon_0 = \frac{hc}{\lambda_0}$$

$\therefore$ the fraction energy lost by the photon in this collision,

$$\frac{\Delta E}{E} = \frac{\Delta\lambda}{\lambda'}$$

$\therefore$
$$\frac{\Delta E}{E} = \frac{0.007}{2.007} = 0.0035$$

Example 21:

For what wavelength photon does Compton scattering result in a photon whose energy is one-half that of the original photon, at scattering angle 45°? In which region of electromagnetic spectrum does such a photon lie? (Compton wavelength of the electron = 0.0242Å).

Solution:

If λ and λ' be the wavelengths of incident photon and scattered photon respectively and θ the scattering angle, then

change in wavelength, $\Delta\lambda = \lambda' - \lambda$

$$= \frac{h}{m_0 c}(1-\cos\theta)$$

where $\frac{h}{m_0 c}$ is the Compton wavelength

Here $h/m_0c = 0.0242Å$ and $\theta = 45°$

$$\Delta\lambda = 0.0242\ (1 - \cos 45°)$$

$$= \frac{0.0242\times(\sqrt{2}-1)}{\sqrt{2}}$$

$$= \frac{0.0242(1.4142-1)}{1.4142}$$

or $$\Delta\lambda = \frac{0.0242\times 0.4142}{1.4142} = 0.007Å$$

It hc/λ be the energy of the incident photon, then the energy of the scattered photon would be

$$\frac{1}{2}(hc/\lambda); \qquad \therefore \qquad \frac{hc}{\lambda'} = \frac{hc}{2\lambda}$$

or $\lambda' = 2\lambda$

Since $\lambda' - \lambda = \Delta\lambda$

$\therefore$ $2\lambda = \Delta\lambda + \lambda$

or $\lambda = \Delta\lambda$

$\therefore$ $\lambda = 0.007$ Å

Example 22:

A beam of γ-radiation having photon energy 510 keV is incident on a foil of aluminium. Calculate the wavelength of the radiation at 90° and also the energy and direction of the emission of the corresponding electron.

Solution:

Energy of each photon = $h\upsilon = \frac{hc}{\lambda} = 510$ keV

or $\frac{hc}{\lambda} = 510 \times 10^3 \times 1.6 \times 10^{-19}$ joule

or $$\lambda = \frac{6.6\times10^{-34}\times3.0\times10^{8}}{510\times10^{3}\times1.6\times10^{-19}}$$

$= 2.246 \times 10^{-12}$ m

According to Compton scattering formula, the wavelength λ' of scattered radiation is given as

$$\lambda' - \lambda = \frac{h}{m_0 c}(1-\cos\theta)$$

or $$\lambda' = \lambda + \frac{h}{m_0 c}(1-\cos\theta)$$

$$\lambda' = 2.426 \times 10^{-12} + \frac{6.6\times10^{-34}}{9.1\times10^{-31}\times3.0\times10^{8}}(1-\cos 90^{\circ})$$

or $\lambda' = 2.426 \times 10^{-12} + 2.418 \times 10^{-12} = 4.844 \times 10/^{-12}$ m

The energy of the recoil electron

$$= h\upsilon - h\upsilon' = \frac{hc}{\lambda} - \frac{hc}{\lambda'}$$

or Energy of recoil electron $= \frac{hc(\lambda' - \lambda)}{\lambda\lambda'}$

or $$E = \frac{6.6\times10^{-34}\times3.0\times10^{8}(4.844\times10^{-12}-2.426\times10^{-12})}{2.426\times10^{-12}\times4.844\times10^{-12}}$$

$$= \frac{6.6\times10^{-34}\times3.0\times10^{8}\times2.418\times10^{-12}}{2.426\times10^{-12}\times4.844\times10^{-12}}$$

$= 4.074 \times 10^{-14}$ joule

The direction of recoil electron φ is given by

$$\tan\phi = \frac{\lambda\sin\theta}{\lambda' - \lambda\cos 90^{\circ}}.$$

Since $\theta = 90^{\circ}$

$$\therefore \quad \tan\phi = \frac{\lambda \sin 90^\circ}{\lambda' - \lambda \cos 90^\circ}$$

$$\text{or} \quad \tan\phi = \frac{\lambda}{\lambda'} = \frac{2.426\times10^{-12}}{4.843\times10^{-12}} = 0.501$$

$$\text{or} \quad \phi = \tan^{-1}(0.501) = 26.61^\circ$$

Example 23:

A beam of gamma-radiation having photon energy 510 KeV is incident on a foil of aluminium. Calculate the wavelength of scattered radiation at 90°.

Solution:

Energy of each photon $= h\upsilon = \frac{hc}{\lambda} = 510$ **KeV**

$$\text{or} \quad \frac{hc}{\lambda} = 510 \times 10^3 \times 1.6 \times 10^{-19} \text{ joule}$$

$$\therefore \quad \lambda = \frac{hc}{510\times10^3\times1.6\times10^{-19}}$$

$$= \frac{6.63\times10^{-34}\times3.0\times10^8}{510\times10^3\times1.6\times10^{-19}}$$

$$= 0.0244 \times 10^{-10} \text{ m}$$

According to the compton scattering formula

$$\lambda' - \lambda = \frac{h}{m_0 c}(1-\cos\theta)$$

$$\text{or} \quad \lambda' = \lambda + \frac{h}{m_0 c}(1-\cos\theta)$$

Wavelength of scattered radiation at 90° is;

$$\lambda' = \lambda + \frac{h}{m_0 c}$$

$$\therefore \quad \lambda' = 0.0244 \times 10^{-10} \text{ m} + \frac{6.63\times10^{-34}}{9.1\times10^{-31}\times3\times10^8}$$

$$= 0.0244 \times 10^{-10} + 0.0243 \times 10^{-10}$$

$$\text{or} \quad \lambda' = 0.0487 \times 10^{-10} \text{ m} = 0.0487 \text{ Å}.$$

2

Photoelectric Effect

INTRODUCTION

Photoelectric effect is the phenomenon of ejection of electrons from a metal plate when light of a suitable wave-length falls on it. The electrons emitted are called photo-electrons to indicate their mode of production.

Laws of Photo-Electric Emission : There are two important laws regarding the photoelectric **Emission.**

Law I: *The velocity of the emitted electrons increases with the increase in frequency of the incident light.* This law means that there is a limiting frequency for a metal, below which no electrons are emitted. This is known as the *threshold frequency.*

Law II: *The number of photo electrons ejected is directly proportional to the intensity of the incident light.* Thus, if the intensity of the incident light is increased, the number of electrons emitted is increased, but their velocity remains constant.

Explanation : The laws of photo-electric emission were explained by Einstein in 1905 on the basis of Planck's quantum theory of radiation. According to the quantum theory, the radiation consists of packets of energy called 'quanta' or 'photons' of energy hv, where h is the Planck's constant and v the frequency of radiation. When a photon of energy hv is incident upon a metal surface, its energy is used up in two ways shown in Fig. 2.1.

(a) A part of its energy is used up in ejecting the electron just out of the surface. The energy depends upon the nature of the metal and is called work-function (ϕ_0).

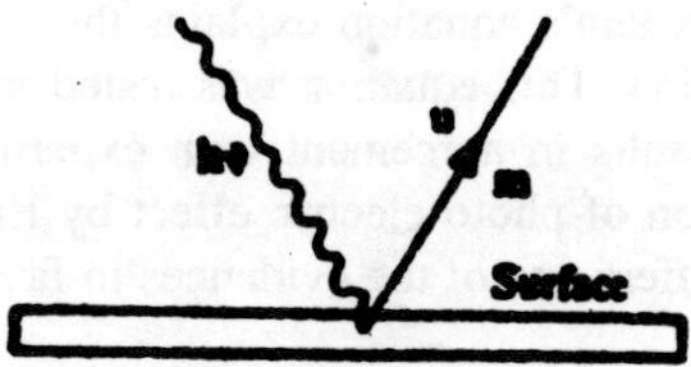

Fig. 2.1

(b) The rest part of the energy of the striking photon is used up in imparting kinetic energy $\frac{1}{2}$ mu² to the ejected electron.

Thus $\quad hv = \phi_0 + \frac{1}{2} mu^2$ ...(1)

But $\quad \phi_0 = hv_0$

where v_0 is the threshold frequency. Therefore, equation (1) becomes as

$$hv - hv_0 + \frac{1}{2} mu^2$$

or $$\frac{1}{2} mu^2 = h(v - v_0)$$ ...(2)

The equation is called Einstein's photo-electric equation.

Explanation of IInd Law : From eq. (2), it follows that the velocity of a photo-electron increases with the increase in frequency. No electron will be emitted when the frequency of the incident radiation v is less than the threshold frequency v_0. This explains the II law of photoelectric emission.

Explanation of 1st Law : If the intensity of light is increased keeping the frequency same, more photons are incident on the metal surface but the energy of each photon will remain the same. Hence the number of electrons emitted per unit area per unit second will increase but their velocities remain the same. This is the I law of photoelectric emission. It is clear that the maximum value of velocity u_{max} for a given frequency v will occur for those electrons for which $\phi = \phi_0$. Hence, equation (2) becomes as

$$1/2\ mu^2_{max} = hv - \phi_0 = hv - hv_0$$ ...(3)

It is clear from the above equation that the maximum velocity of photo- electrons will increase with increase in frequency v of incident radiation and no electrons will be emitted when v is less than v_0.

Testing. The Einstein's equation explains the observed facts about photo-electric emission. This equation was tested in 1915 by Millikan who observed its results in agreement with experimental results. The successful explanation of photo electric effect by Einstein on the basis of quantum theory offers one of the evidences in favour of the quantum theory.

Importance. Photo-electric effect is used in the construction of photo electric cell which converts light energy into electric energy. The photo-electric cells are widely used in various fields.

PHOTOELECTRIC CELLS

It is a device for converting light energy into electric energy. There are *three* main types of photo-cells.

1. *Photo emission type.* It consists of an evacuated glass, or quartz tube which has its inner surface M coated with sodium, potassium or caesium, and this coating is connected to – ve end of a battery. A window is left open through which light can enter the tube, and a wire loop P, or a cylindrical wire in the centre is connected to the +ve end of the battery and it collects the electrons, Fig. 2.2, 2.3 and 2.4. The two parts are insulated from each other and are brought out to two legs at the bottom of the photo cell. On admitting light, which is of a frequency above the threshold frequency, electrons from the photometal are emitted. They travel towards the loop wire and a current is produced in the circuit.

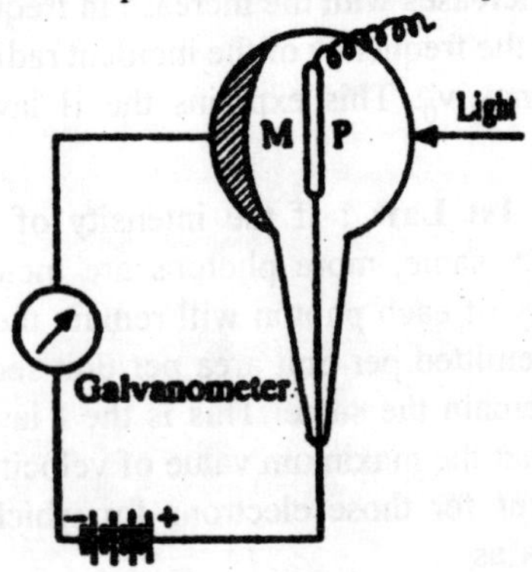

Fig. 2.2

Since a photo cell responds to the light which falls on it, as does the human eye, it is therefore, called an *electric eye.* There are two types of photo emission cells, (1) high vacuum type, and (2) gas filled type.

In vacuum cells, the photoelectric current is very small but these cells keep a strict proportionality between the current and intensity of light. The chief advantages of vacuum type are.

(i) The sensitivity of these cells remain unaltered for a very long time provided the cathode is properly selected.

(ii) There is no time lag between the incident light and the photoelectrons and the photoelectric current is proportional to the intensity of illumination. Thus, they are extremely accurate in response.

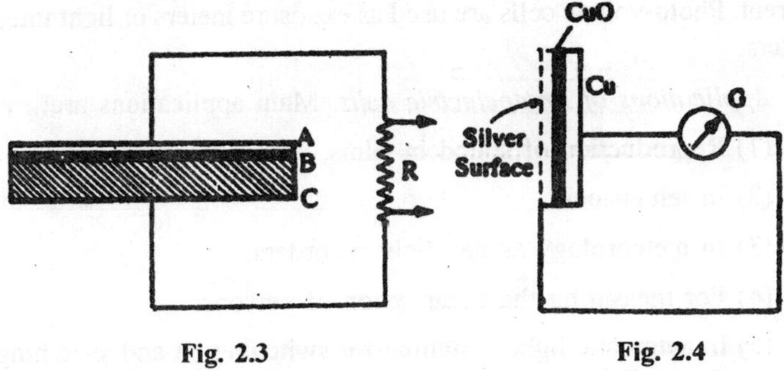

Fig. 2.3 Fig. 2.4

Due to the above mentioned advantages, these cells are being used in television and photometry.

In gas filled cells, the response to light is not so quick but since an inert gas is being used in the cell, ionisation of the gas is produced by the emitted electrons which therefore increase the photo current. Photoelectric current in this case is not proportional to the intensity of illumination and there is some time lag. Such cells are used in cinematography both for recording and reproduction of sound.

2. *Photo-conductive cell.* These cells are based on the property that the resistance of selenium and certain other metals decreases with increase of illumination. No photo-electrons are emitted in this case. A photo-conductive cell consists of a thin film of a semiconductor. *e.g.*, selenium, B, placed below a thin semi-transparent metal film A. The combination is placed on a block of iron C in contact with it (Fig. 2.3). Ordinary selenium is a semi-conductor. When light is incident on A, it is absorbed by the selenium layer, electrons are emitted and if a small external e.m.f, is applied, a current flows in the resistance and can operate a relay when the intensity of incident light is high. These cells are generally connected directly to a micrometer or a low resistance relay and are preferably used

in a Wheatstone bridge arrangement. Their response to light is not as quick as in photo-emission cell, but they are more sensitive to red light.

3. *Photo voltaic cell.* This type of photo-cell consists of a thin-layer of cuprous oxide coated on a disc of copper (Fig 2.4). The oxide film has sputtered silver or gold film on its upper surface. When light falls on the oxide, electrons are emitted from it not into the surrounding air but into the copper. The oxide layer thus becomes + vely charged and copper negatively. An e.m.f. is thus set up. If contacts are made to the sputtered film and to the copper, the photo voltaic e.m.f. Produces a current. Photo-voltaic cells are used as exposure meters or light intensity meters.

Applications of Photoelectric cells. Main applications are:

(1) Reproduction of sound in films,

(2) In televisions

(3) In meteorology as day-light recorders,

(4) For measuring the complexion of persons,

(5) In automatic light switching for switching on and switching off the street light.

(6) In determination of temperature of stars.

(7) Inburglar alarms to detect thieves and in fire alarms to indicate the outbreaks of fire,

(8) In traffic signals.

CHEMILUMINESCENCE

It is the phenomenon of emission of visible light as a result of a chemical change at a temperature at which a black body normally does not emit visible light.

From the above definition, it is evident that chemiluminescence is the reverse of photochemical reaction in which light absorption causes chemical action. Some examples are.

(i) A classical example of chemiluminescence is the phosphorus which produces a greenish-yellow glow. It is probably due to the oxidation of phosphorus vapours by atmospheric oxygen. Phosphorus trioxide also produces glow. This is probably due to the oxidation of P_2O_3 to P_2O_5.

(ii) When a solution of strontium chloride is added to dilute H_2SO_4 in dark, a feeble glow is produced along with the precipitate of $SrSO_4$.

$$SrCl_2 + H_2SO_4 \rightarrow SrSO_4 + 2HCl + h\nu.$$

(iii) It was observed by Evans that a solution of p-bromophenyl magnesium bromide (Grignard reagent) in ether produces greenish- blue glow consisting of a single broad band.

(iv) When alkali metal vapours react with halogens or with mercuric halides, luminescence occurs which consists of the spectrum of the alkali metal. The mechanism of this is based on the fact that sodium metal contains monatomic and diatomic molecules. The mechanism is.

$$Na + Cl_2 \rightarrow NaCl + Cl$$
$$Na + Cl \rightarrow NaCl$$
$$Cl + Na_2 \rightarrow Na + NaCl^*$$
$$NaCl^* + Na \rightarrow NaCl + Na^*$$
$$Na^* \rightarrow Na^* + h\nu \text{ (yellow spectrum)}$$

(v) When the surface of mercury is allowed to come in contact with a stream of atomic hydrogen, a blue fluorescence appears. The spectrum of this blue glow consists of the resonance line 2537 Å and a band of spectral lines from 4520 to 3250 Å due to mercuric hydride. The possible mechanism is due to

$$Hg + H \rightarrow HgH$$
$$H + H \rightarrow H_2 + \text{Energy}$$
$$HgH + \text{Energy} \rightarrow HgH^*$$
$$HgH^* + HgH \rightarrow Hg + Hg^* + H_2$$
$$Hg^* \rightarrow Hg + h\nu$$

Theoretical Explanation of Chemiluminescence : The phenomenon of chemiluminescence can be explained on the basis of quantum theory.- According to this theory, certain chemical reactions result in the formation of products which are in electronically excited states, When these excited products return to their ground levels, they do so by emitting the extra energy in the form of visible light radiations. Thus, by this process, the chemical energy is converted into light energy.

Sensitised Chemiluminescence : In 1925 Kautsky reported that oxidation of unsaturated silicon hydride by $KMnO_4$ is accompanied by a luminescent glow. If the same oxidation is carried out in the presence of contain dyestuffs such as rhodamine-B, a strong red fluorescence characteristics of the dye is observed. This type of change is known as *sensitised chemiluminescence.*

Explanation. During the oxidation of silicon hydride, the energy is given out and is subsequently transferred to the dyestuff, raising it to an excited state. When it returns to normal state, a fluorescence is observed.

PHOTOCHEMICAL KINETICS

The rate laws which photochemical reactions follows are generally more complex than those for thermal reactions because more variables are involved.

Like the kinetics of reactions as discussed in chemical kinetics we will apply the steady state treatment. We will now apply steady state-treatment to reactions which do not involve chains. Examples are :

(1) Dissociation of HI : The dissociation of hydriodic acid was investigated by *Warburg* (1911) in ultraviolet light of wavelength 2070, 2530 and 2820 Å.

$$2HI \xrightarrow{hv} H_2 + I_2$$

The experimental quantum efficiency of this reaction was 2.0. In order to account for this,. Warburg postulated the following mechanism.

(a) $HI + hv$ ¾→ $H + I$ Rate $= I_{abs}$

(b) $H + HI \xrightarrow{k_2} H_2 + I$ Rate $= k_2$ [H] [HI]

(c) $I + \xrightarrow{k_3} I_2$ Rate $= k_3$ $[I]^2$

In the above mechanism, hydrogen iodide is consumed insteps (a) and (b). Thus, its rate of dissociation is given by

$$\frac{-d[HI]}{dt} = I_{abs} + k_2 [H] [HI] \qquad ...(1)$$

Also, the net rate of formation of hydrogen atoms is given by

$$\frac{d[H]}{dt} = I_{abs} - k_3 [H] [HI] \qquad ...(2)$$

As [H] atom is short lived, the steady treatment, therefore, can be applied to it. Thus, equation (2) becomes

$$\frac{d[H]}{dt} = 0\ I_{abs} - k_2\ [H]\ [HI] \text{ or } k_2\ [HI\ [HI] = \text{Iabs}$$

Substituting this value in equation (1), we get

$$\frac{-d[HI]}{dt} = 2I_{abs} \qquad ...(3)$$

Quantum yield. By definition, the quantum yield is given by

$$\phi = \frac{\frac{-d[HI]}{dt}}{I_{abs}} = \frac{2I_{abs}}{I_{abs}} = 2 \qquad \text{[From (3)]}$$

This value of quantum efficiency is quite in agreement with experimental value Table 1.

Table 1 : Photochemical Decomposition of HI

Wavelength	Moles per K cal	Quantum yield
2070 Å	1.44×10^{-2}	1.98
2530 Å	1.85×10^{-2}	2.08
2830 Å	2.09×10^{-2}	2.10

(ii) Dissociation of HBr. The quantum yield for the dissociation of HBr is also two. Spectroscopic studies revealed that the mechanism followed by this reaction is the same as represented above for the dissociation of hydrogen iodide.

We will now apply steady-state to following reaction which involves chain mechanism.

1. Hydrogen and chlorine reaction and
2. Hydrogen and bromine reaction.

For details of chain reactions place see the chapter on 'Chemical Kinetics"

(i) Hydrogen and Chlorine Reaction : It is a classical example of photochemistry which involves chains. The reaction is

$$H_2 + Cl_2 \xrightarrow{hv} 2HCl$$

This reaction was first of all studied by *W. Cruickshank* and later by *J. W. Draper, R Bunsen* and so many other workers.

Facts : Special facts about this reaction are.

(i) The reaction is provoked by the light absorbed by chlorine in the region of 4785 Å.

(ii) An induction period is often observed. This is a period after the start of illumination during which the reaction is either very slow or does not take place at all.

The induction period is due to the presence of impurities like ammonia, NO_2, etc. That is why with pure hydrogen and chlorine no induction period is observed.

(iii) It was found by Draper (1841) that hydrogen-chlorine combination was followed by an increase in volume at constant pressure. This is known as Draper's effect.

(iv) The effect of oxygen is to influence the reaction after it has started. It has been found that the rate of reaction is inversely proportional to the concentration of oxygen.

(v) The quantum yield varies from 10^4 to 10^6 in the absence of oxygen. The value of quantum yield depend upon the impurities and oxygen concentration in the reaction vessel.

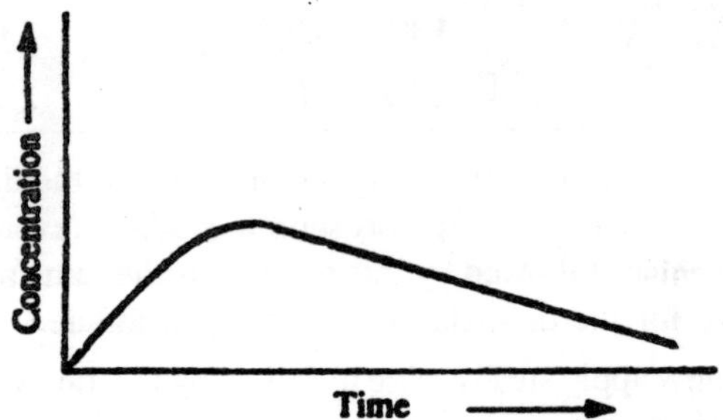

Fig. 2.5 : Combination of H_2 and Cl_2.

On the suggestion of Bodenstein, Nernst (1918) explained the reaction by chain mechanism which is universally accepted.

We will now consider the chain mechanism in two ways to explain. (a) Combination of hydrogen and chlorine in the absence of oxygen and (b) Combination of hydrogen and chlorine in the presence of oxygen.

In the Absence of Oxygen : (a) When a mixture of hydrogen and chlorine is exposed to light in the region of chlorine spectrum, chlorine molecule dissociates into atoms.

Initiation.

(i) $Cl_2 + hv \xrightarrow{k_1} 2Cl]$ Rate = k, I_{abs}

The initiation reaction is followed by the following reactions.

Chain Propagation.

(ii) $Cl + H_2 \xrightarrow{k_2} HCl + H$ Rate = k_2 [Cl] [H_2]

(iii) $H + Cl_2 \xrightarrow{k_3} HCl + Cl$ Rate = k_3 [H] [Cl_2]

Chain Termination.

(iv) $Cl + wall \xrightarrow{k_4} \frac{1}{2}Cl_2$ Rate = k_4 [Cl]

Derivation of Rate Law : The total rate of formation of hydrogen chloride is given by reactions (ii) and (iii). Therefore, its rate is given by

$$\frac{d[HCl]}{dt} = k_2 [Cl] [H_2] + k_3 [H] \{Cl_2] \quad ...(1)$$

The rate of formation of chlorine atoms is given by step (i) and (ii) and the rate of removal of chlorine atoms is given by steps (ii) and (iv). Therefore, the net rate of formation of [Cl] atoms is given by

$$\frac{d[Cl]}{dt} = k_1 I_{abs} + k_3 [H] [Cl_2] - k_2 [Cl] [H_2] - k_4 [Cl] \quad ...(2)$$

Similarly, for hydrogen atoms we can write as

$$\frac{d[H]}{dt} = k_2 [C_l] [H_2] - k_3 [H] [Cl_2] \quad ...(3)$$

When the stationary state is reached, it follows that

$$\frac{d[H]}{dt} = \frac{d[Cl]}{dt} = 0.$$

$$k_1 I_{abs} + k_3 [H] [Cl_2] - k_2 [Cl] [H_2] - k_4 [Cl] = 0 \quad ...(4)$$

and $\quad k_2 [Cl] [H_2] - k_3 [H] [Cl_2] = 0 \quad ...(5)$

By adding Eqs. (4) and (5), we get

$$k_1 \text{ Iabs} - k_4 [Cl] = 0 \text{ or } [Cl] = \frac{k_1}{k_4} I_{abs}. \quad ...(5a)$$

Also, Eq. (5a) gives as

$$k_2 [Cl] [H_2] = k_3 [H] [Cl_2] \quad ...(6)$$

Substituting Eqs. (6) in (1) we get

$$\frac{d[HCl]}{dt} = 2k_2 [H_2] \frac{k_1}{k_4} I_{abs}$$

$$= 2\left(\frac{k_1 k_2}{k_4}\right) I_{abs} [H_2] \quad ...(7)$$

Expression (11) is in perfect agreement with experimental data. In the presence of high chlorine content, step (iv) is replaced by following complicated steps.

(v) $Cl + Cl_2 \rightarrow Cl_3$

(vi) $2Cl_3 \rightarrow 3Cl_2$

If we calculate the rate expression by using steps (i), (ii), (iii), (v) and (vi), we get an expression which is very complex.

(b) In the Presence of Oxygen : When the reaction is carried out in the presence of oxygen, the following mechanism was proposed in 1921 by Gohring.

(i) $Cl_2 + h\nu \xrightarrow{k_1} 2Cl$ Rate $= k_1 I_{abs}$

(ii) $Cl + H_3 \xrightarrow{k_2} HCl + H$ Rate $= k_2 [Cl] [H_2]$

(iii) $H + Cl_2 \xrightarrow{k_3} HCl + Cl$ Rate $= k_3 [H] [Cl_2]$

(iv) $H + O_2 \xrightarrow{k_4} HO$ Rate $= k_4 [H] [O_2]$

(v) $Cl + O_2 \xrightarrow{k_5} ClO_2$ Rate $= k_5 [Cl] [O_2]$

(vi) $Cl + X \xrightarrow{k_6} ClX$ Rate $= k_6 [Cl] [X]$

In step (iv), X is any substance which removes chlorine atom. Applying the steady-state treatment to chlorine atoms, we get

$$\frac{d[Cl]}{dt} = k_1 I_{abs} - k_2 [Cl] [H_2] + k_3 [H] [Cl_2]$$

$$- k_5 [Cl] O_2] - k_6 [Cl] [X] = 0$$

This equation simplifies to

$$[Cl = \frac{k_1 I_{abs} + k_3 [H][Cl_2]}{k_2 [H_2] + k_5 [O_2] + k_6 [X]} \quad ...(8)$$

The rate of formation of hydrogen atoms is given by

$$\frac{d[H]}{dt} = k_2\ [Cl\ [H_2] - k_3\ [H]\ [Cl_2]$$

$$- k_4\ [H]\ [O_2] = 0 \qquad ...(9)$$

This equation on simplification yields.

$$[Cl] = \frac{k_3\,[H][Cl_2] + k_4\,[H][O_2]}{k_2\,[H_2]} \qquad ...(10)$$

By equating Eqs. (8) and (10), we get

$$k_2\ [H_2]\ (k_1\ I_{abs} + k_3\ [H\ [Cl_2]) = (k_2\ [H_2] + k_5\ [O_2]$$

$$+ k_6\ [X]) + (k_3\ [H]\ [Cl_2] + k_4\ [H]\ [O_2])$$

By omitting small term $k_4 k_5 [H] [O_2]^2$, the above equation is simplified to give the value of [H], *i.e.*,

$$[H] = \frac{k_1\ I_{abs}\ k_2\ [H_2]}{k_3k_6\ [Cl_2][X] + [O_2](k_1k_4\ (H_2] + k_3k_5\ [Cl_2] + k_4k_6\ [X])} \qquad ...(11)$$

The rate of formation of hydrogen chloride is given by steps (ii) and (iii). Thus,

$$\frac{d[HCl]}{dt} = k_2\ [Cl]\ [H_2] + k_3\ [H]\ [Cl_2] \qquad ...(12)$$

Reaction (iii) is much faster than (ii), consequently the rate of formation is given by step (iii) only. Therefore, Eq. (12) becomes as

$$\frac{d[HCl]}{dt} = k_3\ [H]\ [Cl_2] \qquad ....(13)$$

Introduction of expression (15) for [H] into this equation, we obtain.

$$\frac{d[HCl]}{dt}$$

$$= \frac{k_1k_2k_3\,[H_2][Cl_2]\,I_{abs}}{k_3k_6\,[Cl_2][X] + [O_2](k_2k_4\,[H_2] + k_3k_5\,[Cl_2] + k_4k_5\,[X])} \qquad ...(14)$$

or

$$\frac{d[HCl]}{dt} = \frac{\left(\dfrac{k_1k_3}{k_4}\right) I_{abs}\ [H_2][Cl_2]}{\dfrac{k_3k_6}{k_2k_4}[Cl_2][X] + [O_2]\left([H_2] + \dfrac{k_3k_5}{k_2k_4}[Cl_2] + \dfrac{k_6}{k_2}[X]\right)}$$

$$= \frac{k\, I_{abs}\, [H_2][Cl_2]}{m\,[Cl_2] + [O_2]\left([H_2] + \frac{[Cl_2]}{10}\right)} \qquad ...(15)$$

where $k = \frac{k_1 k_3}{k_4}$, $m = \frac{[X][k_3 k_6]}{k_2 k_4}$

and $\frac{1}{10} = \frac{k_3 k_5}{k_2 k_4}$ ...(16)

Eq. (15) is in full agreement with the experimental rate equation as given by Bodenstein and Unger.

(2) Reaction Between Hydrogen and Bromine : In 1924, Bodenstein and Lurke-Meyer found that the photochemical reaction proceeds according to the empirical equation

$$\frac{d\,[HBr]}{dt} = \frac{k'[H_2]\, I_{abs}{}^{1/2}}{1 + \frac{[HBr]}{m'\,[Br_2]}} \qquad ...(17)$$

where k′ and m′ are constants, and I_{abs}, is the intensity of the light absorbed.

Mechanism : The photochemical combination between hydrogen and bromine to form hydrogen bromide is an example of a chain reaction. Its mechanism is similar to the thermal reaction with the difference that the initiation is brought about by the absorption of a photon by a bromine molecule.

Chain initiation

(i) $Br_2 \xrightarrow[k_1]{h\nu} 2Br$ Rat = $k_1\, I_{abs}$

The rest part of the mechanism is similar to the thermal reaction

Chain Propagation. Rate = k_2 [Br] $[H_2]$

(ii) $Br + H_2 \xrightarrow{k_2} HBr + H$ Rate = k_2 [Br] $[H_2]$

(iii) $H + Br_2 \xrightarrow{k_3} HBr + Br$ Rate = k_3 [H] $[Br_2]$

Chain inhibition :

(iv) $H + HBr \xrightarrow{k_4} H_2 + Br$ Rate = k_4 [H] [HBr]

Chain breaking : Rate = $k_5\, [Br]^2$

(v) $Br + Br \xrightarrow{k_5} Br_2$

Here k_1, k_2, k_3, k_4, and k_5 represent the specific rates of all the five reactions respectively.

Derivation of Rate Law : Since HBr is produced by reactions (ii) and (iii) and removed by the reaction (iv), the net rate of formation of HBr is given by

$$\frac{d[HBr]}{dt} = k_2\ Br]\ [H_2] + k_2\ [H]\ [Br_2] - k_4\ [H]\ [HBr] \qquad ...(18)$$

As. Eq. (18) involves concentrations of hydrogen and bromine atoms which are too small quantities to be measured directly, it is required that their concentration must be expressed in measurable quantities. This is done by writing down and solving steady-state equations for [H] and [Br].

The Br atoms are formed by (i), (ii) and (iv) and are removed by (ii) and (v), so the net rate of formation is given by

$$\frac{d[Br]}{dt} = k_1\ I_{abs} + k_3\ [H]\ [Br_2] + k_4\ [H]\ [HBr] - k_2\ [H_2]\ [Br] - k_5\ [Br]^2 \qquad ...(19)$$

As [H] atoms are formed by reaction (ii) and removed by (iii) and (iv), (v), so the net rate of formation is given by

$$\frac{d[H]}{dt} = k_2\ [H_2]\ [Br] - k_3\ [H]\ [Br_2] - k_4\ [H]\ [HBr] \qquad ...(20)$$

When the stationary state is reached, it follows that

$$\frac{d[H]}{dt} = 0 \text{ and } \frac{d[Br]}{dt} = 0$$

Therefore, Eqs. (19) and (20) become as

$$k_1\ I_{abs} + k_3\ [H]\ [Br_2] + k_4\ [H]\ [HBr] - k_2\ [H_2]\ [Br] - k_5\ [Br]^2 = 0 \qquad ...(21)$$

$$k_2\ [H_2]\ [Br] - k_3\ [H]\ [Br_2] - k_4\ [H]\ [HBr] = 0 \qquad ...(22)$$

By adding Eqs. (21) and (22), we get

$$k_1 I_{abs} - k_5\ [r]^2 = 0$$

or

$$[Br] = \sqrt{\left(\frac{k_1}{k_5} I_{abs}\right)} \qquad ...(23)$$

Putting the value of Br in Eq. (23). we get

$$k_2\,[H_2]\,\sqrt{\left(\frac{k_1\,I_{abs}}{k_5}\right)} - k_3\,[H]\,[Br_2] - k_4\,[H]\,[HBr] = 0$$

or $$k_3\,[H]\,[Br_2] + k_4\,[H]\,[HBr] = k_2\,[H_2]\,\sqrt{\left(\frac{k_1\,I_{abs}}{k_5}\right)}$$

or $$[H]\,(k_3\,[Br_2] + k_4\,[HBr]) = k_2\,[H_2]) = k_2\,[H_2]\,\sqrt{\left(\frac{k_1\,I_{abs}}{k_5}\right)}$$

$$[H] = \frac{k_2\,[H_2]\sqrt{\left(\frac{k_1\,I_{abs}}{k_5}\right)}}{k_3\,[Br_2] + k_4\,[HBr]} \quad ...(24)$$

On substituting the values of [Br] and [H] from Eqs. (23) and (24) in Eq. (22), we obtain

$$\frac{d\,[HBr]}{dt} = k_2\,[H_2]\sqrt{\left(\frac{k_1\,I_{abs}}{k_5}\right)} + \frac{k_3\,[Br_2]\,k_2\,[H_2]\sqrt{\left(\frac{k_1\,I_{abs}}{k_5}\right)}}{k_3\,[Br_2] + k_4\,[HBr]}$$

$$- \frac{k_4\,[HBr]\,k_2\,[H_2]\sqrt{\left(\frac{k_1\,I_{abs}}{k_5}\right)}}{k_3\,[Br_2] + k_4\,[HBr]}$$

$$\frac{d\,[HBr]}{dt} = k_2\,[H_2]\sqrt{\left(\frac{k_1\,I_{abs}}{k_5}\right)}$$

$$\left(1 + \frac{k_3\,[Br_2]}{k_3\,[Br_2] + k_4\,[HBr]} - \frac{k_4\,[HBr]}{k_3\,[Br_2] + k_4\,[HBr]}\right)$$

or $$\frac{d\,[HBr]}{dt} = \frac{k_2\,[H_2]\sqrt{\left(\frac{k_1\,I_{abs}}{k_5}\right)}}{k_3\,[Br_2] + k_4\,[HBr]}$$

$$(k_3\,[Br_2] + k_4\,[HBr] + k_3\,[Br_2] - k_4\,[HBr])$$

$$= \frac{2k_2\,[H_2]\,k_3\,[Br_2]\sqrt{\left(\frac{k_1\,I_{abs}}{k_5}\right)}}{k_3\,[Br_2] + k_4\,[HBr]} \quad ...(25)$$

Dividing the numerator and denominator by k_3 $[Br_2]$, we get

$$\frac{d[HBr]}{dt} = \frac{2k_2[H_2]\sqrt{\left(\frac{k_1 I_{abs}}{k_5}\right)}}{1+\frac{k_4[HBr]}{k_3[Br_2]}}$$

$$= \frac{k'[H_2]\sqrt{\left(\frac{k_1 I_{abs}}{k_5}\right)}}{1+\frac{HBr}{m'[Br_2]}} \quad ...(26)$$

where $k' = 2k_2$ and $m' = k_3/k_4$ are two constants. Equating (26) in similar to obtained by Bodenstein on the basis of this experimental data. Inspite of the chain mechanism similar to hydrogen and chlorine reaction, the quantum yield of this photochemical reaction is very low, being about 0.01 at ordinary temperatures. This is due to following reasons.

1. Reaction (ii) is highly endothermic and therefore, takes place so slowly at ordinary temperatures that most of the bromine atoms recombine to form bromine molecules. Therefore, the reactions (iii), (iv) and (v) which are due to reaction (i) cannot occur. Hence, the quantum yield is extremely slow.
2. With increasing time, the reversal reaction becomes predominant and hence, the rate of formation of HBr decreases.

(3) Photolysis of Acetaldehyde : The photolysis of acetaldehyde has attracted much attention, in recent years. While discussing this reaction, the following facts must be kept in mind.

(a) The quantum yield of carbon monoxide formation increases with decreasing wave-length.

(b) When the temperature is raised, the quantum yield increases.

(c) The quantum yield of carbon monoxide formation at room temperature increases with decreasing pressure at 3130 A°.

(d) The photolysis of acetaldehyde in the light of 2500=3100 A° wavelength yields methane, carbon monoxide, ethane and a number of other products.

$$CH_3CHO + h\nu \rightarrow CH_4 + C_2H_6 + CO + \text{other products}$$

To account for the above facts, the following mechanism has

been postulated.

(a) $CH_3CHO + hv \longrightarrow CH_3 + CHO$ Rate = Iabs

(b) $CH_3CHO + CH_3 \xrightarrow{k_2} CH_4 + CH_3CO$

Rate = k_2 $[CH_3CHO]$ $[CH_3]$

(c) $CH_3CO \xrightarrow{k_3} CO + CH_3$ Rate = k_3 [CH3Co]

(d) $CH_3CH_3 \xrightarrow{k_4} C_2H_6$ Rate = k_4 $[CH_3]^2$

In the above mechanism, carbon monoxide is formed only in step (c), hence the rate of formation of this substance must be

$$\frac{d[CO]}{dt} = k_3\,[CH_3CO] \qquad ...(27)$$

The CH_3CO radical is formed by step (b) and is used up in the step (c). Therefore, the net rate of its formation is given by

$$\frac{d[CH_3CO]}{dt} = k_2\,[CH_3CHO]\,[CH_3] - k_3\,[CH_3CO] \qquad ...(28)$$

Again, $[CH_3]$ radical is formed by steps (a) and (c) and is consumed in steps (b) and (d). Hence its net rate of formation is given by

$$\frac{d[CH_3]}{dt} = I_{abs} + k_3\,[CH_3CO] - k_2\,[CH_3CHO]\,[CH_3] - k_4\,[CH_3]^3 \qquad ...(29)$$

Applying the steady-state treatment to $[CH_3CO]$ radical, we obtain

$$\frac{d[CH_3CO]}{dt} = k_2\,[CH_3CHO]\,[CH_3] - k_3\,[CH_3CO] = 0$$

or $k_2\,[CH_3CHO]\,[CH_3] - k_3\,[CH_3CO] = 0$ ...(30)

Similarly, the steady-state treatment on applying to $[CH_3]$ radical yields

$$\frac{d[CH_3]}{dt} = I_{abs} - k_2\,[CH_3CHO]\,[CH_3] + k_3\,[CH_3CO] - k_4\,[CH_3]^2 = 0$$

$$I_{abs} - k_2\,[CH_3CHO]\,[CH_3] + k_3\,[CH_3CO] - k_4\,[CH_3]^2 = 0 \qquad ...(31)$$

By adding equations (30) and (31), we obtain

$$I_{abs} - k_4\,[CH_3]^2 = 0$$

or $$k_4\,[CH_3]^2 = I_{abs}$$

or $$[CH_3] = \left[\frac{I_{abs}}{k_4}\right]^{1/2}$$

On substituting the value of $[CH_3]$ in equation (31), we get

$$k_2\,[CH_3CHO]\left(\frac{I_{abs}}{k_4}\right)^{1/2} - k_3\,[CH_3CO] = 0$$

or $$k_3\,[CH_3CO] = k_2\,[CH_3CHO]\left(\frac{I_{abs}}{k_4}\right)^{1/2} \qquad ...(32)$$

Substituting this value of k_3 $[CH_3CO]$ in equation (32), we obtain

$$\frac{d[CO]}{dt} = k_2\,[CH_3CHO]\left(\frac{I_{abs}}{k_4}\right)^{1/2}$$

$$= \frac{k_2}{(k_4)^{1/2}}\,[CH_3CHO]\,I_{abs})^{1/2}$$

$$= I_{abs}{}^{1/2}\,[CH_3CHO] \qquad ...(33)$$

where $$k = \frac{k_2}{(k_4)^{1/2}}$$

Quantum Yield : If the rate of formation of carbon monoxide is divided by the intensity of absorbed radiation, the value of quantum yield is obtained. Thus,

$$\text{Quantum yield} = \frac{d[CO]/dt}{I_{abs}}$$

$$= \frac{k\,I_{abs}{}^{1/2}\,[CH_3CHO]}{I_{abs}} \qquad \text{[From eq. (33)]}$$

$$= \frac{k\,[CH_3CHO]}{I_{abs}} \qquad ...(34)$$

General Discussion

1. From equation (34), it follows that the quantum yield of carbon monoxide formation increases as the intensity of light is decreased.
2. The activation energy of chain carrying process (b) has an appreciable energy, *e.g.*, about 10 K cal.

(b) $CH_3CHO + CH_3 \rightarrow CH_4 + CH_3CO$

This high activation energy accounts for the fact that quantum yield decreases as the temperature is lowered.

3. The effect of pressure on quantum yield may be explained by assuming that primary process

(a) $CH_3CHO + h\nu \rightarrow CH_3 + CHO$

is the main one at 3130 Å. At this wavelength, the CHO radical does not possess enough energy to decompose spontaneously. At high pressure, the CHO radicals may combine to form glyoxal. At low pressure, the CHO radicals may diffuse to the walls to form carbon monoxide.

LASER TECHNIQUES

When a large percentage of an ensemble of molecules can be brought into an excited state, a way frequently can be found by which large numbers of these excited molecules can be triggered into an almost simultaneous joint transition back to the inactive state, with the resulting emission of a beam of intense coherent radiation. Such a device is now known as laser, an acronym for light amplification by stimulated emission of radiation.

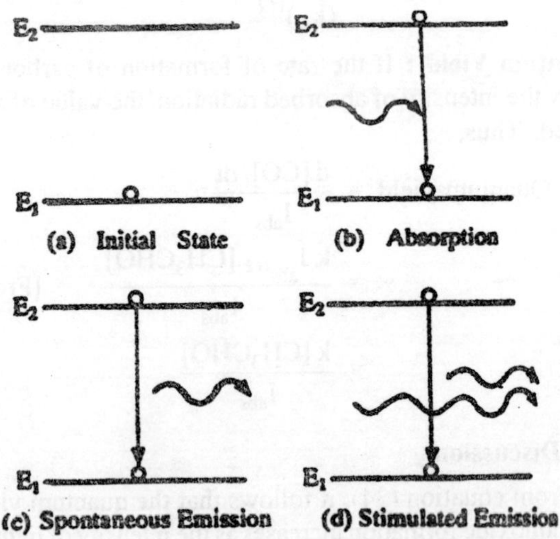

Fig. 2.6 : Four stage os laser action.

The four steps associated with such a device are illustrated in Fig. 2.6. In (a) the molecule is inactive, in (b) the molecule gets activated by a photon, in (c) the molecule returns spontaneously to its initial state, emitting radiation, and in (d) the transition to the ground state is triggered by an incoming beam, which is then amplified by the emitted radiation. For effective laser operation the amount of emission form (c) must be small compared with (d).

In a typical solid laser like a ruby crystal, the active atoms are those of chromium, which are held in a lattice of aluminium oxide. A flash of intense light raises a large percentage of the chromium atoms to the excited state.

Then radiation can trigger the return transition, resulting in the emission of light which is coherent both in space and in time because the stimulating photon on entering the atom causes it to emit a photon which is precisely "in step" with it.

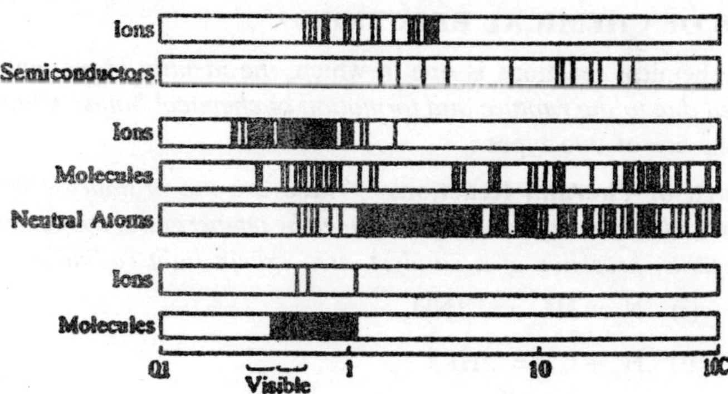

Fig. 2.7 : Laser and maser radiation sources available.

In this way, monochromatic beams os great intensity can be produced which travel great distances without spreading. There are many important applications possible for such beams both in science and technology. They are valuable in studying Raman scattering and are certainly going to be an asset in communication. As the laser principle applies over the entire range of the spectrum, it is used extensively in the generation of microwaves. The device is then called a *maser,* microwave amplification by simulated emission of radiation. In Fig. 2.7, some of the wavelengths of various laser and maser sources are depicted.

PHOTOCHEMICAL REACTIONS

Photochemistry is the branch of chemistry which is mainly concerned with rates and mechanisms of reactions resulting from the exposure of reactants to light radiations.

This board definition of photochemistry would include reactions produced by all radiations of wave lengths ranging from those of radio waves to chose γ rays. But for practical purposes, the light radiations of the visible and ultraviolet regions lying between 2000 to 8000 Å are mainly concerned in bringing about such reactions which are termed as *Photochemical Reactions*. Therefore, the definition of photochemistry may be summarised as follows.

"It is the study of chemical effects produced by light radiations ranging from 2000 to 8000 Å wave-length"

The study of photochemistry helps us to know the changes when a molecule absorbs radiations.

TYPES OF CHEMICAL REACTIONS

A chemical reactions is one in which, *the identity of molecules is changed due to the rupture and formation of chemical bonds.* Chemical reactions are of two types.

Dark of Thermal Reactions : *These are the ordinary chemical reactions which are influenced or induced by temperature, concentration of reactants, presence of a catalyst, etc., except light radiations.*

(i) $N_2 + 3H_2 \rightleftharpoons 2NH_3$

(ii) $H_2 + I_2 \rightleftharpoons 2HI$

(iii) $PCl_5 \rightleftharpoons PCl_3 + Cl_2$

(b) Photochemical Reactions : *A photochemical reaction may be defined as any reaction which is induced or influenced by the action of light on the system.*

Some of the photochemical reactions are accompanied by increase in free energy unlike dark reactions. Examples of some photochemical reactions are.

(a) *Dissociation.*

$$2HBr \rightarrow H_2 + Br_2$$

(b) *Rearrangement.* Fumaric acid → Maleic acid

(c) *Addition Reaction.*

$$Br_2 + (C_6H_5)_2C = C(C_6H_5) \rightarrow (C_6H_5)_2\ CBr - CBr(C_6H_5)_2$$

(d) *Polymerisation.*

$$nCH_2 \rightarrow (C_2H_5)Nx$$

(e) Photo-catalytic *Reaction*

$$CO_2 + H_2O + \text{chlorophyll} \xrightarrow{\text{light}} \frac{1}{n}(H_2CO)_n + O_2 + \text{Chlorophyll}$$

(f) *Combination.*

$$H_2 + Cl_2 \rightarrow 2HCl$$

(g) *Decomposition.*

$$2O_3 \rightleftharpoons 3O_2$$

(h) *Double decomposition.*

$$C_6H_{12} + Br_2 \rightarrow C_6H_{11}Br + HBr$$

DIFFERENCES BETWEEN DARK AND PHOTOCHEMICAL REACTIONS

Photochemical reactions differ from ordinary dark or thermal reactions in certain aspects which are as follows.

(i) In ordinary thermal reactions the energy of activation is provided by the collisions. On the other hand in photochemical reactions the energy required is gained through the absorption of quanta of visible or ultraviolet light.

(ii) All ordinary chemical reactions are always accompanied by a decrease in free energy. On the other hand, the free energy of the light or photo-chemical reactions increases as some of the light energy is convered into free chemical energy of the products. Some examples of such photochemical reactions are ozonisation of oxygen, polymerisation of anthracene, and photosynthesis occurring in plants.

In these photochemical reactions, the energy of absorbed radiation gets transformed into free chemical energy of the products. However, when the source of radiations is removed, the system tends to return to its original state, though at a very slow rate.

A feature of photochemical activation is its selectivity. The absorbed quanta of light excite and thus activate a separate atom or group of atoms in a given molecule. This is a great advantage of activating molecules with light in comparison with thermal activation.

(iii) The difference between photo excitation and thermal excitation is that while in the former a few molecules succeeding in absorbing photons are strongly excited, in the case of latter a significant rise in temperature increases the average energy of all the molecules by a small amount. Thus, light absorption can bring about reactions at room temperature which might otherwise require a raising of temperature by several tens or hundreds of degrees centigrade.

While the rate of thermal reactions depends upon the temperature, the photochemical reaction rate is independent of temperature. But the reaction rate does depend upon the intensity of the radiation used. In certain photochemical reactions, the rate is seen to vary with the temperature, but this is due to the temperature dependence of thermal reactions which follow the light absorption step.

PHOTOSENSITISATION

Certain reactions are known which are not sensitive to light. These reactions can be made sensitive by adding a small amount of foreign material which can absorb light and stimulate the reaction without itself taking part in the reaction. Such an added material is known as photo-senistiser and the phenomenon as photosensitisation.

Formerly, it was thought that photosensitisation might be akin to catalysis. It has now been confirmed that photosensitisation is different in nature from ordinary catalysis. Photosensitized reactions are spontaneous involving an increase in free energy of the system.

Role Played by a Photosensitiser : The formation of a photosensitiser is to absorb light, become excited and then pass on this energy to one of the reactants and thereby activate them for reaction, without itself taking part in the reaction. Thus, a photosensitiser acts as a carrier of energy. Some examples are.

REACTIONS SENSITISED BY MERCURY ATOMS

(i) When a mixture of mercury vapour and hydrogen gas is illuminated by light of wavelength 2537 Å, the dissociation of molecular hydrogen into atomic hydrogen takes place.

$$Hg + H_2 \rightarrow 2H + Hg$$

The *mechanism* which is now suggested is the excitation of mercury atom and then to transfer the energy to hydrogen molecule which will dissociated to form atoms.

$$Hg + hv \rightarrow Hg^*$$

$$Hg^* + H_2^* \rightarrow H_2^* + Hg$$

$$H_2^* \rightarrow 2H$$

(ii) When the reaction is carried out between hydrogen and oxygen in the presence of mercury vapour under the influence of light radiations, the reaction leads to the formation of H_2O_2 and H_2O.

$$3H_2 + 2O_2 \xrightarrow[\text{Light}]{\text{Hg}} 2H_2O + H_2O_2$$

The accepted *mechanism* of the above reaction is that the first stage is the formation of hydrogen atoms by collision between excited mercury atom and hydrogen molecule.

$$Hg + hv \rightarrow Hg^*$$

$$Hg^* + H_2 \rightarrow 2H + Hg$$

The above reaction may be followed by reactions such as

(1) $H + O_2 + X \rightarrow HO_2 + X$

(2) $HO_2 + HO_2 \rightarrow H_2O + O_2$

(3) $HO_2 + H_2 \rightarrow H_2O_2 + H$

Now H_2O_2 may either be isolated as such or further decomposed to form H_2O and O_2

$$H_2O_2 \rightarrow H_2O + \frac{1}{2}O_2$$

(iii) The reaction between hydrogen and carbon monoxide is photosensitised by mercury atoms. The main products are formaldehyde and glyoxal in similar amounts. The quantum yield for this reaction in nearly 2. The accepted *mechanism* is that the primary process is the formation of hydrogen atom which further carries out chain reactions as given below.

$H + CO + X \rightarrow HCO + X$ (where X is the foreign body)

Either $2HCO \rightarrow HCHO + CO$ or $2HCO \rightarrow (CHO)_2$

(iv) Mercury photosensitises the decomposition of ammonia in the presence of light of wavelength 2537 Å. Quantum efficiency of this reaction is 7 and the rate of this photosensitised reaction is, about 200 times as great as the photolysis of ammonia (in the absence of mercury vapour), if both the reactions are carried out at the same wavelength.

The possible mechanism is

$$NH_3 + Hg^* \rightarrow NH_3^* + Hg$$
$$NH_3^* \rightarrow NH_2 + H$$
$$H + H \rightarrow H_2$$
$$NH_2 + NH_2 \rightarrow N_2H_4$$
$$N_2H_4 + H \rightarrow NH_3 + NH_2$$
$$NH_2 + NH_3 \rightarrow N_2 + 2H_2$$

Other example which are photosensitised by mercury atoms are. (a) Decomposition of phosphine, (b) Ozonolysis of oxygen, (c) Decomposition of acetone, (d) Decomposition of water vapour, (e) Decomposition of ethyl alcohol.

(b) Chlorine as A photosensitiser : In many reactions, chlorine acts as a photosensitiser. Let us consider the decomposition of ozone in the presence of chlorine (as a photosensitiser) under the influence of ultra-violet light.

$$2O_3 \xrightarrow[hv]{Cl_2} 2O_2 .$$

The rate of above reaction is independent of concentration, but is proportional to the intensity of absorbed light. Many mechanisms have been proposed from time to time. The most satisfactory mechanism is given below.

1. $Cl_2 + hv \rightarrow 2Cl$
2. $Cl + O_3 + X \rightarrow ClO_3^* + X.$

   The excited ClO3* radical may be absorbed on the walls of the containing vessel to form a mixture of Cl_2O_6, Cl_2 and O_2.
3. $ClO_3^* + ClO_3^* + M \rightarrow Cl_2 + 3O_2 + M.$

   In addition to the above reactions, $ClO_3^* + M \rightarrow Cl_2 + 3O_2 + M$

   In addition to the above reactions, ClO_3^* in the gas phase may react to form ClO_3 and O_2.
4. $ClO_3^* + O_2 \rightarrow ClO_2 + 2O_2$
5. $ClO_2 + O_3 \rightarrow ClO_3 + O_2$

(b) *Bromine as a photosensitiser.* Bromine acts as photosensitiser in the conversion of maleic acid into fumaric acid.

```
H—C—COOH                 H—C—COOH
  ||        hv'            ||
            ——→
  ||        Br2            ||
H—C—COOH              HOOC—C—H
Maleic acid             Fumaric acid
```

(c) *Cadmium vapour as a photosensitiser.* Cadmium vapour acts as a photosensitiser for the polymerisation of ethylene, for the decomposition of ethane and propane to hydrogen, methane and higher hydrocarbons.

$$nC_2H_4 \xrightarrow[hv]{Cd} (C_2H_4)_n$$

Photosensitisation in Solid Phases : When silver halides in photographic plates are exposed to red (7500 Å) and yellow (5900 Å) light, they are not appreciably affected. But addition of certain photosensitisers like red dye to the emulsion makes the photographic plate to respond not only to red but even to far infrared radiations.

$$AgBr + hv \xrightarrow{\text{Red dye}} Ag + Br$$

Photosensitisation in Solution : When the decomposition of oxalic acid is carried out by light of shorter wavelength in presence of uranyl ions, the quantum yield for this reactions is 0.5 or more. The mechanism of this reaction is not clear.

$$UO_2^{2+}\ hv \rightarrow [UO_2{}^{+2}]$$

$$[UO^{2+}]^{2*} + \begin{array}{c} COOH \\ | \\ COOH \end{array} \longrightarrow CO_2 + CO + H_2O + UO_2^{+2}$$

Uranyl ions also act as a photosenitizer in the photolysis of formic acid.

Chlorophyll as a photosensitiser : Chlorophyll acts as a photosenitiser in the photosynthesis of carbohydrates from CO_2 and H_2O.

$$\text{Chlorophyll} + hv \rightarrow [\text{Chlorophyll}]^*$$

$$6CO_2 + 6H_2O + [\text{Chlorophyll}]^* \rightarrow C_6H_{12}O_6 + 6O_2 + [\text{chlorophyll}]$$

sunlight

ABSORPTION OF LIGHT

Introduction : *When light* (monochromatic or heterogeneous) *is incident upon a homogeneous medium, a part of the incident light is reflected, a part is absorbed by the medium and the remainder is allowed to transmit as such.* If I_0 denotes the incident light, I, the reflected light, I_r the absorbed light and I_a, the transmitted light, then one can write

$$I_0 = I_a + I_t + I_r \qquad ...(1)$$

It a comparison cell is used, the value of I_r which is very small (about 4 per cent), can be *eliminated for air-glass interfaces.* Under the condition, equation (1) becomes as

$$I_0 = I_a + I_t \quad ...(2)$$

Bouguer actually investigated the range of absorption of light with the thickness of medium. But the credit was enjoyed by *Lambert* who simply extended the concepts developed by *Bouguer*. Beer later applied Lambert's concept to solution of different concentrations and reported his results just prior to those of *Bernard.* However, the two separate laws governing, absorption are generally known as *Lambert's* Law and *Beer's Law*. We will now discuss these one by one.

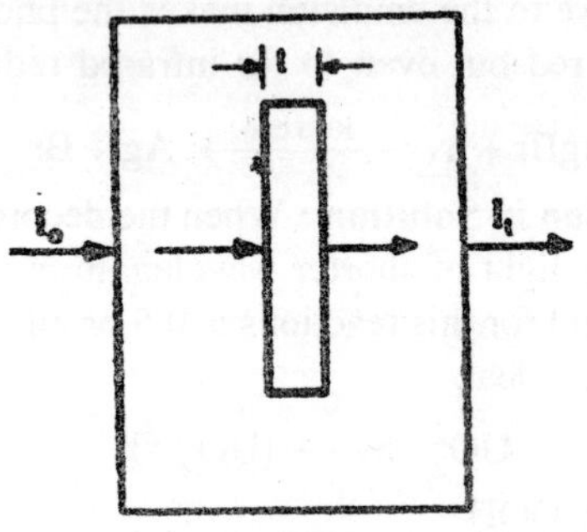

Fig. 2.8

Lambert's Law : This law can be stated as follows. "*When a beam of light is allowed to pass through a transparent medium, the rate of decrease of intensity with the thickness of medium is directly proportional to the intensity of the light.*"

Mathematically the Lambert's law may be stated as follows.

$$-\frac{dI}{dt} \propto I \text{ or } -\frac{dI}{dt} = kI \quad ...(3)$$

where I denotes the intensity of incident light of wavelength λ, *t* denotes the thickness of the medium and *k* denotes the proportionality factor. On integrating equation (3) and putting $I = I_0$, when $t = 0$, we get

$$\ln \frac{I_0}{I_t} = kt \text{ or } I_t = I_0 e^{-kt} \quad ...(4)$$

where I_0 denotes the intensity of the incident light, I_t denotes the intensity of the transmitted light and k is a constant which depends upon the wavelength and absorbing medium used. On changing equation (4) from natural to common logarithms, we get

$$\ln \frac{I_0}{I_t} = kt \text{ or } I_t = I_0 e^{-Kt} \quad ...(5)$$

where $\qquad K = k/2.3026 \qquad$ ...(6)

In equation (6) K is the *absorption coefficient* which is defined as. *"It is the reciprocal of the thickness which is required to reduce the light to 1/10 of its intensity."*

The above definition follows from equation (5),

$$It/I_0 = 0.1 = 10^{-Kt}$$

or $\qquad Kt = 1 \quad \text{or} \quad K = K \propto \frac{1}{t}$

The ratio I_t/I_0 is termed as the *transmittance,* T and the ratio log I_0/I_t is termed as the *absorbance* A, of the medium. Formerly absorbance was termed as *optical density* D or *extinction coefficient* ≤. The ratio I_0/I_t, is termed as *opacity,*

$$A = \log \frac{I_0}{I_t} \qquad ...(7)$$

Lambert's law is very rigid and like Faraday's laws of electrolysis, it has no exception.

Beer's Law : Lambert's law shows that there exists a logarithmic relationship between the transmittance and the length of the optical path through the sample. Beer observed that a similar relationship holds between transmittance and the concentration of a solution, *i.e., the intensity of a beam of monochromatic light decreases exponentially with the increase in concentration of the absorbing substance arithmetically.*

Thus, equation (4) becomes as

$$I_0 = I_0\, e^{-k'0} \qquad ...(8)$$

$$= I_0 \cdot 10^{-0.4348k'c} = I_0\, 10^{-K'2} \qquad ...(9)$$

where k′ and K′ are constants and *c* is the concentration of the absorbing substance. On combining equations (5) and (9), we get

$$I_t = I_0 . 10^{-act}$$

or $\qquad \log (I_0/I_t) = act \qquad$ ..(10)

where a is the new constant.

Equation (10) is termed as mathematical statement of *Beer-Lambert law*. This is also the fundamental equation of colorimetry and spectrophotometry..

In equation (10), the value of a depends upon the units of concentration. If c is expressed in mole dm^{-3} and *t* in centimeters, then *a* is replaced by the symbol ε and is termed as the *molar absorption coefficient* or *molar absorptivity* (formerly the molar extinction coefficient). It is important to remark here that there exists a relationship between the absorbance A, the transmittance T and the molar absorption coefficient ε, *i.e.*,

$$A = \varepsilon\ ct = \log\frac{I_0}{I_t} = \log\frac{1}{T} = -\log T \qquad ...(11)$$

In spectrophotometers, the scales are calibrated to read directly absorbances. In colorimeters, I_0 is considered to be the light transmitted by the pure solvent whereas I_t is considered to be the light transmitted by the solution.

Equation (11) may be put as

$$\varepsilon = A/ct \qquad ...(12)$$

If c = 1 mole dm^{-3} and T = 1 cm, equation (12) becomes as

$$\varepsilon = A \qquad ...(13)$$

From equation (13) it follows that the molar absorption coefficient is the specific absorption coefficient for a concentration of 1 mole dm^{-3} and a path length of 1 cm. When a system contains several absorbing substances, each of these contributes to the rate of absorption of light.

$$A = \varepsilon_1 c_1 t + \varepsilon_2 c_2 t + ... = \sum_i \varepsilon_t\ c_i\ t$$

Nature of Molar Absorptivity and Absorbance : Absorbance is an *extensive property* of a substance whereas absorptivity is its *intensive property.*

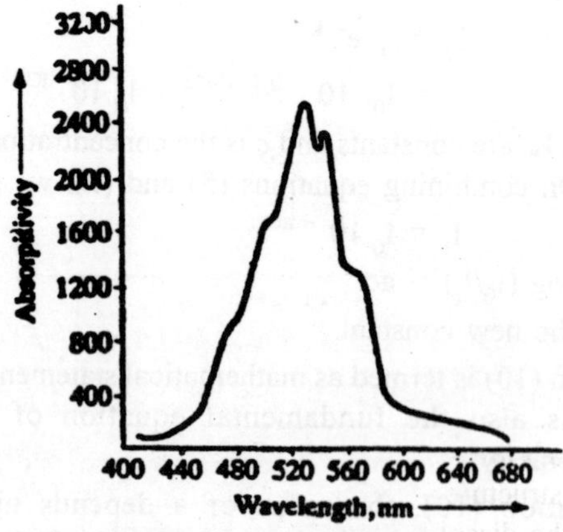

Fig. 2.9 : Molar absorption at different wave lengths.

If there is a change in concentration and in the thickness of the container, the value of molar absorptivity will remain constant within the Beer's law range but the absorbance will change significantly.

The value of molar absorptivity will be different at different wavelengths.

This is shown in Fig. 4.28. From this figure, it is alto evident why a monochromatic radiation is used in spectrophotometry and colorimetry. Similarly absorbance also varies with wavelength.

VALIDITY OF BEER'S LAW

When Beer's law is obeyed by a solution, it means that.

(a) There are no interactions between molecules of the solution.

(b) No new types of molecules will be formed as the concentration is changed.

Deviation from Beer's Law : From Beer's law it follows that if we plot absorbance A against concentration, a straight line passing through the origin should be obtained Fig. 2.10. But there is usually a deviation from a linear relationship between concentration and absorbance and an apparent failure of Beer's law may ensue. Deviations from the law are reported as *positive* or *negative* according to whether the resultant curve is concave upwards or concave downwards.

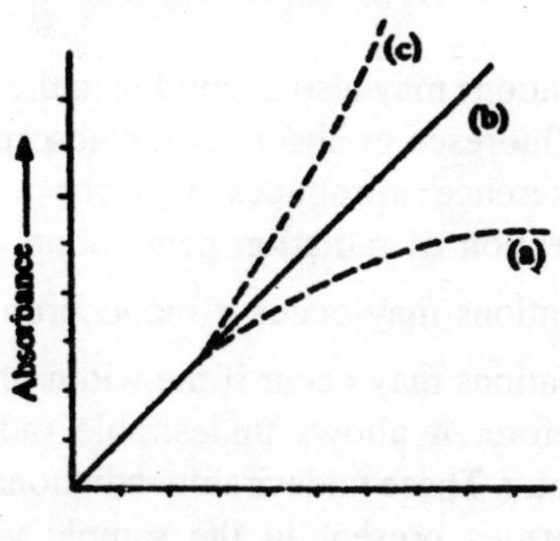

Fig. 2.10

Deviation from Beer's law can arise due to following factors.

(i) Beer's law will hold over a wide range of concentration provided the structure of the coloured ion or of the coloured non-electrolyte in the dissolved state does not change with concentration. If a coloured solution is having a foreign substance whose ions do not react chemically with the coloured components, its small concentration (foreign substance) does not affect the light absorption whereas its large concentration may affect light absorption and may also alter the value of the extinction coefficient.

(ii) Deviation may also occur if the coloured solute ionises, dissociates or associates in solution.

For example, benzyl alcohol, in chloroform exists in a polymeric equilibrium.

$$4C_6H_5CH_2OH \rightleftharpoons (C_6H_5CH_2OH)_4$$

Dissociation of the polymer increases with dilution. The monomer absorbs at 2.0 to 2.765 μ whereas the polymer absorbs at 3.000 μ. Hence, absorption at 2.050 μ shows negative deviation whereas at 3.000 μ *positive deviation.*

Another example of deviation is the change of colour of dichromate ion on dilution. The effect can be represented by the equilibrium shown below.

$$\underset{\text{(Orange)}}{Cr_2O_7^{2-}} + H_2O \rightleftharpoons 2HCrO_4^- \rightleftharpoons 2H^+ + \underset{\text{(Yellow)}}{CrO_4^{2-}}$$

It can be seen from above that a solution of dichromate is converted gradually into chromate. The absorption at 450 μ, is positive and at 350 μ is negative.

(iii) Deviations may also occur due to the presence of impurities that fluoresce or absorb at the absorption wavelength. This interference introduces an error in the measurement of absorption of radiation penetrating the sample.

(iv) Deviations may occur if monochromatic light is not used.

(v) Deviations may occur if the width of slit is not proper and, therefore, it allows undesirable radiations to fall on the detector. These undesirable radiations might be absorbed by impurities present in the sample which would cause an apparent change in the absorbance of the sample. The magnitude of this deviation becomes appreciable at higher concentrations.

(vi) Deviation may occur if the solution species undergoes polymerisation. For example, benzyl alcohol in carbon tetrachloride in high concentration exists in polymeric form

$[(C_6H_5CH_2OH)_4]$, association of this polymer increases with dilution. Due to this change, absorbance will change.

(vii) Beer's law cannot be applied to suspensions but the latter can be estimated colorimetrically after preparing a reference curve of known concentrations.

Illustration : From the Beer's law it follows that if the absorption remains constant, then the concentration (c) should be related to the thickness of the absorbing layer (t) in a reciprocal manner. This fact was demonstrated by *Halbon* (1922). He observed that a layer of chlorine 20 cm long at a pressure of 0.1 atmosphere would have the same transmission at 10 cm layer at a pressure of 0.2 atmosphere

Applications of Beer's Law : Beer-Lambert's law can be used for the determination of an unknown concentration by comparison with a solution of known concentration by using colorimeter or spectrophotometer, the principle of which is based upon Beer's law.

LAWS OF PHOTOCHEMISTRY

There are three laws which govern the effect of radiation on chemical reactions. These laws are.

I. Grotthuss–Draper Law : This is sometimes referred to as the first law of photochemistry. On the basis of certain theoretical considerations, Grotthuss in 1818 discovered this law. This law was reaffirmed in 1841 by J.W. Draper as a result of his experimental researches on photochemical reaction between hydrogen and chlorine. It may be stated as follows.

"When light falls on any substance, only the fraction of incident light which is absorbed by the substance can bring about a chemical change', reflected and transmitted light do not produce any such effect."

It is important to remark that all light radiations which are absorbed by reacting system are not effective in producing desired chemical reactions. When conditions are not favourable for the molecules to react, some portion or the whole light absorbed by reacting substances is converted into heat in some cases. While in some other cases, the absorbed light is re-emitted as radiations of the same or another frequency.

Grotthuss-Draper law is purely qualitative. It does not give any relationship between the amount of light absorbed by a system and the number of molecules which have reacted,.

II. Law of Photochemical Equivalence : It is one of the most important laws in photochemistry. According to quantum theory of light, energy is absorbed or emitted only in small packets of quanta of magnitude $\varepsilon = hv$ where ε is the energy, v is the frequency W light and h is the Planck's constant. In 1905, Einstein applied quantum theory to photochemical reactions and enunciated the law of Photochemical Equivalence which states that

"When an atom or molecule absorbs light of a given frequency, it absorbs one quantum only,"

The photochemical importance of the above law emphasized in 1909 by Stark and other. In 1913, Einstein considered the work of Stark and restated his law as.

"Each molecule which takes part in a chemical reaction absorbs one quantum of light which induces the reaction " or briefly "one molecule one quantum."

The term one photon (or one quantum) means energy equal to hv, where h is the Planck's constant and v is the frequency of radiation

$$AB + hv \rightarrow AB^*$$

Thus, in the primary process, the number of molecules that are activated is equal to the number of quantas absorbed. Therefore, amount of energy E, for activation of 1 mole will be N hv where N is the Avogadro's number and is equal to 1 mole, *i.e.,*

$$E = Nhv$$

This quantity of energy 'E" *absorbed per mole of the substance is called an Einstein.*

But $\quad N = 6.023 \times 10^{28}$ molecules

$$h = 6.624 \times 10^{-27} \text{ erg sec}$$

$$\therefore \quad E = 6.023 \times 10^{28} \times 6.624 \times 10^{-27} \times v \text{ ergs}$$

$$= \underline{6.1023 \times 6.624 \times 10^{-27}} \; \underline{\times v} \text{ calories} \quad ...(1)$$

$$E = 9.53 \times 11-11 \times v \text{ calories}$$

$$[\because 1 \text{ cal} = 4.186 \times 10^{7} \text{ ergs}]$$

But $\quad v = \dfrac{c}{\lambda} = \dfrac{\text{velocity of light in cm/sec}}{\text{wave length in cm}} = \dfrac{3 \times 10^{10}}{\lambda}$

Substituting the value of v in equation (1), we obtain

$$E = \frac{9.53 \times 10^{-11} \times 3 \times 10^{10}}{\lambda} \text{ calories}$$

where λ is expressed in cm. But 1 A° = 10^{-8} cm. Therefore,

$$E = \frac{9.53 \times 10^{-11} \times 3 \times 10^{10}}{\lambda \times 10^{-8}} = \frac{2.859}{\lambda} \times 10^{8} \text{ cal}$$

$$= \frac{2.859}{\lambda} \times \frac{10^{8}}{10^{3}} \text{ K cal} \quad [\because 1 \text{ K cal} = 1000 \text{ cal}]$$

$$= \frac{2.859}{\lambda} \times 10^{5} \text{ K cal}$$

Thus, the value of one Einstein of radiation of given frequency or wavelength is given by the relation (2). It is evident from equation (2) that the energy absorbed per mole decreases with increasing wavelength. In other words, the value of E is inversely proportional to the wavelength of light absorbed, *i.e.*,

$$E \propto \frac{1}{\lambda}$$

From the above facts it is evident that shorter the wavelength of light, the greater the energy. It has been verified that.

1. For the wavelength 4000Å (ultraviolet light), the energy is 71 K cal;
2. For the wavelength 7500Å (red light), the corresponding energy is 38 K, cal.

From the above results, it follows that in ultraviolet and violet portions, the radiations will be more active chemically than those of longer wavelengths.

QUANTUM YIELD OR QUANTUM EFFICIENCY

The photochemical equivalence law applies to the primary photochemical process, *i.e.,* as a result of primary absorption of one quantum, only one molecule undergoes dissociation and the products enter no further reaction. In such cases, there will be 1 : 1 relationship between the number of quantas absorbed and the number of reacting molecules.

In most cases, a molecule activated photochemically initiates a series of thermal reactions called secondary reactions. As a result of such reactions many reactant molecules may undergo chemical change by absorbing one quantum only. Under such conditions, there will be no. 1 : 1 relationship between the number of quanta absorbed and the number of reacting molecules.

In some cases, a molecule activated photochemically undergoes deactivation. Thus, less than one molecule may react per quantum

To describe the deviation from 1 : 1 relationship, the idea of quantum yield or efficiency of a process 'ϕ' was introduced. This can be defined as follows :

"It is the number of molecules which undergo chemical transformation per quantum of absorbed energy."

Mathematically, this quantity is defined as follows.

or $$\phi = \frac{\text{No. of molecules reacting in a given time}}{\text{No. of quanta absorbed in t hesame time}}$$

or $$\phi = \frac{\text{No. of molecules reacting in a given time}}{\text{No. of einsteions absorbed in t hesame time}}$$

or $$\phi = \frac{\text{Rate of chemical reaction}}{\text{No. of einsteins absorbed}} \qquad ...(1)$$

The number of quanta absorbed (n_a) in a unit time i

$$n_a = \frac{Q}{hv} \qquad ...(2)$$

and the number of molecules transformed under the action of light (n_r) is

$$n_r = \frac{Q}{hv} \qquad ...(3)$$

On substituting equation (2) in (1), we get

$$\phi = \frac{n_r}{n_a} = \frac{n_r}{Q/hv} \qquad ...(4)$$

The rate of a chemical reaction; according to equation (4), is as follows.

$$-\frac{dn}{dt} = \frac{dn_r}{dt} = \phi\frac{dn_a}{dt} = \phi\,\frac{Q}{hv}$$

The concept of quantum yield or quantum efficiency was first introduced by Einstein. Because of the frequent complexity of photochemical reactions, quantum yields as observed vary from a million to a very small fraction of unity.

The concept of quantum yield can be extended to any act, physical or chemical, following light absorption. The quantum yields are dependent on the light intensity.

The concept of quantum yield provides a mode of account keeping for partition of absorbed quanta into various pathways.

If Stark-Einstein law is correct, then value of ϕ should always be unity. However, later on, it was realised that the law of photochemical equivalence is applicable to only primary processes.

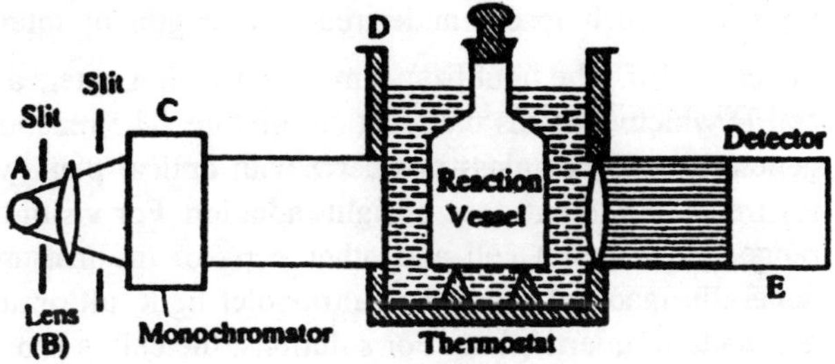

Fig. 2.11

Experimental Determination of Quantum Yields : Knowledge of quantum efficiency of a photochemical reaction offers valuable information about the mechanism of photochemical reactions. In order to determine the value of quantum yield of photochemical reaction, it is essential to know (a) Number of moles reacting, and (b) Number of Einsteins absorbed. For this purpose, an arrangement such as shown in Fig. 2.12 is required.

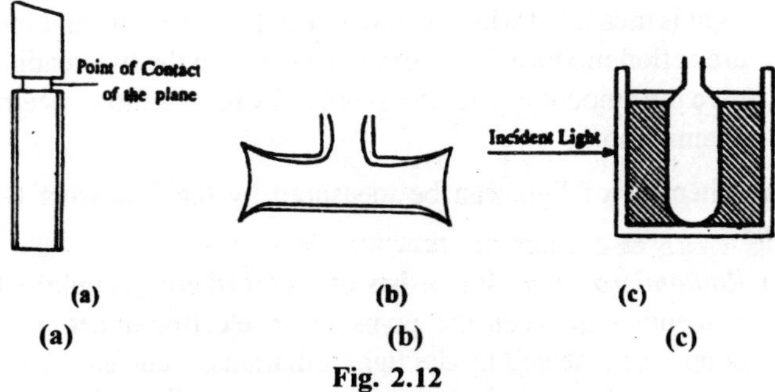

(a) (b) (c)

Fig. 2.12

1. *Light source.* 'A' is the source of light which may be sunlight, arc lamp, mercury vapour lamp, discharge tube, etc. For a quantitative type of work, mercury vapour lamp is the best source of ultra-violet radiations because it emits radiation of

suitable intensity in the desired spectral range., For visible or ultraviolet regions, the tungsten lamp is often used in place of mercury vapour lamp.

2. *Monochromator.* C is a monochromator or a filter. When light from source, A is passed through the lens B and then allowed to pass through, C_0 the monochromator (C) will absorbed that these colour filters transmit light within certain range of wavelengths. Monochromators are generally made of gelatin or coloured glass or transparent plates with metal films of suitable thickness which absorb undesired wavelengths by interference.
3. *Reaction cell.* The light from a monochromator enters a reaction cell D which contains the reaction mixture. The reaction cell is generally made of glass or quartz with optical plane windows for free exit and entrance of light radiation. For visible spectral range, the reaction cell and other parts of the instrument are generally made of glass. For ultraviolet light, all optical parts are made of quartz glass. For solutions, the cell is provided for stirring.

The actual design of the cell depends upon the nature of the reaction, but the front and the back should be plane parallel. The types of cells generally used. These represent the various types of cells which are generally used. In sealing plane window absorption cell; there is a tubular absorption cell, there is a absorption cell with in blow window.

4. *Detector.* D is the detector which is used for determining the intensity of light coming from a reaction cell. The intensity of light is measured with the reaction cell when empty and then with are action mixture. The difference between the two readings will give the amount of energy absorbed by the reaction system under examination.

The intensity of light can be measured by the following types of detectors :

(a) *Radiomicrometer:* It consists of a coil (thermocouple) which is suspended between the poles of an electromagnet. When the couple gets heated by the light radiations, a current flows in the circuit and it is deflected in the magnetic field. The deflection can be measured by lamp and scale arrangement. This deflection is thus, proportional to the intensity of light radiations.

(b) *Photo-electric cell.* The principle of photo-electric cell is based upon the photo-electric effect. It is a device which converts light energy into electrical energy, A photo-electric cell consists of lithium or sodium metal which is enclosed in a vessel. The metal is generally connected to gold leaves of an electroscope or to an electrometer. When the light radiations are incident upon lithium metal, it loses electrons and develops a positive charge on itself whereby the gold leaves diverge. This divergence of the gold leaves is proportional tc the intensity of light radiations. Photo-electric cells are generally used for measuring very low intensities of light shown in Fig. 2.13.

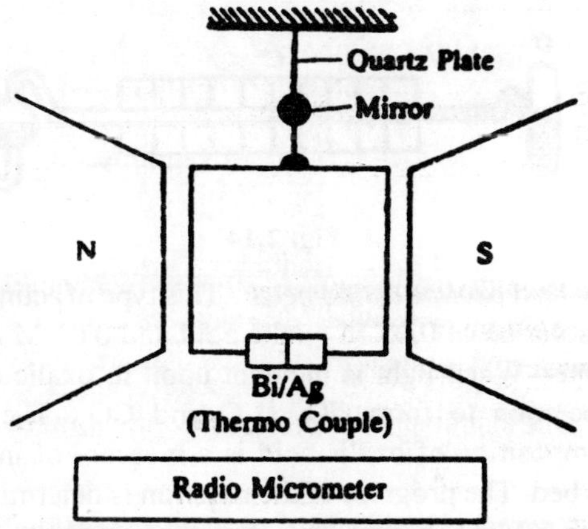

Fig. 2.13

(c) *Chemical actinometers. When* extreme accuracy is not desired, a chemical actinometer can be used to measure the intensity of light radiation. A chemical actinometer generally consists of gas mixtures or "solutions which are sensitive to light. When radiations Fall upon these substances, a chemical reaction will take place and the extent of which is a direct measure of energy absorbed. Following are the main types of actinometers.

(i) *Bunsen and Roscoe's actinometer.* It is illustrated.

The bulb B′ contains a mixture of hydrogen and chlorine over water. The bulb D′ also contains water. The two bulbs B′ and D′ are connected through a graduated tube C′. When the light is incident on B′, some of the gases in A′ react to form hydrochloric acid.

This is absorbed in water of B' and the water moves from C' to B'. The motion of the water thread will be proportional to the amount of hydrochloric acid formed and hence to the intensity of light absorbed.

(ii) *Eder's actinometer : J. M. Eder (1979)* utilized the following reaction to determine the absorbed radiations

$$2HgCl_2 + (NH_4)_2C_2O_4 \rightarrow 2NH_4Cl + 2CO_2 + Hg_2Cl_2$$

The system was exposed to radiations and the mercurous chloride formed in the reaction was weighed. From the weight of mercurous chloride or CO_2 evolved, the absorbed radiations may be determined quantitatively.

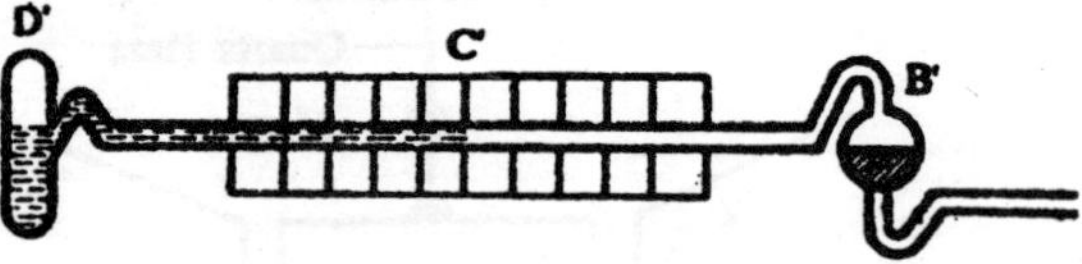

Fig. 2.14

(iii) *The uranyl oxalate actinometer :* This type of actinometer consists of a solution of 0.05 M oxalic acid and 0.01 M uranyl sulphate in water. When light is incident upon it, oxalic acid undergoes dissociation to form CO, H_2O and CO_2. The extent of the decomposition of oxalic acid is a measure of intensity of light absorbed. The progress of the reaction is determined by titrating it with $KMnO_4$ before and after the exposure.

$$UO_2^{2+} + \begin{matrix} COOH \\ | \\ COCH \end{matrix} \xrightarrow{hv} CO + CO_2 + H_2O + UO_2^{2+}$$

(d) *Thermopile :* It consists of a series of unlike metals which are connected, to a galvanometer. One end of it is coated with lamp black, or platinum black and the other end "is kept as such, The thermopile is kept in, glass or quartz vessel.

When the light from a reaction cell is incident upon the black surface, it gets heated up, resulting a difference in temperatures of the black and the other cold end. Thus, a current flows in the circuit. The current produced is proportional to the intensity of light absorbed. Thermopiles are generally calibrated with standard light sources.

(e) *Geiger Muller counter :* It is a modification of the ionisation chamber and is generally used for measuring very low intensities. The apparatus consists of an ionisation chamber with a thin wire of steel or tungsten. When a charged particle enters the chamber the air present in it gets ionised and discharge is initiated in the gas. A current then .flows between the electrodes which can be detected in the usual manner,.

Process For the Determination of Quantum Yield : Set up the apparatus. The choice of the detector depends upon the nature of the experiment. First keep the reaction vessel empty. Allow the light to pass through it. Determine the light intensity as such. In case of solution, it is filled with the solvent.

Now, fill the reaction vessel with reaction mixture. Allow the light to pass through it for a known time. Then note the intensity of light after the reaction is over. The difference between the two readings will give the amount of energy absorbed by the reaction system under examination. Analyse the contents of reaction mixture. Then, determine the no, of moles reacted in given time. Then, apply the following equation to get the value of quantum yield.

$$\text{Quantum yield} = \frac{\text{No. of moles reacting in a given time}}{\text{No. of einsteins absorbed in t he same time}}$$

DEVIATIONS IN THE LAW OF PHOTOCHEMICAL EQUIVALENCE

The photochemical equivalence law applies to the primary photochemical process. As a result of primary absorption of one quanta, only one molecule undergoes dissociation and the products enter no further reaction. In such cases, there will be 1 : 1 relationship between the number of quantas absorbed and the number of reaction molecules. In most cases, a molecule activated photochemically initiates a series of thermal reactions called secondary reactions. As a result of such reactions, many reactant molecules may undergo chemical change by absorbing one quanta only under such conditions.

Class I : High Quantum Yield Reactions : These are the photochemical reactions in which the quantum yield is greater than unity. Examples are :

(i) The quantum yield of the combination of carbon monoxide and chlorine by a light of wavelength ranging from 4000 to 4360 Å is 10^3.

$$CO + Cl_2 \rightarrow COCl_2$$

(ii) Another example is the combination of hydrogen and chlorine to form hydrogen chloride by a wavelength less than 4500°. The quantum yield of this reaction has been found to vary from 10^4 to 10^6, *i.e.,* one quantum brings about the combination of 10^4 to 10^6 molecules.

$$H_2 + Cl_2 \rightarrow 2HCl$$

(iii) Another example is the photochemical decomposition of H_2O_2, by a wavelength of about 3100Å". The quantum yield for this reaction is more than 7.

$$2H_2O_2 \rightarrow 2H_2O + O_2$$

Class II. Low Quantum Yield Reactions : This type includes such photochemical reactions in which the quantum yield is less than unity. Examples are.

(i) Combination of hydrogen and bromine to form HBr is an example of this type. The quantum yield of this reaction is about 0.01.

$$H_2 + Br_2 \rightarrow 2HBr$$

(ii) Photolysis of ammonia by a wavelength 2100Å is another example.

$$2NH_3 \rightarrow N_2 + 3H_2,$$

The quantum yield of this is nearly 0.2.

(iii) The quantum yield for the dissociation of acetone to form carbon monoxide and ethane by a light of wavelength 3000Å is 0.1.

$$CH_3COCH_3 \rightarrow CO + C_2H_6$$

(iv) The quantum yield of the concentration of maleic acid into fumaric acid by a light of wave-length ranging from 2000 to 1800 Å is 0.04.

```
H—C—COOH                 H—C—COOH
  ‖          ———→           ‖
H—C—COOH              HOOC—C—H
 Maleic Acid             Fumaric Acid
```

Class III. Small Integer Quantum Yield Reaction : This type includes the photochemical reaction in which the quantum yield is a small integer like 1, 2, 3, etc. Examples are.

(i) The quantum yield for the dissociation of hydriodic acid by a light of wave length 2070 – 2823Å is 2.

$$2HI \longrightarrow H_2 + I_2$$

(ii) The quantum yield of the following reaction is unity.

$$2Fe^{3+} + I_2 \longrightarrow 2Fe^{2+} + 2I^-$$

This reaction is carried out in liquid phase by a wave-length 5790Å.

(iii) The quantum yield of the combination of SO_2 and chlorine to form SO_2Cl_2 by a light of wavelength 4200 Å is one.

$$SO_2 + Cl_2 \longrightarrow SO_2Cl_2.$$

(iv) The quantum yield of the ozonisation of oxygen by a light of wave-length 1700 – 1900 Å is three.

$$3O_2 \longrightarrow 2O_3$$

REASONS OF HIGH AND LOW QUANTUM YIELD

In order to explain the exceptions to Einstein's Law of photochemical equivalence, *Bodenstein and others* pointed out that photochemical reactions involve two distinct processes I

(a) *Primary process.* It is the process in which each molecule capable of entering into chemical reaction absorbs one quantum of radiation (hv). The absorbed energy may give rise to the formation of excited molecule, *i.e.,*

$$AB + h\nu \rightarrow AB^*$$

or the molecule which absorbs light may undergo dissociation to yield free atoms (some in excited state) or free radicals.

$$AB + h\nu \rightarrow A. + B.$$

(b) *Secondary process.* It is the process which involves the excited atoms or molecules or free radicals produced in the primary process. A secondary process can take place in the dark as well. Due to this process, a different number of molecules may undergo reaction, by absorbing one quanta only. As a result, the quantum efficiency of the reaction, as a whole, will differ from unity.

From the above discussion it follows that the value of quantum efficiency or yield will depend upon secondary processes whereas the primary process remains the same for every reaction. We will now discuss the reasons for different quantum yields for different reactions on the basis of secondary process.

Reasons of High Quantum Yield : These are outlined below:

(a) When the excited atoms or molecules or free radicals produced in the primary state undergo secondary processes, each secondary process gives rise to some further excited particles or radicals; this process continues unless it is checked due to one or the other reason. Thus, by absorbing one quanta only, a large number of reactant molecules undergo reaction. Hence, the quantum efficiency of this type of reaction will be greater than unity.

(b) In some reactions, such as the. Combination of hydrogen and chlorine, a chain is set up and quantum yield becomes as high as 10s. These reactions are known as chain reactions. Some evidences in favour of chain reactions are given below.

(i) The chain is broken when a foreign substance is added in the system to remove H_2 or Cl_2.

(ii) **Marshall and Taylor** found that at least one step in the chain reaction can be carried out experimentally.

Reasons of Low Quantum Yield : The low quantum yield is explained on the basis of following reactions.

(a) If the excited particles formed in the primary process are such that they cannot react due to their deactivation by collisions, by fluorescence or by internal arrangements or by coming in contact with inert molecules the quantum yield will be extremely low.

(b) The excited particles produced in the primary process may recombine to form the reactants so as to give low quantum yield.

To illustrate the above mentioned reasons, a few examples will now be taken.

(i) **Dissociation of HI :** In the primary process, a molecule of hydrogen iodide absorbs light of wavelength less than 4000 Å AND dissociates to form an atom of hydrogen and an atom of iodine. The primary process may be represented as.

$$HI + h\nu \rightarrow H + I$$

The primary process has been established from spectroscopic studies. The possible secondary reactions are.

$$H + HI \rightarrow H_2, + I \quad ...(1)$$

$$I + HI \rightarrow I_2 + H \quad ...(2)$$

$$H + H \rightarrow H_2 \quad ...(3)$$

$$I + I \rightarrow I_2 \quad ...(4)$$

$$I + H \rightarrow HI \quad ...(5)$$

Reaction (2) is endothermic and, thus, it cannot take place at ordinary temperatures. Reactions (3) and (5) are highly exothermic. The heat produced in these reactions is so high that the products formed in these reactions undergo dissociations. Thus, the occurrence of reactions (3) and (5) is highly unlikely. If we eliminate reactions (2), (3) and (5), the reactions (1) and (4) are the only two likely secondary reactions. On this basis, the overall process may be written as.

$$HI + hv \rightarrow H + I \quad \text{(Primary)}$$

$$H + HI \rightarrow H_2 + I \quad \text{(Secondary)}$$

$$I + I \rightarrow I_2 \quad \text{(Secondary)}$$

$$2HI + hv \rightarrow H_2 + I_2$$

Thus, two molecules of hydrogen iodide can undergo dissociation by absorbing one quanta only. Thus, the quantum efficiency of the reaction is 2 which has been found experimentally.

(ii) **Dissociation of HBr :** The mechanism of the reaction is analogous to that proposed above for HI.

(iii) **Combination of Hydrogen and Bromine :** The quantum efficiency of the reaction is about 0.01. The primary process involves the dissociation of bromine molecule into bromine atoms.

$$Br_2 + hv \rightarrow 2Br$$

The expected subsequent secondary processes are.

$$Br + H_2 \rightarrow HBr + H \quad ...(1)$$

$$H + Br_2 \rightarrow HBr + Br \quad ...(2)$$

$$H + HBr \rightarrow H_2 + Br \quad ...(3)$$

$$Br + Br \rightarrow Br_2 \quad ...(4)$$

Reaction (1) is highly endothermic and is, therefore, taking place very slowly at ordinary temperature. At this slow speed, most of the bromine atoms recombine to give the bromine molecules. Hence the reactions (2), (3) and (4) which are due to reaction (1) cannot occur. Therefore, the quantum yield will be extremely low.

(iv) Polymerisation of Anthracene : This reaction can be represented as

$$2C_{14}H_{10} \underset{\text{dark}}{\overset{\text{light}}{\rightleftharpoons}} C_{28}H_{20}$$

The quantum yield of this reaction should be 2 but actually it is only 0.5. It is due to the reason that this reaction involves a back thermal reaction. The low quantum yield of isomeric transformation of maleic acid into fumaric acid can be explained on the basis that a state of equilibrium is reached in this reaction which will lower the quantum yield.

FACTORS AFFECTING QUANTUM YIELD

Except those reactions which obey Einstein's law of photochemical equivalence, it is found that the quantum yield depends upon the following factors.

1. Temperature : The change of quantum yield with temperature is given by the relation.

$$\frac{d \log \phi}{dT} = \frac{Q}{RT^2}$$

where f = quantum yield.

Q = amount of heat evolved by the formation of one gm mole of the substance.

Kuhn found that the quantum yield of photochemical decomposition of ammonia increases by about 15% for every 100° temperature increase. It reaches seven fold at 500°.

For some photochemical reactions the values of temperature coefficient and activation energy are given in Table 2.

Table 2 : Temperature Coefficient and Activation Energy of Some Photochemical Reactions.

Reactions	Temperature coefficient	ΔE (K cal).
$H_2 + Br_2 \rightarrow 2HBr$	1 – 1.02	40.7
$H_2 + Cl_2 \rightarrow 2HCl$	1 – 1.04	44.9
$H_2 + O_2 \rightarrow H_2O_2$	1.09	47.6
$2O_3 \rightarrow 3O_2$	1.27	80.5
$2C1_4H_{10} \rightarrow C_28H_20$	1.04	44.8

2. **Wave-length :** It has been found experimentally that the power of a quantum or photon of light energy to induce photochemical change is greater, the higher the frequency of incident radiation. In other words, the quantum yield will be lower at the higher-wave lengths.

Bonhoeffer and *Harteck* (1930) suggested that the lower quantum yield at longer wavelength is due to a low efficiency of the primary dissociation process in that region. Generally, the quantum yield is always found to rise over a range of frequencies, as in the case of decomposition of nitrogen' dioxide.

Wavelength	310	365	405	436
Quantum yield	2.97	1.54	0.74	0.009

Henri, Wurmser and Lasareff have obtained a proportionality between the amount of light of different wavelengths absorbed and the photochemical effect produced.

3. **Light intensity :** It was shown by Dhar and his coworkers that decrease in light intensity results in an increase of quantum yield. The light intensity was found to change by varying the iris of the aperture, through which light was allowed to pass into the reacting system. These workers found that.
   (i) The velocity of photochemical reactions is found to be directly proportional to the intensity of light.
   (ii) The quantum yield decreases as the amount of light falling on the reacting substance is increased.
4. **Inert Gases :** It was found by Hartel that in most of the photochemical reactions, the addition of inert gases increases the quantum yield. This may be explained due to the occurrence of an induced pre-dissociation which will increase the number of starting chains as it may retard the diffusion of the atoms to the walls. This latter effect would decrease the rate of chain terminating step thereby increasing the speed of reaction.

LUMINESCENCE

We know that the usual methods of obtaining light are involving the heating of solids, or solid particles to sufficiently high temperatures. For example, by heating a tungsten, platinum wire or a carbon filament electrically or in a flame, bright light can be obtained. These are examples of *incandescence* in which thermal energy is converted into light. If, however, light is produced by processes other than those involving heating, such light is called *luminescence light or cold light.*

The above discussion may be summarised as follows.

"When the emission of visible radiation occurs due to some cause other than temperature, the phenomenon is known as luminescence.

As light is produced at low temperatures, the term luminescence may be regarded as 'light' produced without 'heat' or 'cold light'

But in incandescence and luminescence, emission of light occurs due to the return of electrons from excited outer position to lesser excitation position or ground state. Luminescence is of the following types.

1. **Photoluminescence :** Luminescence caused by light is called photoluminescence. Photoluminescence that ceases immediately after the cause of exciation is cut off is called *fluorescence.* Photoluminescence that persists for an appreciable time after the stimulating process is cut off is called *phosphorescence.*
2. **Chemiluminescence :** Luminescence resulting from chemical reactions is called *cathodoluminescence.*
3. **Cathodoluminescence :** Luminescence caused by bombardment of electrons is called *cathodoluminescence.*

   Let us discuss these one by one.
4. **Electroluminescence :** Luminescence resulting from the application of an electric field to matter is called *electroluminescence.*

FLUORESCENCE AND PHOSPHORESCENCE

Introduction to Fluorescence : *When a beam of light is incident on certain substances, they emit visible light or radiations and they stop emitting light or radiation as soon as the incident light is cut off. This phenomenon is known as fluorescence.*

Such substances which emit radiations during the action of stimulating light are called fluorescent substances.

When a substance absorbs light energy, it will result in the excited state of an atom or molecule. But if the light absorbed is not sufficient energetic to eject an electron, it will cause the electrons to move from inner orbits to outer orbits. When these excited electrons return to the ground state, the light is emitted which may possess different frequency than the incident light. The successive stages may be put as follows.

$$\underset{\text{Normal state}}{X} + \underset{\text{photon}}{h\nu} \longrightarrow \underset{\text{Excited state}}{X^*}$$

$$X \longrightarrow X^{**} + h\nu'$$

$$X^{**} \longrightarrow X + h\nu''$$

where X** represents an energy state between X and X*. The emitted radiation frequencies ν' and ν'' are different from ν, *i.e.*, the frequency of incident radiations.

Some characteristics of the phenomenon of fluorescence are as follows

(i) This phenomenon is instantaneous and starts immediately after the absorption of light and stops as soon as the incident light is cut off.

Fluorescence is stimulated by light of the visible or ultraviolet regions of the spectrum. Line, band and continuous spectra of emitted light of fluorescence are observed. The character of the spectrum depends essentially on the state of aggregation of the substance.

(ii) It is a general phenomenon and is exhibited by gases, liquids and solids. No fluorescence will be observed in gases, unless the pressure is low.

(iii) Different substances fluoresce with light of different wavelengths. 'Thus fluorspar fluoresces with blue light, chlorophyll with red light, uranium glass with green light and so on.

(iv) The fluorescent light from solutions is polarised and the degree of polarisation depends in some cases upon the concentration of the solution.

(v) The extent of fluorescence depends upon the nature of the solvent and the presence of certain anions in solution. Thus the thiocyanate, iodide and bromide ions show a marked quenching effect.

(vi) According to *Stoke's law,* during fluorescence light is absorbed at a certain wavelength and should be emitted at a greater wavelength.

The above principle was recognised before the quantum theory was proposed.

(vii) The quantum yield in fluorescence is the ratio of the number of photons of luminescent radiation to the number of photons absorbed from the stimulating light upon a fixed wavelength of the latter. The quantum efficiency of fluorescence increases in proportional to the wavelength l of absorbed radiation. Then, after reaching its maximum value in a certain interval of ~ λ_{max} the efficiency drops rapidly to zero upon a further increase in l.

(viii) Fluorescence may be regarded as a secondary effect resulting from the primary process of absorption of a quantum of light by an atom or molecule.

Introduction to Phosphorescence : *When light radiation is incident on certain substances, they emit light continuously even after the incident light is cut off. This type of delayed fluorescence is called phosphorescence and the substances are called phosphorescent substances.*

Some characteristics of the phenomenon of phosphorescence are as follows.

(i) Materials exhibiting fluorescence generally re-emit excess radiation within 10^{-6} to 10^{-4} second of absorption. On the other hand, materials exhibiting phosphorescence re-emit excess radiation within 10^{-4} to 20 seconds or longer. Thus, the life-time of phosphorescence is much longer than fluorescence.

(ii) The phenomenon of phosphorescence is caused chiefly by the ultraviolet and violet parts of the spectrum.

(iii) The phenomenon of phosphorescence is shown mainly by solids.

(iv) The magnetic and dielectric properties of phosphorescent substances are different before and after illumination.

(v) The time for which the light is emitted from phosphorescent substance defends upon the nature of substance and sometimes on the temperature changes.

(vi) Different colours may be obtained by mixing different phosphorescent substances. *Examples of Fluorescent Substances.*

Among the naturally occurring substances, chlorophyll present in green leaves show the phenomenon of fluorescence, *i.e.*, when leaves are strongly illuminated in presence of oxygen, they emit fluorescent light whose intensity changes with time of irradiation. Franck and Herzfeld explained this phenomenon on the basis of complex formation.

Petroleum, vapours of sodium, iodine, acetone and hydrocarbons (paraffins and olefins) have been found to fluoresce in ultraviolet light.

For example, acetone absorbs at 2700Å corresponding in > C = O group and emits blue fluorescence. Same thing happens with other aldehydes and ketones.

We know that all molecules are capable of absorption. But fluorescence is not exhibited by a large number of compounds. Fluorescence is generally observed in those organic molecules which have rigid framework and not many loosely coupled substituents through which vibrionic energy can flow out. In analogy with chromophores, following structures are termed as *fluorophores.*

C=C, N=O, —N=N, —C=O, —C=N, —C=S

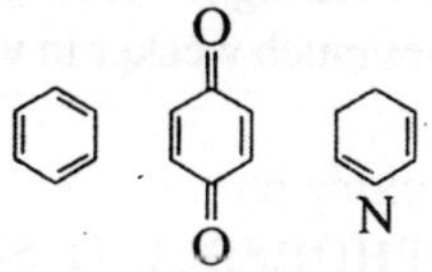

A large number of substances enhance fluorescence. These are known as fluorochromes in the same analogy as *auxochromes*. Generally electron donors act as auxochromes.

–OH

—NH_2

—CH_2

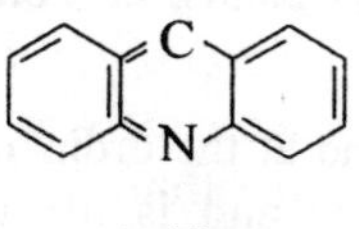

Acridine
(non-fluorescent)

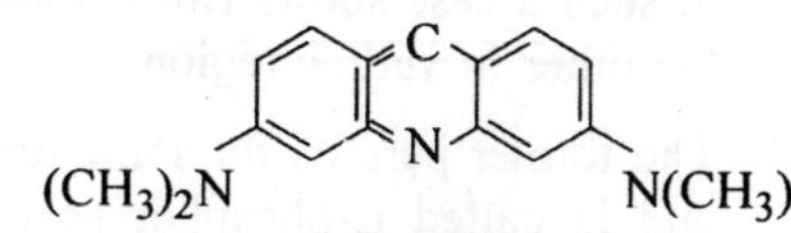

$(CH_3)_2N$ $N(CH_3)_2$

Acridine Derivative
(fluorescent)

Electron-withdrawing substituents like – COOH tend to diminish or inhibit fluorescence completely.

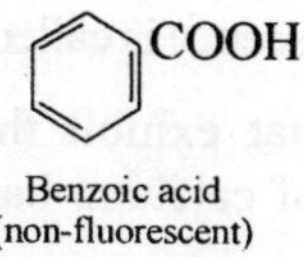

Benzoic acid
(non-fluorescent)

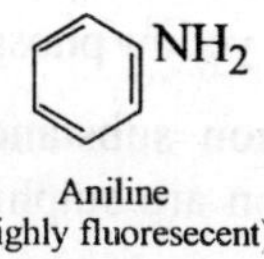

Aniline
(Highly fluoresecent)

Some examples of fluorescent substances are as follows:

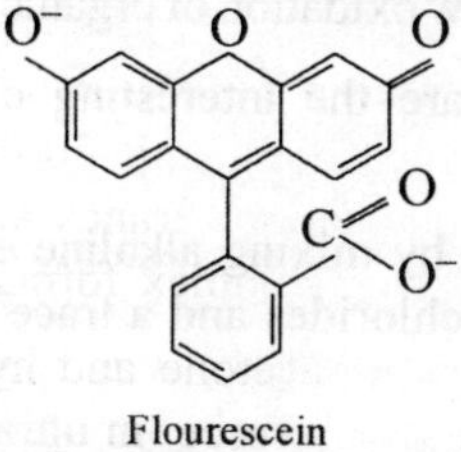

Flourescein
(fluorescent)

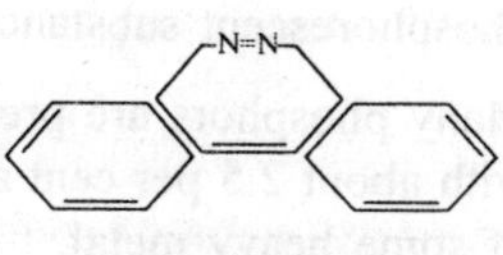

Azophenanthrene
(fluorescent)

Among inorganic substances there are only few examples such as fluorite (the name fluorescence originates from it), uranium compounds and rare earths which exhibit fluorescence.

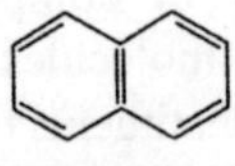

Naphthalene (fluorescent)

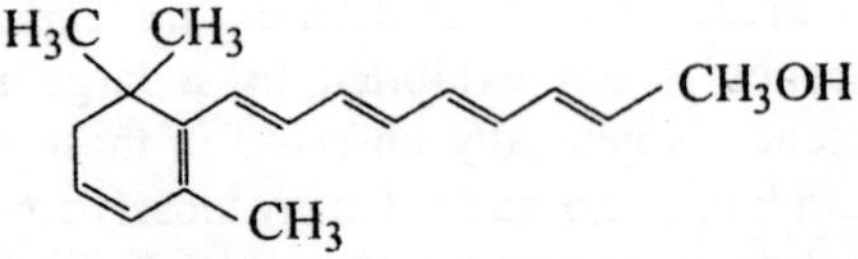

Vitamin A (fluorescence as that of naphthalen)

Nitrogen peroxide in blue light (4920 to 4550A°) shows intense fluorescence which becomes much weaker in violet light (4550 to 3650Å) and vanishes at 3650Å).

EXAMPLES OF PHOSPHORESCENT SUBSTANCES

(i) Many dyes which fluoresce in ordinary light in aqueous solution exhibit phosphorescence when dissolved in fused boric acid or glycerol and cooled to form a glassy solid. The phosphorescence in such a case shows two distinct bands, *i.e.,* one in blue and the other in yellow region.

The former persists for some time at the ordinary temperature and is called α-phosphorescence and is identical with the fluorescence. If the temperature is lowered, α–phosphorescence becomes less marked and disappears altogether at 0°C. The yellow phosphorescence, on the other hand, is practically independent of temperature and has been observed down to – 250°C. The yellow phosphorescence is called β-*phosphorescence.*

(ii) The common substances that exhibit this phosphorescence phenomenon are sulphides of calcium, barium, strontium.

(iii) Many organic substances and certain fungi phosphoresce to emit a faint light This is due to the slow oxidation of organic substances.

(iv) Minerals, *e.g.,* ruby, emarld are the interesting examples of phosphorescent substances.

(v) Many phosphors are prepared by mixing alkaline earth metals with about 2.5 per cent alkali chlorides and a trace of sulphide of some heavy metal.

THEORY OF FLUORESCE.\CE AND PHOSPHORESCENCE

1. Singlet and Triplet States : In order to understand the theory of fluorescence and phosphorescence, one has to understand the meaning of singlet and triplet states. These terms arise from multiplicity considerations to atomic spectroscopy and simply define the *number of unpaired electrons in the absence of magnetic field.* If there are *n* number of unpaired electrons, it means that (n + 1) -fold degeneracy (equal energy states) will be associated with the electron spin, regardless of the molecular orbital occupied.

Thus, if no unpaired electrons are present (n = 0), where there is only n + 1 or 0 + 1 or 1 spin state. Such a state is a called a *singlet state.* Similarly, systems having 1, 2, 3, 4, ... unpaired electrons refer to doublet, triplet, quartet, etc. respectively.

Most of the molecules in their ground state do not have unpaired electrons *(singlet state).* When such a molecule absorbs ultraviolet or visible radiation of the proper frequency one or more of the paired electrons (generally a re-electron) get raised to an *excited singlet state.* In this excited state, the spin of the electron does not undergo any change and the net spin is still zero.

One more possibility is that one set of electron spins may have undergone unpairing, resulting in two unpaired electrons which make an *excited triplet state.* Fig. Fig. 2.15(a) represents a molecule in the ground singlet state, Fig. 2.15(b) exhibits a molecule in the excited singlet state. Fig. 2.15(c) represents a molecule in an excited triplet state.

One should remember that there is very small probability of a direct transition from the singlet ground state to the triplet excited state.

2. Excited-State Processes in Molecules : When molecules are irradiated with light of the appropriate frequency, it will be absorbed in about 10^{-15} second. In the process of absorption the molecules may move from the ground to the first excited singlet electronic state. Although at room temperature molecules may be present in their ground vibrational level, after absorption the excitational molecules can end up in any one of the vibrational levels in the first excited electronic state. From the excited singlet state, one of the following three phenomena will probably occur, depending on the molecules involved and the conditions.

(a) The *first possibility* is that the excited singlet state is relatively *unstable.* In such a situation, the excited molecule will return to the ground state by *collisional deactivation without emitting any radiation.*

(b) The *second possibility* is that the molecule in the excited singlet state may emit an ultraviolet or visible light photon. This process is known as *fluorescence.*

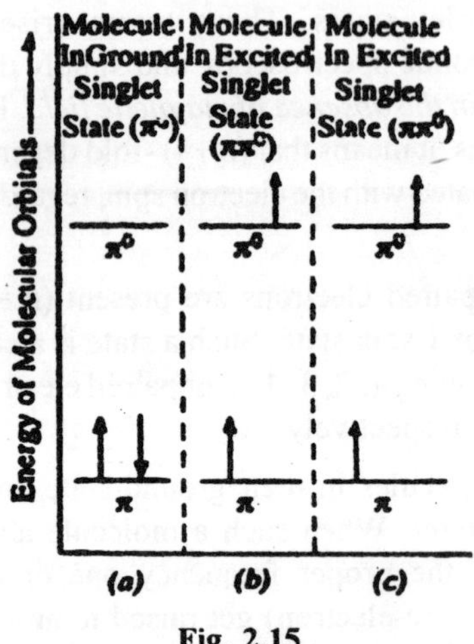

Fig. 2.15

If we compare the absorption and fluórescence spectrum of the same compound, they do not superimpose on each other as expected but they are mirror images of each other with the fluorescence spectrum shifted to longer wavelengths. The reason for this is that as the time required to execute a vibration is about 10^{-13} second which is much shorter than the decay or mean life of 10^{-9} second, most of the excess vibrational energy will be given to the surroundings and the excited molecules will decay in their ground vibrational levels. This is illustrate in Fig. 2.16.

(c) The third possibility is that the molecule with relatively stable *excited singlet state* may undergo transition to a *metastable triplet state* and some time thereafter returns to the ground states, usually by emission of an ultraviolet or visible light photon. This is known as *phosphorescence emission* Fig. 2.16, and the process of crossing from a singlet state (no unpaired electron) to a triplet state (two unpaired electrons) is termed as *intersystem crossing.*

The decay from the triplet to the ground state singlet is *forbidden by spin symmetry* and is therefore *slow.* Thus, the life-time of phosphorescence is much longer than flour-essence. The above mechanism of phosphorescence involving singlet-triplet decay scheme has been confirmed by the magnetic susceptibility and ESR measurements.

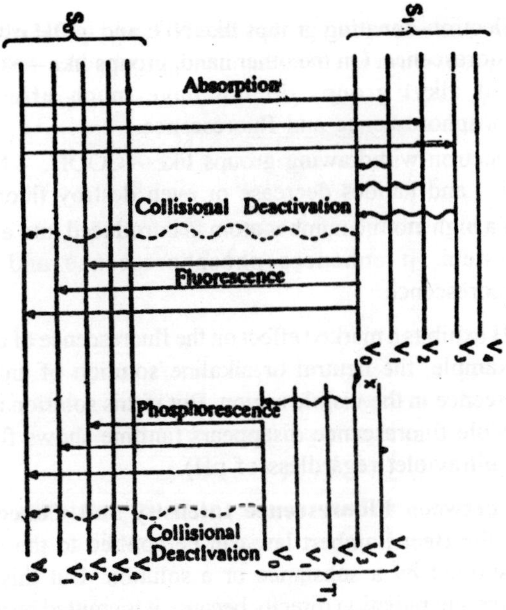

Fig. 2.16

According to Hund's rule, the triplet level always lies *lower* than the corresponding singlet level and for this reason phosphorescence spectrum is *not the mirror image* of the absorption spectrum and it always occurs at *longer wavelengths* compared with the absorption and flourescence spectrum. Phosphorescence is rarely observed in gases It is almost never observed at room temperature (biacetyl is one exception).

Factors Affecting Fluorescence and Phosphorescence : The various factors are as follows.

(a) All the molecules cannot show the phenomena of fluorescence and phosphorescence. Only such molecules show these phenomena that are able to absorb ultraviolet or visible radiation. In general, the greater the absorbancy of a molecule, the more intense its luminescence. This requirement means that molecules having conjugated double bonds (π bonds) are particularly suitable for this study. On the other hand, aliphatic and saturated cyclic organic compounds, are not suitable.

(b) Substituents often exhibit a marked effect on the fluorescence and phosphorescence of molecules. There are no rigid rules but a few generalities may be useful. These are as follows.

(i) Electron-donating groups like NH_2 and – OH often enhance fluorescence. On the other hand, groups like $– SO_2H$, $– NH_4^+$ and alkyl groups do not have much effect on both phosphorescence and fluorescence.

(ii) Electron-withdrawing groups like – COOH, $– NO_2$. – N = N – and halides decrease or even destroy fluorescence.

(iii) If a high atomic number atom is introduced into a re-electron system, it enhances phosphorescence and decreases fluorescence.

(c) The pH exhibits a marked effect on the fluorescence of compounds. For example, the neutral or alkaline solution of aniline shows fluorescence in the visible region. But if this solution is acidified, the visible fluorescence disappears (aniline shows fluorescence in the ultraviolet regardless of pH).

Relation between Fluorescence Intensity and Concentration : We know that the Beer-Lambert law can be applied to the intensity of radiation transmitted by a substance or a solution. But this cannot be applied to fluorescent radiation directly because it is emitted by a substance. However, the following relation has been developed.

$$F \propto I_D - I$$

or $$F = K (I_0 - I) \quad ...(1)$$

where F is the intensity of fluorescent radiation, K is a proportionality constant. I_0 is the intensity of incident radiation and I is the intensity of transmitted radiation. We know that the Beer-Lambert law is

$$I = I_0 \, 10^{-abc}$$

or $$I_0 - I = I_0 - I_0 \, 10^{-abc} \quad ...(2)$$

$$= I (1 - 10^{-abc}) \quad ...(3)$$

On substituting equation (3) in (1), we get

$$F = KI_0 (1 - 10^{-abc)} \quad ...(4)$$

On rearranging equation (4), we get

$$\log \frac{KI_0}{KI_0 - F} = abc \quad ...(5)$$

In the above equation, K is the fraction of the incident radiation that is adsorbed, *a* is the absorptivity, *b* is the length of cell path and *c* is the concentration of absorbing substance.

For very dilute solutions, equation (4) simplifies to the following equation

$$F = 2.303\ KI_0\ abc$$

or

$$F = K'c$$

For the above equation it follows that the fluorescent intensity is practically proportional to the concentration of the fluorescent substance. Equation (6) holds good for solutions of few parts per million, *i.e.*, dilute solution. For connected solutions, fluorescence concentration curve will bend towards the concentration axis. Fig. 2.17 shows typical curves at – 196°C, and 35°C.

From Fig. 4.40. it follows that the curve levels off at high concentration at optimum temperature. Factors such as association, dissociation, or solvation which are responsible for deviations in Beer-Lambert law would be expected to show a similar effect in flourescence.

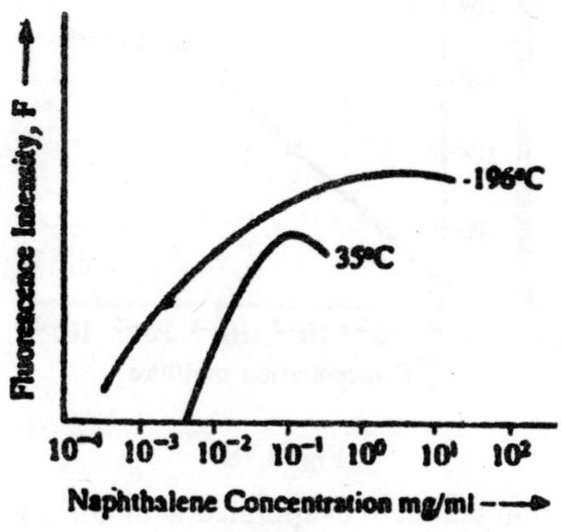

Fig. 2.17

Another important factor in fluorescence is temperature. In most of the instruments, sources of the very high intensity are used. This causes the heating of the sample or sample chamber. An increasing temperature usually decreases flourescence intensity. In order to minimise temperature effects in fluorescence instruments, shutter mechanisms are provided to minimise exposure time and thermally insulating the sample holder.

Relation between Phosphorescence Intensity and Concentration : Exactly the same considerations as discussed for fluorometry are important in relating phosphorescence intensity P to concentration *c*, and thus the final equation is as follows.

$$\log \frac{K'I_0}{K'I_0 - P} = abc$$

where K′ depends on the instrument used. A typical curve or phosphorimetry is shown in Fig. 2.18.

Applications of Fluorescene : We know that the phenomenon of fluorescence is a well established analytical tool. A large number of applications are known. But we will describe some of these as follows.

(a) An interesting example is the determination of uranium in salts by fluorescence. This is used extensively in the field of nuclear research.

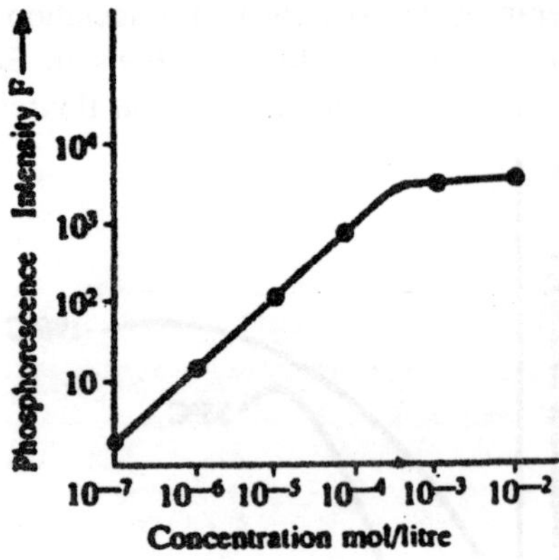

Fig. 2.18

The uranium sample is evaporated with nitric acid to bring about oxidation. Then the sample is fused with sodium fluoride to a melt having fluorides of sodium and uranium. On cooling, this solidifies to a glass which is examined in a specially designed fluorimeter. By this method, one can determine uranium of the order of 5×10^{-9} g in a 1 g of solid sample.

(b) In general, inorganic ions do not exhibit fluorescence. However, some of these inorganic ions form fluorescent chelates with non-fluorescent organic molecules. This has provided the basis for very sensitive analysis of many elements including most of the transition elements. Some interesting examples are as follows.

(i) An interesting example is the determination of ruthenium ion in the presence of other platinum metals. With 5-methyl-

1, 0-phenanthroline ruthenium forms the complex ion which fluoresces strongly at pH 6. By this method, the determination of ruthenium can be carried out in the range of 0.3 to 2.0 μg/ml in the presence of interfering elements of platinum group which may be present to the extent of at least 30μ/ml.

(ii) Another interesting example is the determination of aluminium (III) in alloys. Aluminium (III) forms a complex with dye Pantachrome Blue-black RM at a pH of 4.8 which fluoresces strongly. One can determine aluminium in the range of 0.2 to 25μg in a volume of 50 ml with a sensitivity of 1 part 10^8.

(iii) Another inorganic fluorimetric determination is the estimation of traces of boron in steel by means of the complex formed with benzoin. The boron present in the acid solution of steel is first converted into boric acid which is separated from the other constituents by distillation with methyl alcohol.

The resulting distillate having boric acid is neutralised with sodium hydroxide. Then, the methyl alcohol is distilled off leaving the sodium salt of boric acid in the distillation flask which on treatment with an alcoholic solution of benzoin yields a fluorescent solution. It has been found that the fluorescent power is linear with the concentration upon 100 *ug* of boron in 50 ml volume but drops off at higher concentrations. This method for its determination is superior to old techniques in speed and sensitivity.

(iv) Cadmium can be estimated by precipitating it with 2-(2- hydroxyphenyl)-benzoxazole in the presence of tartrate. The complex on dissolving in glacial acetic acid yields a solution with an orange tint and a bright blue fluorescence in ultraviolet light. The acetic acid solution forms the basis for the determination of cadmium.

(v) Similarly, calcium can be estimated by fluorimetry with calcein solution.

(c) Fluorescent indicators : The intensity and colour of the fluorescence of many substances depend upon the *pH* of the solution, *i.e.,* their colours depend upon the pH range. These are termed *as flourescent indicators.*

These are mainly used in acid-base titrations. These can be employed in the titration of coloured solutions in which the changes in colour of indicators get masked. Some examples of fluorescent indicators are given in Table 3.

Table 3 : Some Fluorescent Indicators.

Name of Indicators	approx. pH	Colour Change
Eosin	3.4–4.0	Colourless to green
Fluorescein	4.0–6.0	Colourless to green
Quinine sulphate	3.0–5.0	Blue to violet
Acridine	5.2–6.6	Green to violet blue
2 – Naphthaquinone	4.4–6.3	Blew to colourless
2 – Hydroxycinnamic acid	7.2–9.0	Colourless to green

(d) **Determination of vitamin B_1 :** Vitamin B_1 (thiamine) is non-fluorescent whereas its oxidation product, thiochrome fluoresces with blue colour. This property is used for the determination of vitamin B_1 in the food samples like meat, cereal, etc.

The food sample is treated with phosphatase which brings about hydrolysis of the phosphate esters of thiamine present in the sample. The solution on filtration removes phosphatase and other insoluble matter. Then, the filtrate is diluted to a known volume. From the filtrate, two equal aliquots are taken, one for analysis and other for a blank.

To both aliquots, equal quantities of sodium hydroxide and isobutyl alcohol are added. To the first oxidising agent like potassium ferricyanide is added. After shaking, the alcoholic solution is separated from the aqueous solution. Then, the alcoholic solution is examined in the fluorimeter. The whole procedure, including a blank, is repeat with a standard thiamine solution.

(e) **Determination of Vitamin B_2 (Riboflavin) :** Determination of vitamin B_2 is done by a fluorescence method because the fluorescent power depends upon the reaction conditions and upon the nature and amount of impurities.

Generally, the method of standard increment is used to become sure about the fact that impurities have the same effect upon the standard and unknown. In the method of standard increment, one measures fluorescence of a portion of the standard in the same solution with the unknown.

The procedure also considers the simple fact that riboflavin on oxidation yields a non-fluorescent substance.

An acid solution of the sample (a food stuff) is treated with different reagents to precipitate various interfering ions. This solution is then oxidised with dilute permanganate, Then, the residual fluorescence is measured as a blank. Now, a slight excess of solid sodium dithionite ($NagSaO_4$) is added.

Fluorescence is again determined. Then, a known volume of standard is added and fluorescence is again measured. Finally, the results are computed by the following scheme.

Table 4

Fluorescence of solution	Designation
10 ml oxidised sample + 1 ml water	F_A
Same + dithionite	F_B
Same + 1 ml standard	F_C

From the above results, the concentration of fluorescing material can be calculated by using the following formula.

$$\frac{F_B - F_A}{F_C - F_A} = \frac{m_x}{m_x + m_s}$$

In the above equation, m_x and m_s are the masses of riboflavin from sample and standard in the cuvette respectively.

(f) **Food-Stuffs :** The phenomenon of fluorescence is being utilised in examining conditions of food-stuffs.

When ultraviolet light is incident on newly laid eggs, they fluoresce with rosy colour, while bad eggs appear blue. Similarly butter, lard and different kinds of honey can be readily distinguished.

(g) **Police Work :** The difference in the fluorescence caused by ultraviolet rays in different types of inks enables police to detect forged documents.

(h) **Medicine** : Ringworm may be detected by the fluorescence caused by ultraviolet radiations.

When fluorescent liquids are injected into an animal body, the internal organs will fluoresce and can be observed by a microscope of special type called fluorescent microscope. In this way the internal organs of human or animal body may be diagnosed by doctors without any difficulty.

When, in fluoroscope. X-rays are allowed to fall on a screen of barium platinocyanide or other materials, there will be characteristic fluorescence which may be used for diagnosis by doctors. This is the principle of fluoroscope used in X-ray diagnosis.

(i) **Analysis** : The characteristic fluorescence of various substances when exposed to ultraviolet light offers a good method of quantitative analysis. This has been used in the analysis for the following.

(a) Drugs and dyes

(b) Textile and paper industry.

(c) Medicine and bacteriology,

(d) Fuels and chemicals.

(j) **Lamps** : Fluorescent lamps which are now widely used for lighting depend upon the fluorescence caused by ultraviolet light on phosphorus coated inside the fluorescent tubes.

(k) **Science** : Use of the phenomenon may be made in seeing the invisible radiations like ultraviolet rays.

(l) **Organic Analysis** : Fluorescence has been used to carry out qualitative as well as quantitative analysis for a great many aromatic compounds present in cigarette smoke, air-pollutant concentrates and automobile exhausts. A specific example is the determination of benzopyrene in the nanogram range.

APPLICATIONS OF PHOSPHORESCENCE

Similar to fluorescence, phosphorescence finds applications in biology and medicine. But these applications are not accepted for routine analysis. Some interesting applications are as follows.

(a) One can carry out the determination of aspirin (acetylsalicylic acid) in blood serum with high sensitivity by phosphorimetry at liquid nitrogen temperatures. By this method, 0.02 – 1.00 mg aspirin per ml of serum can be analysed. The metabolic product of aspirin, *i.e.,* salicylic acid, does not exhibit appreciable phosphorescence but is quite strongly fluorescent whereas aspirin is exhibiting strong phosphorescence but not fluorescence. Therefore, it becomes possible to carry out a simple chloroform extraction of the serum and then analysis is done for both the drug (aspirin) and its metabolite (salicylic acid) in the presence of one another by using these two complementary techniques (fluorimetry and phosphorimetry).

(b) Low concentrations of procaine, cocaine, phenobarbital and chloropramazine in blood serum have been determined by phosphorimetry in combination with extraction procedures.

(c) Cocaine and atropine in urine have been determined by employing phosphorimetry in combination with extraction procedures.

(d) Phosphorimetry has been employed in combination with thin layer or paper chromatography. An interesting example is given by Winefordner and Moye. They have separated three tobacco alkaloids (nicotine, nornicotine and anabasine) from crude tobacco by thin layer chromatography on alumina. The R_f values of three alkaloids are 0.80, 0.26 and 0.48 respectively. The separated drugs were scrapped from the plate and determined by phosphorimetry. This method is quite rapid.

COMPARISON OF FLUORESCENCE AND PHOSPHORESCENCE

As phosphorescence is more complicated experimentally than fluorescence, far fewer applications have been realised for phosphorimetry than for fluorimetry. But the potential of phosphorimetry is great and it is hopped that it will find valid applications in chemistry in the near future. However, it is clear that the two techniques are complementary.

Whenever we analyse complex samples having much interfering substances, we prefer phosphorimetry because it is more selective than fluorimetry.

One more advantage of phosphorimetry is that it is more sensitive than fluorimetry due to the following reasons.

(i) Scattering problems do not exist in phosphorimetry whereas these become severe in fluorimetry.

(ii) Quantum efficiencies are much more for phosphorescence than for fluorescence. The reason for this is that phosphorimetric determinations are carried out at – 196°C whereas fluorimetric determinations are carried out at room temperature. One more advantage for the low temperatures in phosphorimetry is that the quenching does not pose a serious problem, than in fluorimetry.

From the above discussion it follows that whenever fluorimetry and phosphorimetry are of equal sensitivity, fluorimetry is preferred because it is experimentally less complicated. However, phosphorimetry is preferred when fluorimetry is not sensitive or selective.

Antistoke's Behaviour : According to G.G. Stoke, the emitted radiation in fluorescence has greater wave-length than the absorbed radiations. There are however, cases where Stoke's law is violated. Two exceptions are.

(a) **Resonance Fluorescence :** *In certain cases, the fluorescent light has the same frequency as that of incident light. This fluorescent light is called resonance radiation and the phenomenon is known* as resonance fluorescence. This phenomenon is analogous to resonance of sound. For example, when mercury vapour with atoms in normal state ($^1S_0$) is exposed to ultraviolet radiations, it was found to absorb the radiation of wave-length 2537 Å. It then converts to $^3P_1$. If excited electrons return to the ground-state, they emit the radiations of the same wavelength, *i.e.,* 2537 Å.

$$\text{Hg}\ (^1S_0) + \text{HV} \rightarrow \text{Hg}^*\ (^3P_1)$$

$$\text{Hg}^*\ (3P_1) \rightarrow \text{Hg}\ (^1S_0) + \text{hv}.$$

(b) **Sensitised Fluorescence :** *A substance which is normally non-fluorescent may be made fluorescent in the presence of other fluorescent substances. The phenomenon is known as* Sensitised fluorescence. If the vapour of thallium is added to mercury vapour and then exposed to radiations of wave-length, 2537 Å, the vapour of thallium fluoresces. Thus the thallium which is a non-fluorescent substance can be made fluorescent by mercury vapour. This can be explained by saying that in sensitised fluorescence, mercury vapour gets energy from incident light and thus its atoms get excited to higher energy levels.

$$\text{Hg}\ (^1S_0) + \text{hv} \rightarrow \text{Hg}^*(^3P_1).$$

Then, thallium atoms undergo collision with excited mercury atoms and thus a part of energy of excited mercury atoms is transferred to thallium atoms which are raised to higher energy levels.

$$Hg^* + T_l \rightarrow T_l^* + Hg$$

These excited atoms of thallium emit their own characteristic fluorescence on reverting to the ground state

$$T_l^* \rightarrow T_l + h\nu'$$

Quenching of Fluorescence : *When a photochemically excited atom has a chance to undergo collision with another atom or a molecule before it fluoresces, the intensity of the fluorescent radiation may be diminished or stopped. This phenomenon is known as quenching of fluorescence.* Some facts about quenching of fluorescence are as follows.

(i) Quenching of fluorescence depends greatly on the concentration of the fluorescent atoms and of the quenching substance.

(ii) At low pressures, very little quenching will occur. At high pressures, appreciable quenching takes place.

(iii) Quenching of fluorescence takes place appreciably in a liquid medium because collisions are very frequent.

Explanation. The quenching of fluorescence is due to transfer of energy from the photochemically excited atom to the molecule or atom with which it undergoes collisions. As a result of transfer of energy, the following changes may be possible.

(i) A photochemically excited atom may activate the atom with which it undergoes collision.

$$Hg^* + T_l \rightarrow Hg + T_l^*.$$

This results in sensitized fluorescence.

(ii) The photochemically excited atom may activate a molecule by collision.

$$Cd^* + H_2 \rightarrow Cd + H_2^*$$

(iii) A photochemically excited atom may react chemically with a molecule to form the products. Example is

$$Hg^* + H_2 \rightarrow HgH + H$$

(iv) An excited atom may undergo collision with another molecule and, thus, resulting its dissociation. Example is

$$Hg^* + H_2 \rightarrow Hg + 2H.$$

This phenomenon is known as photosensitisation.

Importance. The photochemical importance of the quenching of fluorescence is that the excited molecules or atoms obtained in the quenching process may undergo further reaction so as to continue the photochemical changes such as (a) photosensitisation (b) sensitised fluorescence, (c) photochemical reactions and (d) dissociation of molecules to form atoms.

SPECTRUM

When a beam of white light is passed through a prism (or grating) it breaks up into the beams of constituent colours. The different coloured beams, when focussed on a screen by a converging lens, from an array of colours on the screen. This is called the 'spectrum' of white light.

The are two main types of spectra : (i) 'emission spectra' and (ii) 'absorption spectra'.

When light coming directly from a source is allowed to enter a prism or grating spectroscope, and 'emission spectrum' of the source is observed in the spectroscope. On the other hand, when light from a source showing a continuous emission spectrum is passed through an absorbing material and then into the spectroscope, and 'absorption spectrum' of the material is observed.

Emission and absorption spectra are further classified according to their appearance. There are three classes.

(i) Continuous Spectra

When the emitting source is an incandescent solid, or liquid, such as a lamp filament or a gas at high pressure, the spectrum is continuous containing all colours (wavelengths) from red to violet. Its appearance is like an unbroken luminous band of light in which the colour changes gradually from point to point but without any sharp boundary. In this spectrum, the intensity is maximum at a certain point and decreases on both sides of it. The point of maximum intensity shifts towards the violet end of the spectrum as the temperature of the source increases.

(ii) Line Spectra

If the light source is a low-pressure gas (as in a discharge tube), flame, arc or spark, the spectrum is discontinuous, showing a number of sharp bright coloured lines. These lines are the images of the slit of the spectroscope formed by lights of different colours. The entire series of images is called a 'line spectrum'. The different lines differ in intensity and nature. Some are sharp, some are sharp on one side and diffused on the other, and others are diffused on both sides.

The line spectrum is the characteristic of the atom or the ion. It means that a particular atom or ion always gives the characteristic set of spectral lines, and no two atoms or ions can give the same spectral line. For example, sodium atom gives two intense yellow lines called D_1 and D_2 lines.

(iii) Band Spectra

Such spectra are obtained by the radiation from gas molecules such as oxygen (O_2), nitrogen (N_2), cyanogen (CN), etc. They consist of illuminated regions, called 'bands', separated by dark spaces. With a high-resolving instrument each band is seen to consist of very fine lines which become closer and closer on one side of the band until they coincide. This side thus has a sharp and bright edge called the 'head' of the band.

Absorption Spectra

The absorption spectra are obtained when the absorbing substance is placed between a source emitting a continuous spectrum and the slit of the spectroscope. In such cases, certain colours (wavelengths) are absorbed by the substance. Hence the spectrum is found to consist of dark lines or bands against a bright background. An example of such spectra is sun's spectrum. It is a line absorption spectrum. It was studied in detail by Fraunhoffer offer who named the absorption lines as A, B, C, D... These lines inform us about the elements which are present around the sun.

Similarly, when an intense beam of continuous white light is passed through sodium vapour and then sent into a spectroscope, we obtain two dark lines on a continuous background in the same positions as the yellow D_1 and D_2 lines in the sodium, emission spectrum. If the sodium vapour is replaced by Iodine vapour (I2), an absorption band spectrum of I2 molecule is obtained.

Series Relationship in Atomic Spectra

The atom spectra consist of a large number of lines. A quantitative experimental study of these spectral lines was made in the second half of the 19th century when several regularities were observed in the spacings of the lines. For example, in 1870. Lveing and Dewar noticed that the spectral lines of various elements could be grouped into distinct 'series'. *In each series the spacing and intensity of lines decrease regularly towards shorter wavelengths,* until it becomes impossible to distinguish the individual lines. The point at which the lines of the series finally converge is called the 'series limit'. The various series in complicated spectra overlap.

The simplest atomic spectrum is that of hydrogen. Its visible part consists of a single series which was first observed by Balmer in 1885. This is called the 'Balmer series' of hydrogen. Its first line having longest wavelength of 6563 Å (Fig. 2.19), is named Ha, the next Hb, and so on, the series limit reaching at 3646 Å (Fig. 2.19) Besides this, there is a series of lines in the ultraviolet part of the hydrogen spectrum which is known as 'Lyman series, and three series in the infra-red part which are known as 'Paschan series', 'brackett series' and 'Pfund series'.

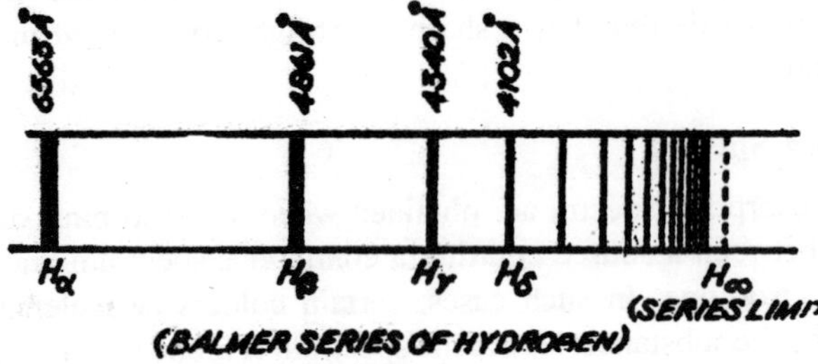

Fig. 2.19

Balmer discovered a formula for the wavelengths of all the lines of the Balmer series. His formula is

$$\lambda = 3646 \frac{n^2}{n^2 - 4} \text{ Å}, \ n = 3, 4, 5, \ldots$$

n = 3 gives the wavelength of H_α line, n = 4 gives H_β line, ... and n = ¥ gives the series limit.

Subsequently, Rydberg found that Balmer's formula was a special case of a more general formula which is as follows.

$$\bar{v} = \frac{1}{\lambda} = R_H \left(\frac{1}{m^2} - \frac{1}{n^2} \right),$$

where $\bar{v}$ is wave number (reciprocal of wavelength), m and n are positive integers (n > m) and RH is the Rydberg constant for hydrogen. Its value is 1.097 × 107 m^{-1}. To obtain formula for Lyman series we set m = 1 and n = 2, 3, 4, ..., for Balmer series m = 2 and m = 3, 4, 5, ... and so on. Thus.

$$\bar{v} = R_H \left(\frac{1}{1^2} - \frac{1}{n^2}\right), n = 2, 3, 4, ... \quad \text{(Lyman)}$$

$$\bar{v} = R_H \left(\frac{1}{2^2} - \frac{1}{n^2}\right), n = 3, 4, 5,... \quad \text{(Balmer)}$$

$$\bar{v} = R_H \left(\frac{1}{3^2} - \frac{1}{n^2}\right), n = 4, 5, 6, ... \quad \text{(Paschen)}$$

$$\bar{v} = R_H \left(\frac{1}{4^2} - \frac{1}{n^2}\right), n = 5, 6, 7, ... \quad \text{(Brackett)}$$

$$\bar{v} = R_H \left(\frac{1}{5^2} - \frac{1}{n^2}\right), n = 6, 7, 8... \quad \text{(Pfund)}$$

We see that the wave numbers of hydrogen lines can be expressed as differences of two terms of the form $\frac{R_H}{n^2}$.

The next simplest spectra are of the 'monovalent' atoms of alkali metals Li, Na, K, etc. The lines in the spectrum of an alkali atom can be grouped unto four distinct series, a principal series' of intense lines a 'sharp series' of fine lines, a diffuse series' of comparatively broader lines and a 'fundamental series' which lies in the infra-red region.

The sharp and diffuse series line in the visible part and converge to a common limit. Rydberg represented the lines of a particular series by the formula.

$$\bar{v} = Z^2 R_A \left[\frac{1}{(m - \Delta_1)} - \frac{1}{(n - \Delta_2)^2}\right],$$

where R_A is the Rydberg constant for a particular element A, Z is the atomic number, m and n are positive integers, and Δ_1 and Δ_2 are constants for the particular series. Actually each line of an alkali spectrum is a close doublet. Again, we find that the wave numbers of the lines can be expressed as differences of two terms like $\frac{Z^2 R_A}{(n - \Delta)^2}$.

After alkali spectra, next in complexity are the spectra of 'divalent' atoms of alkaline-earths Be, Mg, Ca. In a typical alkaline-earth spectrum, we can distinguish two distinct systems of lines, a system of singlets and a system of distinguish two distinct systems of lines, a system of singlets and a system of triplets. Each system has a principal series, a sharp series, a diffuse series and a fundamental series.

As we proceed to atoms having several valence electrons, the spectra becomes more complex and the groupings of lines into series becomes less pronounced. Still regularities can be observed in complex spectra, and it is possible to express the wave number of any spectral line as the difference of two terms.

Rydberg-Ritz Combination Principle

The principle states *the wave numbers of spectral lines can be expressed by the difference of spectroscopic terms in such a way that other difference of those terms give also the wave numbers of lines in the same spectrum.*

Suppose in a spectrum the wave numbers of two lines are given as

$$\bar{v}_a = T_2 - T_3 \text{ and } \bar{v}_d = T_1 - T_4,$$

Then lines of the following wave numbers are also expected in the same spectrum,

$$\bar{v}_b = T_2 - T4 \text{ and } \bar{v}_c = T_1 - T_3.$$

This means that constant differences exist between the wave numbers.

$$\bar{v}_b - \bar{v}_a = \bar{v}_d - \bar{v}_c$$

$$\bar{v}_c - \bar{v}_a = \bar{v}_d - \bar{v}_b.$$

Displacement Law

According to this law *the spectrum of any neutral atom of atomic number Z closely resembles the spectrum of the singly ionised atom of atomic number Z + 1*. For example, the spectrum of H^+ (Z = 2) closely resembles the spectrum of H (Z = 1) and can be represented by similar formula,

$$\bar{v} = 4 R_{He} \left[\frac{1}{m^2} - \frac{1}{n^2}\right].$$

Other elements deprived of all but one electron, also produce hydrogen-like spectra which an be represented by the general formula

$$\bar{v} = Z^2 R_A \left[\frac{1}{m^2} - \frac{1}{n^2}\right].$$

In a similar way, the spectrum of a singly ionised alkaline-earth atom resembles the spectrum of an alkali atom. From this we conclude that it is the number of valence electrons in an atom which determines the qualitative character of the spectrum of that atom.

Hydrogen Spectrum

The spectrum of hydrogen atom consists of a number of lines. These lines have been grouped into a number of 'series'. The lines in each series are such that *their separation and intensity decrease regularly towards shorter wavelengths,* converging to a limit called the 'series' limit. The wavelengths in each series can be given by a simple empirical formula. The first such spectral series was observed by balmer in 1885 and is called the Balmer series of hydrogen. The first line with the longest wavelength (6563 Å) is named Ha. The next Hb, and so on. The series limit lies at 3646 Å, beyond which is a faint continuous spectrum. Balmer's formula for the wavelengths of the series is

$$\frac{1}{\lambda} = R\left(\frac{1}{2^2} - \frac{1}{n^2}\right), \; n = 3, 4, 5, \ldots \qquad \text{(Balmer)}$$

The quantity R is called the 'Rydberg constant' and has the value

$$R = 1.097 \times 10^7 \text{ meter}^{-1}.$$

The Ha line corresponds to n = 3, the H_β line to n = 4, and so on. The series limit experiment.

The Balmer series contains only those spectral lines which fall in the visible part of the hydrogen spectrum. The lines falling in the ultraviolet and infrared parts form other series. The lines in the ultraviolet form the Lyman series whose wavelengths are give by

$$\frac{1}{\lambda} = R\left(\frac{1}{1^2} - \frac{1}{n^2}\right), \; n = 2, 3, 4, \ldots \qquad \text{(Lyman)}$$

In the infrared, three spectral series have been observed whose lines have the wavelengths given by the formulas

$$\frac{1}{\lambda} = R\left(\frac{1}{3^2} - \frac{1}{n^2}\right), \; n = 3, 5, 6, \ldots \qquad \text{(Paschen)}$$

$$\frac{1}{\lambda} = R\left(\frac{1}{4^2} - \frac{1}{n^2}\right), \; n = 5, 6, 7, \ldots \qquad \text{(Brackett)}$$

$$\frac{1}{\lambda} = R\left(\frac{1}{5^2} - \frac{1}{n^2}\right), \; n = 6, 7, 8, \ldots \qquad \text{(Pfund)}$$

Bohr's Theory of Hydrogen Spectrum

The existence of sharp spectral lines cannot be explained from the classical electro-magnetic theory. Bohr explained it by applying Planck's quantum hypothesis to the Rutherford's atomic model. The Rutherfod's atom consists of a central massive nucleus containing the positive charge of the atom, and the electrons move round the nucleus in circular planetary orbits. The centripetal force required for the orbital motion is provided by the electrostatic attraction between the positively-charge nucleus and the negatively-charged electron.

Bohr proposed two postulates.

(i) An electron can move only in those orbits for which the angular momentum L of the electron is an integral multiple of $h/2\pi$ where h is Planck's constant. (Thus, Bohr quantised toe angular momentum of the electron.). The electron moving in any of the permitted orbits does not radiated energy in spite of its acceleration towards the centre of the orbit. The atom, therefore, is said to exist in a *stationary* state.

(ii) The emission (or absorption) of radiation by the atom takes place when an electron jumps for one permitted orbit to another. The radiation is emitted (or absorbed) as a single quantum (photon) whose energy hv is equal to the difference in energies of the electron in the two orbits involved. Thus, if E_i be the energy of the initial orbit of the electron and Z_f that of the final orbit then we have

$$hv = E_i - E_f,$$

where v is the frequency of the emitted (or absorbed) radiation.

Let e, m and v be the charge, mass and velocity of the electron (measured in column, kg and meter/sec respectively) and r the radius of the orbit measured in meter. The positive charge on the nucleus is Z e, where Z is the atomic number (Fig. 2.20) In case of hydrogen Z = 1, so that positive charge on the nucleus is e. As the centripetal force is provided by the electrostatic attraction, we have

$$\frac{mv^2}{r} = \frac{1}{4\pi\varepsilon_0}\frac{e^2}{r^2}$$

or $$m\,v^2 = \frac{e^2}{4\pi\varepsilon_0 r}. \qquad ...(1)$$

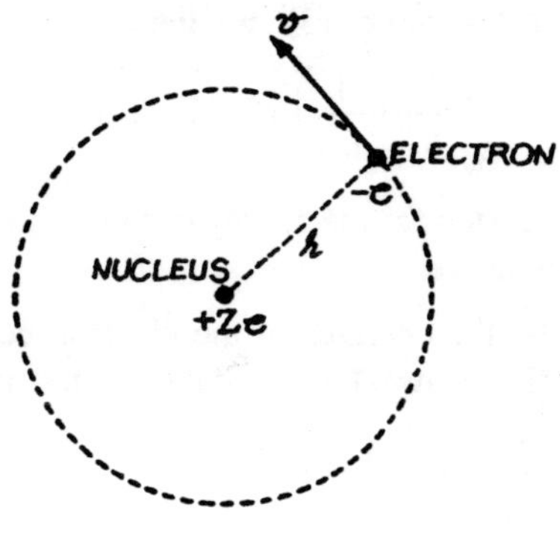

Fig. 2.20

From the first postulate, the angular momentum of the electron is given by

$$L = m\,v\,r = n\,\frac{h}{2\pi}, \qquad ...(2)$$

where n is called as 'quantum number' having values 1, 2, 3, ...

Squaring eq. (2) and dividing by eq. (1) we get

$$r = n^2\,\frac{h^2\varepsilon_0}{\pi m e^2},\ n = 1, 2, 3, ... \qquad ...(3)$$

This is the expression for the radius of the permitted orbits.

Now, the energy E of the electron in an orbit is the sum of kinetic and potential energies. The kinetic energy of the electron is

$$K = \frac{1}{2}\,mv^2 = \frac{e^2}{8\pi\varepsilon_0 r}. \qquad \text{[from eq. (i)]}$$

The potential energy at a distance r to infinity against the electrostatic attraction $\left(-\frac{e^2}{4\pi\varepsilon_0 r^2}\right)$, and is given by

$$U = \int_r^\infty -\frac{e^2}{4\pi\varepsilon_0 r^2}\,dr = \frac{1}{4\pi\varepsilon_0}\left[\frac{e^2}{r}\right]_r^\infty = -\frac{e^2}{4\pi\varepsilon_0 r}.$$

Hence the total energy of the electron is

$$E = K + U = \frac{e^2}{8\pi\varepsilon_0 r} - \frac{e^2}{4\pi\varepsilon_0 r} = -\frac{e^2}{8\pi\varepsilon_0 r}.$$

Substituting for r from eq. (3), we get

$$E = -\frac{me^4}{8\varepsilon_0{}^2h^2}\left(\frac{1}{n^2}\right) \quad n = 1, 2, 3... \qquad ...(4)$$

This is the expression for the energy of the electron in the n th orbit. We see that it in negative.

Let E_i and E_f be the energies of the electron corresponding to the initial (higher) and final (lower) orbits of the excited atom. Then, we have

$$E_i = \frac{me^4}{8\varepsilon_0{}^2h^2}\left(\frac{1}{n_i{}^2}\right)$$

and
$$E_f = -\frac{me^4}{8\varepsilon_0{}^2h^2}\left(\frac{1}{n_f{}^2}\right),$$

where n_i and n_f are the corresponding quantum numbers. The energy difference between these states is

$$E_i - E_f = \frac{me^4}{8\varepsilon_0{}^2h^2}\left(\frac{1}{n_f{}^2} - \frac{1}{n_i{}^2}\right).$$

Hence, from Bohr's second postulate, the frequency v of the emitted photon is

$$v = \frac{E_i - E_f}{h}.$$

$$= \frac{me^4}{8\varepsilon_0{}^2h^2}\left(\frac{1}{n_f{}^2} - \frac{1}{n_i{}^2}\right)$$

The corresponding wavelength l is given by

$$\frac{1}{\lambda} = \frac{v}{c} = \frac{me^4}{8\varepsilon_0{}^2ch^3}\left(\frac{1}{n_f{}^2} - \frac{1}{n_i{}^2}\right).$$

This equation indicates that, *Since ni and nf can take only integral values, the radiation emitted by excited hydrogen atoms should contain certain discarte wavelengths only.*

The value of the constant *term* $\frac{me^4}{8\varepsilon_0{}^2ch^3}$ comes out to be the same as the Rydberg constant R in the Balmer's empirical formula. Thus we have

$$\frac{1}{\lambda} = R\left(\frac{1}{n_f^2} - \frac{1}{n_i^2}\right).$$

Emission of Spectrum

When the hydrogen atom gets sufficient energy from outside by some means, the electron from an inner orbit of lower energy goes up to an outer orbit of higher energy. This excited state of the atom lasts for jumping down, it emits the difference in energy between the two orbits as electromagentic radiation. If the electron jumps from an orbit ni to an orbit nf, the wavelength of the emitted radiation will be

$$\frac{1}{\lambda} = R\left(\frac{1}{n_f^2} - \frac{1}{n_i^2}\right).$$

It is found that for

$n_f = 1,$ $n_i = 2, 3, 4, \ldots$ we obtain Lyman series,
$n_f = 2,$ $n_i = 3, 4, 5, \ldots$ we obtain Balmer series,
$n_f = 3,$ $n_i = 4, 5, 6$ we obtain Paschen series,
$n_f = 4,$ $n_i = 5, 6, 7, \ldots$ we obtain Brackett series,
$n_f = 5,$ $n_i = 6, 7, 8, \ldots$ we obtain Pfund series,

The corresponding energy level diagram is shown in Fig. 2.21. The top horizontal line represents zero energy, that is, energy of the electron outside the atom ($n = \infty$). The other horizontal lines represent energies of different orbits given by the formula

$$E = -\frac{me^4}{8\varepsilon_0^2 h^2}\left(\frac{1}{n^2}\right).$$

The arrows ending at the lines n = 1, 2, 3, 4, and 5 represent the transitions responsible for the Lyman, Balmer, Paschen, Brackett and Pfund series respectively.

Shortcomings of Bohr's Theory

Bohr's theory although very successful l in explaining the hydrogen spectrum and giving valuable information about atomic structure, has the following shortcomings.

(i) An individual line of hydrogen spectrum, when examined under a high resolving spectroscope, is found to be accompanied by a number of theory as such. It can, however, be explained when the relativistic variation in the mass of the electron and the electron 'spin' are taken into account.

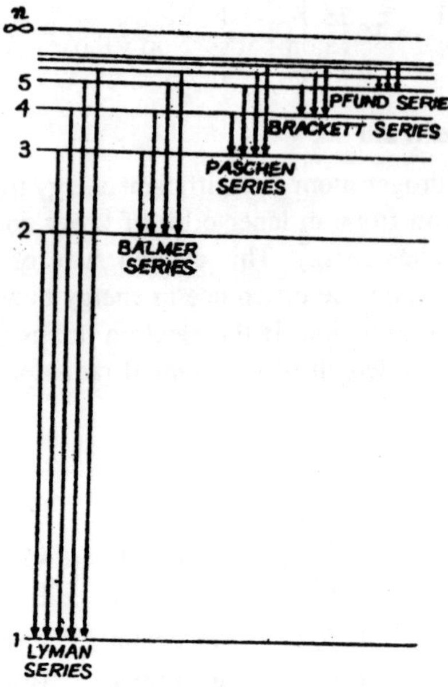

Fig. 2.21

(ii) Bohr's theory cannot explain the variation in intensity of the spectral lines of an element. The intensity can be explained by quantum mechanics.

(iii) The theory is only applicable to one-electron atoms such as hydrogen isotopes, singly-ionized helium, doubly-ionised lithium, etc. It does not explain the spectra of complex atoms.

(iv) The success of Bohr's theory in explaining the effect of magnetic field on spectral lines is only partial. The theory cannot explain the 'anomalous' Zeeman effect.

(v) The theory does not satisfactorily explain the distribution of electrons in atoms.

No Balmer Lines in Absorption Spectrum of Hydrogen

The Bohr theory also explains the absorption line spectrum of hydrogen. When a beam of - continuous light (containing all wavelengths) is passed through hydrogen and then sent into a spectrograph, a set of dark lines is obtained. In terms of quantum theory the incident light is a beam of quanta (photons) of all sorts of energies. Now according to

Bohr theory the hydrogen atoms absorb only those quanta whose energies correspond to transitions between its discrete energy levels.

The resulting excited hydrogen atoms re-radiate the absorbed energy almost atones but these photons come off in random directions with only a few in the same direction as the original beam of continuous light. The dark lines in the absorption spectrum are therefore never completely black. Obviously the absorption lines will have exactly the same frequencies as the emission lines.

Now it is found that all the emission lines of hydrogen spectrum do not appear in the absorption spectrum. The reason is that normally the atom is always in the ground state n = 1. Therefore, absorption transitions can only occur from n = 1 to n > 1. Hence lines of only the Lyman series can appear in absorption spectrum.

To obtain Balmer series in absorption, the atom must initially be in the state n = 2, because Balmer lines require transitions from n = 2 to n > 2. Since atoms are usually in the ground state. Balmer lines are not obtained in absorption.

DIFFERENT SPECTRAL LINES OF HYDROGEN

Every atom when excited emits radiations. The radiations form a line spectrum which is the characteristic of the emitter. Each atom has its own particular line spectrum which is regarded as the characteristic of that element to which the atom belongs. Like other elements, hydrogen possesses its own characteristic line spectrum. Hydrogen spectrum consists of a number of lines. These have been grouped into five series which are named after their discoverers, Fig 2.22. Many attempts were made by various workers to find a rule for underlying relationship which governed the wave lengths of these lines. We will discuss these by one.

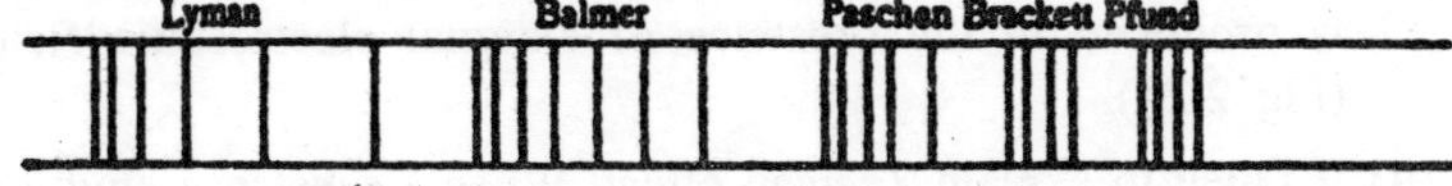

Fig. 2.22

Calculation of Rydberg's Constant

The energy of the hydrogen atom when the electron is in the n_1th orbit.

$$E_{n_1} = -\frac{me^4}{8\varepsilon_0^2h^2} \cdot \frac{1}{n_1^2} Z^2$$

and atom the energy of the atom when the electron is in the n_2th orbit

$$En_2 = -\frac{me^4}{8\varepsilon_0{}^2h^2}\cdot\frac{1}{n_2{}^2}Z^2$$

and the frequency of the photon emitted, when an electron jumps from n_2th to n_2th orbit is given by Bohr's third postulate, *i.e.*,

$$hv = En_1 - En_2 = -\frac{me^4}{8\varepsilon_0{}^2h^2}\left(\frac{1}{n_1{}^2}-\frac{1}{n_2{}^2}\right)Z^3$$

or
$$v = \frac{me^4}{8\varepsilon_0{}^2h^2}\left(\frac{1}{n_2{}^2}-\frac{1}{n_1{}^2}\right)Z^2.$$

Since the velocity of light $c = v\lambda$, we have

$$\frac{v}{c} = \frac{1}{\lambda} = \frac{me^4}{8\varepsilon_0{}^2ch^2}\left(\frac{1}{n_2{}^2}-\frac{1}{n_1{}^2}\right)Z^2 \quad ...(1)$$

where
$$R = \frac{me^4}{8\varepsilon_0{}^2ch^2}.$$

On substituting the values, we get $R = 109737.302\ cm^{-1}$. Here ε_0 is a constant whose numerical value is equal to $8.854. \times 10^{-12}$.

This value is found to be in agreement with the value obtained from the spectroscopic data of the Balmer's series which is $109677.67 cm^{-1}$.

Success of Bohr's Theory : This theory explains the following facts about atomic spectra.

FAILURES OF BOHR'S THEORY

(i) It failed to explain *Stark effect.* It is similar to Zeeman effect and is produced in the presence of external electrostatic field (Fig. 2.23).

(ii) It failed to explain Zeeman effect. When a substance emitting a line space trum is placed in a magnetic field, its lines would split up into a number of closely spaced lines. This is known as *Zeeman effect* (Fig. 2.23)

(iii) When spectral lines of hydrogen are observed very closely, each line is further made up of much closely spaced lines.

(iv) It failed to explain spectra of atoms other than hydrogen.

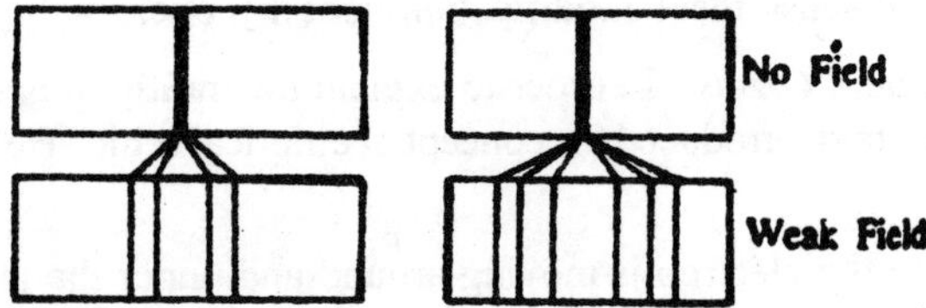

Fig. 2.23

(v) Another fundamental objection against Bohr's theory is that is used two theories which are opposed to each other, *i.e.*, *quantum theory* was used to account for the existence of stationary orbits and for frequencies of radiations emitted while motion of electron in its orbit obeyed *the law of classical mechanics.*

(vi) In the light of the Heisenberg's uncertainty principle, both the velocity and the position of the electron cannot be specified at a given time as Bohr did. Thus, it is another weakness of Bohr's theory.

(vii) Another weakness of Bohr's theory is that it did not throw light on the distribution and arrangement of electrons in atoms.

(viii) Bohr assumed that the nucleus is stationary and only electrons are revolving around it. Detailed facts have revealed that both the nucleus and the electrons move in closed orbits around their centre of mass. Therefore, it is major weakness of Bohr's theory.

THE SOMMERFELD MODEL

Inspite of many success, Bohr's theory was found to be inadequate to explain certain details in the spectrum of hydrogen. For instance, the Hx line of the Balmer series was found to contain several components. This fine structure of spectral lines could not be explained by Bohr's theory which assumed that there was only one orbit for each quantum number, whereas the observed fine structure suggested that for any given quantum number n there might be several orbits of slightly different energies.

Sommerfeld, in 1915, guided by the above suggestion modified Bohr's theory by introducing the following modifications.

(a) Concept of elliptical orbits, and

(b) Relativistic variation of the mass of electron.

We will discuss these modifications one by one.

(a) *Elliptical Orbits :* In order to explain the multiplicity of spectral lines. Sommerfeld introduced the concept of elliptical orbits. He postulated that :

(i) Since the electron is moving around and under the influence of a massive nucleus, like a planet around the central massive sun, it might describe elliptical orbits as well.

(ii) While retaining the first circular orbit suggested by Bohr, Sommerfeld assumed one additional elliptical orbit in the case of Bohr's second orbit and added two additional elliptical orbits to Bohr's third orbit and so on (Fig. 2.24).

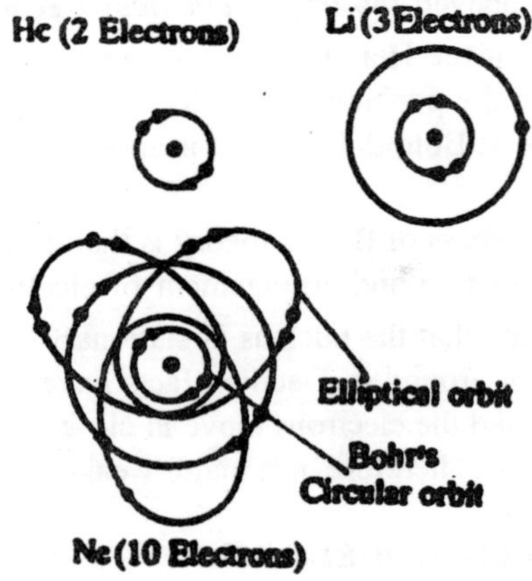

Fig. 2.24

(iii) The nucleus is one of the foci for all these orbits.

(iv) In an elliptical orbit, the major and minoraxes will differ in lengths, but as the orbit broadens, they will approach each other and becomes equal when the orbit becomes circular. Circular orbit is only a special case of the elliptical orbit. The orbit is defined by two quantum numbers, *i.e.*, n and k which correspond to the major and minor axes of the ellipse. They are related as follows.

$$\frac{\text{Principal quantum number}}{\text{Azimuthal quantum number}} = \frac{n}{k} = \frac{\text{length of major axis}}{\text{length of minor axis}}.$$

It is clear from the above that for any given value of n, k cannot be zero as in that case the ellipse would degenerate into a straight line passing through the nucleus. Further, k cannot be more than n since b is always less than a. When n = k the path becomes circular. This is a limit to the number of different orbits that an electron may have in any energy level.

Such possible paths are referred to as sub-levels. For a given value of n, k can have only n different values which mean that there may be only n elliptical orbits or sub levels with different eccentricity. For elliptical orbits the values of k would be (n – 1), (n – 2) etc., down to k = 1. When k = 0, the ellipse would be a straight line. When n = 3, k = 3, (circular), 2, 1, two of sommerfeld model lines in its subdivision of the original Bohr stationary levels into various sub-levels of slightly differing energies as given by difference in orbit shapes.

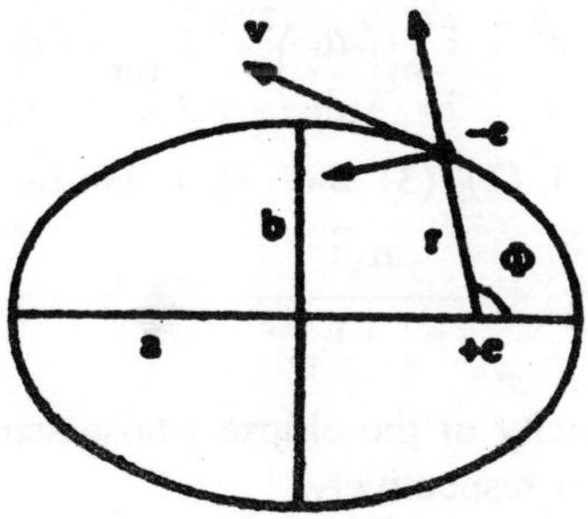

Fig. 2.25

Simple mathematical approach to Sommerfeld model : Let us consider the motion of an electron (– c) in an elliptical orbit as shown in (Fig. 2.25). Its position at any instant can be fixed in terms of polar coordinates, r and f where r is the distance of the electron from the nucleus (+ e) at one of the foci of the ellipse and f is the angle which the radius vector makes with the major axis of the ellipse. The tangential velocity v of the moving electron at any instant can be resolved into two components.

(i) One radial, *i.e., dr/dt* along the radius vector. Corresponding to this there will be a radial momentum Pr equal to m (dr/dt).

(ii) Other transverse, *i.e.*, m at right angles to the radius vector equal to r (df/dt). Corresponding to this there is angular or azimuthal momentum of equal to mr^2 (df/dt), where m is the mass of the electron.

As Sommerfeld considered the circular orbits to be special cases of elliptical orbits, he assumed that the elliptical orbits should satisfy the quantum condition of Bohr just as the circular orbits, *i.e.*,

$$\oint p_r\, dr = n_r h \qquad ...(1)$$

$$\oint p_f\, d\phi = n_\phi l_t \qquad ...(2)$$

Thus, the single n of Bohr's theory has been replaced by the two new quantum numbers n_r and n_ϕ, *i.e.*,

$$n = n_r + n_\phi \qquad ...(3)$$

where n_r is the radial quantum number and nf is the angular or azimuthal quantum number. The total energy E is given by

$$E = \text{P.E.} + \text{K.E.}$$

$$= \text{P.E.} + \text{Radial K.E.} + \text{Angular K.E.}$$

$$= -\frac{e^2}{r} + \frac{1}{2}m\left(\frac{dr}{dt}\right)^2 + \frac{1}{2}mr^2\left(\frac{d\phi}{dt}\right)^2 \qquad ...(4)$$

From equations (1), (2), (3) and (4), it can be shown that

$$1 - e^2 = \frac{b^2}{a^2} = \frac{n_\phi^{\ 2}}{(n_\phi + n_r)^2}$$

where e is the eccentricity of the ellipse whose semi-major and semi-minor axes are a and b respectively.

and $$E = -\frac{2\pi^2 mZ^2e^4}{h^2}\left(\frac{1}{n_\phi + n_r}\right)^2 \qquad ...(6)$$

From equation (5), we get

$$(1 - e^2)^{1/2n} = \frac{b}{a} = \frac{n_\phi}{n_\phi + n_r} = \frac{n_\phi}{n}.$$

Thus, equation (6) can be written as

$$E = \frac{2\pi^2 mZ^2e^4}{n^2h^2} \qquad ...(7)$$

From equation (7), it follows that

(i) The elliptical orbits which have the same value of n, though of different eccentricities, have the same energy. It is not correct.

(ii) All orbits having the same values of the semi major axis possess the same energy, since the length for the semi-major axis is determined solely by the total quantum number n (= $n_f + n_r$). Hence the energy of any of these permitted elliptical orbits is identical with that of a circular Bohr orbit, whose radius is equal to the semi-major axis of the ellipse.

From the above it follows that the theory of elliptical orbits in spite of two new quantising conditions involved, introduces not new energy levels other than those given by Bohr's theory of circular orbits. No new spectral lines, which would explain the fine structure, are there fore predicted. Thus, Sommerfeld has to modify his own model by suggesting the variation of mass of electron with velocity called relativistic variation.

Relativistic Variation of the Mass of Electron : The velocity if an electron moving in an es path of the electron is found to be no longer a simple ellipse, it is indeed, no longer a closed figure, but is transformed into a complicated curve known as rosette –a processing ellipse (Fig. 2.26). The total energy E of the system corrected by the relativistic variation of the mass of electron can be shown to be

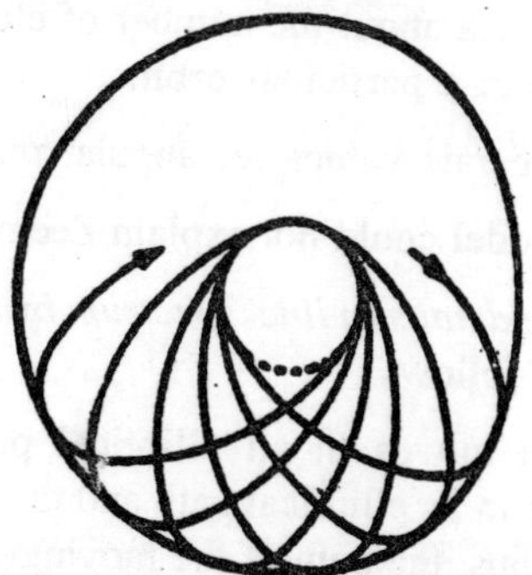

Fig. 2.26 : Sommerfeld elliptical orbit precessing about an axis through one of its focci.

$$E = -\frac{2\pi^2 m Z^2 e^4}{h^2}\left[\frac{1}{n^2} + \frac{4\pi^2 e^4 Z^2}{c^2 h^2}\left(\frac{n}{n\phi} - \frac{3}{4}\right)\frac{1}{n^4} + \ldots\right] \quad \ldots(8)$$

The relativistic correction, therefore, results in splitting up a given energy level E_α into n levels differing slightly from one another in energy. The splitting up of each energy levels gives rise to a fine structure of single spectral line, one application of the usual Bohr frequency conditions, When one explains the fine structure of lines, one should not take into account all theoretical possible transitions of the electron from one elliptical orbit to another that actually occur.

According to a principal known as the selection rule, transition can take place only between orbits for which the azimuthal quantum number changes by + 1 or – 1, *i.e.*, nf = ± 1.

Limitations of Sommerfeld model : (i) Though Sommerfeld's theory is fairly well verified yet it leads only to three components for the structure of Hx line, while there should be really five. Thus the relativistic atom model has met with a partial success.

(i) The concept of elliptical orbits due to Sommerfeld gives the correct total (n) of possible azimuthal quantum number, but the actual values are not correct. The experimental studies as well as theoretical treatment based on wave mechanics show that azimuthal quantum number can be zero, so that the values can be 0, 1, 2,... –1, thus making a total of n possibilities. The correct and new azimuthal quantum number is denoted by 'l' to avoid the confusion. Thus, l is equal to K – 1.

(ii) It provides no idea about the number of electron which can be accommodated in a particular orbit.

(iii) It provides macerate values for angular momentum.

(iv) Sommerfeld model could not explain Zeeman and Stark effect.

Explanation of fine details in line spectrum by Sommerfeld model : It can be explained as follows.

When an electron moves in an elliptical path there will be a displacement each time in its elliptical path and thus resulting in a small difference in energy. Thus, the path of the moving electron is no longer a simple closed ellipse but a complicated curve, known as rosette, which is made up of different elliptical orbits of slightly different energies. This *difference in energies of elliptical orbits explains the existence of components of spectral lines in the spectrum of higher elements.*

Criticism of the theory : (1) Suppose there are two orbits with principal quantum numbers, n = 2 and n = 3, (Fig. 2.27) respectively. Consider an electron which falls from third to the second orbit. There may be six possible transitions, each giving one fine line. But actual observations yield only five lines (Fig. 2.27).

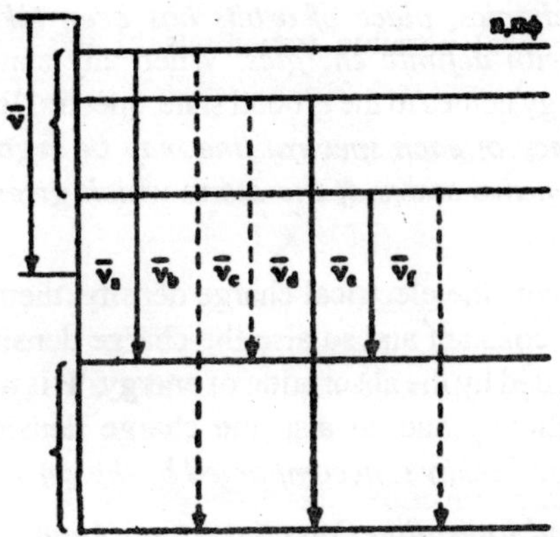

Fig. 2.27

Possibilities.

(i) $3K_3 \rightarrow 2K_2$ *i.e.*, $\Delta K = 3 - 2 = +1$ (iv) $3K_2 \rightarrow 2K_1$ *i.e.*, $\Delta K = 2 - 1 = +1$

(ii) $3K_2 \rightarrow 2K_1$ *i.e.*, $\Delta K = 3 - 1 = +2$ (v) $3K_1 \rightarrow 3K_2$ *i.e.*, $\Delta K = 1 - 2 = -1$

(iii) $3K_2 \rightarrow 2K_3$ *i.e.*, $\Delta K = 2 - 2 = 0$ (vi) $3K_1 \rightarrow 2K_1$ *i.e.*, $\Delta K = 1 - 1 = 0$

where ΔK indicates the change in azimuthal quantum number. Sommerfeld applied *selection rule* which limited the number of transitions. According to such rule only those transitions are possible for which the quantum number K changes by -1 or $+1$ *i.e.*, $\Delta K = \pm 1$. Hence transitions like $(3K_1 \rightarrow 2K_1)$, $(3K_3 \rightarrow 2K_1)$ and $(3K_2 \rightarrow 2K_2)$ are forbidden. Therefore, according to this theory, the fine structure of H_α lines should made only of three lines. But actually it splits up into five lines. Sommerfeld theory fails to explain complicated systems.

(2) Sommerfeld's model could not explain Zeeman and Stark effect.

(3) The model gives no information regarding the relative intensities of lines, whose frequencies alone are predicted.

Wave mechanics and spectral lines : According to wave mechanics orbits do not exist in the atom. Thus, the interpretation of the emission of radiation due to a jump of the electron from outer to the inner orbits does not hold good.

In wave mechanics, place of orbits has been taken by stationary states of atoms with definite energies. When any state excited by the absorption of energy comes to the ground state, spectral lines are produced. Thus, *the frequency of each spectral line may be regarded as a 'beat' frequency between two states of the atom, which gives the same result as that of Bohr.*

As Ψ^2 represents the electrical charge density, then in the stationary state this remains constant and so also the charge density. But when any state becomes excited by the absorption of energy, Ψ is no longer constant and varies periodically and so also the charge density. *This periodic variation in charge density is accompanied by the emission of radiation.*

Excitation and Ionisation Potentials of an Atom : According to the Bohr's theory, in an atom there are certain discrete orbits only in which an electron can revolve without radiating energy. Each of these orbits of a given atom is characterised by a certain (quantised) energy. Hence we say that there are certain discrete energy levels in an atom.

When an electron in an atom absorbs sufficient energy from an outside source, it rises from its present energy level to an higher level. The atom is than said to be 'excited'. This excited state lasts only for about 10^{-8} second after which the electron jumps back to the inner level, emitting the absorbed energy is emitted as electromagnetic radiations.

The energy required to excite or to ionise a given atom is perfectly definite.

For example, in case of hydrogen atom, the energy of the nth orbit is

$$E_n = -\frac{me^4}{8\varepsilon_0^2 h^2}\left(\frac{1}{n^2}\right).$$

Substituting the known values of

$$m\ (= 9.1 \times 10^{-31}\ \text{kg}),$$
$$e\ (= 1.6 \times 10^{-19}\text{C}),$$
$$h\ (= 6.62 \times 10^{-34}\ \text{J-s})$$

and ε_0 ($= 8.85 \times 10^{-12}\text{C}^2/\text{N m}^2$), we get

$$E_n = -2.17 \times 10^{-18}\left(\frac{1}{n^2}\right)\ \text{joule (J)}$$

$$= -\frac{2.17 \times 10^{-18}}{1.6 \times 10^{-19}} \left(\frac{1}{n^2} \right)$$

$$= -\frac{13.6}{n^2} \text{ eV.} \qquad [1 \text{ eV} = 1.6 \times 10^{-19} \text{ J}]$$

Putting n = 1, 2, 3, ... ∞, the energies of the 1st, 2nd, 3rd, ... ∞ energy levels come out to be – 13.6, – 3.4 – 1.51, ... 0 electron volts respectively. Hence the energy to be supplied to the atom to raise an electron from the first to the second orbit is (13.6 – 3.4) = 10.2 electron-volts, from first to the third orbit is (13.6 – 1.51) = 12.09 electron volts,* and to the ionized state is (13.6 – 0) = 13.6 electron volts.

Now, the atoms are usually excited or ionised by bombarding them with electrons accelerated through a potential. But the excitation or ionisation of an atom can take place only when thc bombarding electron has the required amount of energy. *The minimum accelerating potential which imparts to the bombarding electron an energy sufficient to excite a given atom is called the 'excitation potential' of the atom. Similarly, the minimum accelerating potential which imparts to the bombarding electron an energy sufficient to ionise the atom is called the 'ionisation potential' of the atom.*

Franck-Hertz Experiment

Franck and Hertz, in 1914, performed a series of experiments to measure the excitation potentials of atoms of different elements. These experiments showed directly that in an atom discrete energy levels do exist.

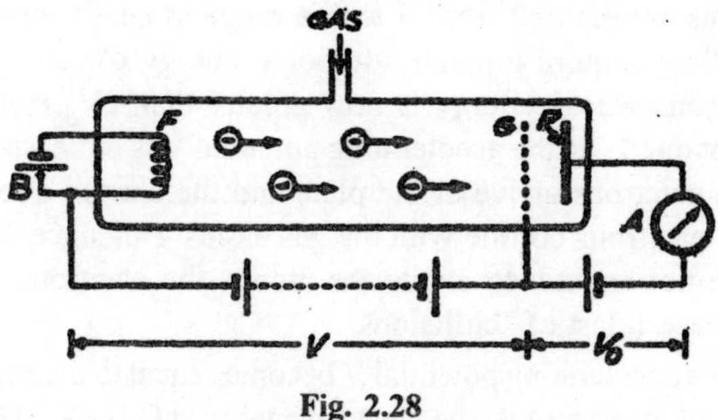

Fig. 2.28

Their apparatus is shown in Fig. 2.28. It consists of a glass tube in which are mounted a filament F, a grid G, and a plate P, as shown. The filament F is heated by a small battery B. An accelerating potential V is applied between F and G, and *a small fixed* retarding potential V_0 (about 0.5 volt) between G and P. The gas of the element whose atoms are to be studied is introduced in the tube at about 1 mm of mercury pressure.

The electrons emitted from the hot filament F are accelerated between F and G by the potential V, and retarded between G and P by the potential V_0. Thus, only electrons having energies greater than eV_0 at G are able to reach P. The current to P is recorded by an ammeter A.

The current is plotted against the *accelerating* potential V which is gradually increased from zero. The curve obtained shows a series of regularly spaced peaks (Fig. 2.29).

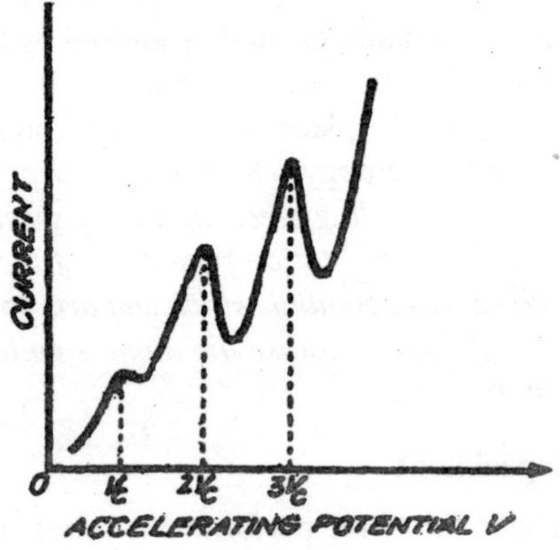

Fig. 2.29

Interpretation of the Curve

Electrons are emitted from F with a range of small energies. In the beginning, they acquire a small additional energy eV on reaching G. Those electrons whose energy is now greater than eV_0 reach P, and a current is obtained. As the accelerating potential V is increased, more and more of the electrons arrive at the plate and the current rises. Between F and G the electrons collide with the gas atoms. But since the electrons energies are not enough to excite the atoms, the electrons do not lose energy in these (elastic) collisions.

When the accelerating potential V becomes equal to the first excitation potential V_c of the as atoms, the electron energy at G is eV_c. The electrons can now suffer *inelastic* collisions with gas atoms near G and excite them to an energy level above their ground state. The electrons which do so, lose their energy and are unable to reach P against the retarding potential.

Hence the current drops sharply. Thus, the position of the first peak gives the excitation potential V_c of the gas atoms.

As the accelerating potential V is increased further, the electrons suffer inelastic collisions nearer and nearer to the filament, F, so that when they reach the grid G, once again they acquire enough energy to reach the plate P. The current thus begins to rise again. When V becomes equal to $2V_c$, a second inelastic collision occurs near the grid and the current drops again. Thus, the second peak gives $2V_c$. This process repeats as V is further increased.

The first peak occurs at a potential slightly less than V_c. This is because electrons are emitted from the filament with a finite velocity and hence with some energy. The true excitation potential V_c is obtained by measuring the difference between two successive peaks.

Limitations : Atoms have more than one excitation potentials and also an ionisation potential. Therefore, in an actual experiment the curve obtained is quite complicated and we cannot distinguish between which are excitation and which are ionisation potentials.

Demonstration of the Existence of Discrete Energy Levels : The experiment shows that electrons transfer energy to the atoms indiscrete amounts, and that they cannot excite atoms if their energy is less than eV_c. Franck and Hertz demonstrated it directly by observing the spectrum of the gas during electrons collisions. They showed that a particular spectral line does not appear until the electron energy reaches a *threshold* value. For example, in the case of mercury vapour, they found that minimum electron energy of 4.9 eV was required to obtain the 2536 Å line of mercury, and a photon of 2536 Å light has an energy of just 4.9 eV. This shows that discrete energy levels do exist in atom and the electrons of the atom can exist only in these levels.

Bohr's Correspondence Principle : The quantum theory gives results which are altogether different from those given by the classical theory in the microscopic world. Yet the two theories approach each other as the quantum number in question increases. This fact was pointed out by Bohr in 1932 who enunciated the following principle.

The predictions of the quantum theory for the behaviour of any physical system must coincide with the predictions of the classical theory in the limit in which the quantum numbers specifying the state of the system becomes very large. This is knows as 'Bohr's correspondence principle'. We establish it by means of an example

According to classical theory, an electron moving in a circular orbit must emit radiation only of a particular frequency, namely, the frequency of revolution of the electron itself (or its integral multiples). According to quantum theory, on the other hand, radiation is emitted when the electron jumps from one orbit to the other and the frequency of radiation is determined by the difference in energy between the two orbits. Let us try to find a relation between the two frequencies.

The basic equations in the Bohr's theory of the hydrogen atom are the following.

$$\frac{mv^2}{r} = \frac{1}{4\pi\varepsilon_0}\frac{e^2}{r^2}$$

and $$mvr = n\frac{h}{2\pi}, \qquad n = 1, 2, 3, ...$$

where v is the velocity of the electron of mass m in the orbit of radius r. These equations give

$$v = \frac{nh}{2\pi mr}$$

and $$r = \frac{n^2h^2\varepsilon_0}{\pi me^2}.$$

Hence the classical frequency of revolution of the electron is

$$f = \frac{\text{electron speed}}{\text{orbit circumference}} = \frac{v}{2\pi r}$$

$$= \frac{nh}{4\pi^2 mr^2} = \frac{nh}{4\pi^2 m\left(\dfrac{n^2h^2\varepsilon_0}{\pi me^2}\right)^2}$$

$$= \frac{me^2}{4\varepsilon_0{}^2n^3h^3} = \frac{me^4}{8\varepsilon_0{}^2ch^3}\frac{2c}{n^3}$$

$$= \frac{2Rc}{n^3} \qquad ...(i)$$

where $R\left(=\dfrac{me^4}{8\varepsilon_0{}^2ch^3}\right)$ is the Rydberg constant. This is the frequency which must be radiated by the moving electron classically.

Now, the frequency radiated on the basis of quantum theory is given by

$$v = \frac{E_i - E_f}{h} = \left(= \frac{me^4}{8\varepsilon_0{}^2 h^3} \right)\left(\frac{1}{n_f{}^2} - \frac{1}{n_i{}^2} \right)$$, when the electron drops

from the ni th orbit the nf the orbit. Again, introducing R, we have

$$v = Rc\left(\frac{1}{n_f{}^2} - \frac{1}{n_i{}^2} \right).$$

This may be written in the form

$$v = Rc\,\frac{(n_i - n_f)(n_i + n_f)}{n_f{}^2 n_i{}^2} \qquad \text{...(ii)}$$

Let us write $n_f = n$ and $n_i = n + 1$. Then the frequency of the emitted radiation for the transition $n + 1 \rightarrow$ (so that $\Delta n = 1$) is given by

$$v = Rc\,\frac{2n+1}{n^2\,(n+1)^2}.$$

When n is very large, we can write $\frac{2n+1}{n^2\,(n+1)^2} \simeq \frac{2}{n^3}$.

Under this condition, the emitted frequency is

$$v = \frac{2Rc}{n^3}. \qquad \text{...(iii)}$$

Equations (i) and (iii) yield

$$v = f.$$

If we consider transitions Dn = 2, 3, 4, ... we shall have

$$v = 2f, 3f, 4f, ...$$

Thus, for very large quantum numbers, the quantum theory frequency v of the radiation is identical with the classical frequency of the revolution (or its harmonics) of the electron in the orbits. This is in accordance with Bohr's correspondence principle.

SPECTRA OF ALKALI METALS

Like the hydrogen atom, the spectra of the alkali metals consist of a series of lines with regularly decreasing separation and decreasing intensity. One important feature of the alkali-spectra is that the lines are mostly doublets though each and every line has not been resolved into two in the spectroscope

Quantum defect : The spectral lines obtained from alkali metals cannot be represented by a formula exactly analogous to Balmer's formula. As the lines converge to a limit, one is able to represent them by the difference of two spectral terms. Thus the values of the terms are not of Balmer's form *i.e.*, R/n^2 but of the form R/n^{*2}, where n is the principal quantum number and n* is the effective quantum number and its value is given by $(n + \alpha)$, where a is a small quantity and is known as the quantum defect.

The *Magnitude* of the *quantum defect* depends upon the penetration of the core by the orbit of the emitting electron. Increased penetration of the core depends upon the *shape of the orbit*. It is, therefore, expected that a series of different values of a will occur for different values of azimuthal quantum number.

There are *four* types of quantum defect (α) designated by S, P, D, and F. These words stand for the first letter of sharp, principal, diffuse and fundamental. The names of the series are attributed according to the order, as to which of them appears in the quantum defect of the running term. The principal series is most easily excited and is hence so calculate. The lines in the sharp series are well defined and those of the diffuse series are blurred.

In emission spectra, other series in addition to the principal series may be observed for the alkali metals. These series partly overlap each other. The last named series viz., the fundamental of Bergmann series, occur in the infra-red.

However, the frequencies of the lines of the four series may consequently be represented by the following general expression.

Sharp series $\bar{v} = aP - nS$, Principal series $\bar{v} = aS - nP$,

Diffuse series $\bar{v} = aP - nD$, Fundamental series $\bar{v} = aD - nF$,

where a is a constant, and n is the principal quantum number.

***Explanation*:**

As earlier stated, the volume of n* is given by $(n + \alpha)$. Ωηεν we calculate the values of n and n* for the S, P, D and F terms of the different alkali metals, we get the following values.

	Terms Series	*n*(= n + α)*	*n*	*l*
Lithium	S	1.59	2	0
	P	1.96	2	0
	D	3.00	3	2
	F	4.00	4	3
	Terms Series	n* (= n + α)	*n*	*l*
Sodium	S	1.63	3	0
	P	2.12	3	1
	D	2.99	3	2
	F	4.00	4	3.

From the above table that the discrepancies between n* and n decrease in the order of S, P, D, and F of terms series. The discrepancies [n* – n (n + α) – n = α], also called quantum defect have been explained as follows.

If we compare the structure of an alkali metal with hydrogen, it may be regarded that its nucleus is a single unit and around which the additional electron rotates. The additional electron is known as *optical electron.*

In hydrogen, the electron remains in its fixed orbits and therefore, definite spectral lines will be observed.

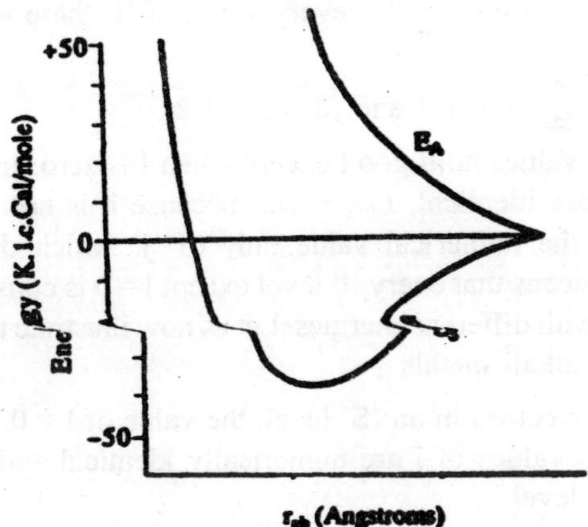

Fig. 2.30

In the case of an alkali metal, the optical electron does not always rotate in its own orbits. It has a tendency to penetrate into the other inner orbits and this penetration depends upon the eccentricity of the orbit. The greater the eccentricity of the orbit, the more closely the electron may approach the nucleus in the course of its rotation.

It is, therefore, more likely that greater the penetration of the inner levels, the greater will be divergence from hydrogen like behaviour. A simplified energy level diagram for sodium is shown in (Fig. 2.30), those for the other alkali metals are similar, except for the difference in the principal quantum number of the optical electrons. The integers indicate the various values of the principal quantum number n in each spectral state, represented by S, P,D and ZF. It may be noted that the S, P, D and F terms are actually doublets but this is not shown in the Fig. 2.30.

EXPLANATION OF THE DOUBLET STRUCTURE

When we study the spectra of elements other than hydrogen, we observe that many lines are actually multiples consisting of two, three or more lines close together.

In order to explain Uhlenbeck and Goldsmit (9135) suggested that this multiplicity of spectral lines is due to the spin of the electron. The contribution due to the spin is quantised and is expressed in terms of spin quantum numbers. The value of this number is + 1/2 and – 1/2. The resultant of the spin and azimuthal quantum number is given by j = 1 + s, where 'j' is called the *inner quantum number.* As 's' can have + 1/2 and + 1/2, it follows that for every value of 'l' there are two values of 'j' viz.,

$$j_1 = l + 1/2 \text{ and } j2 = l - 1/2.$$

The above values hold good except when l is zero. In that case the two 'j' values are identical, *i.e.*, + 1/2, because it is not the + ve or – ve sign but is the numerical value only of 'j' which determines the momentum. It means that every 'l' level except l = 0 is consequently split into two levels with different energies. Let us now illustrate this discussion by applying to alkali metals.

(i) For an electron in an 'S' level, the value of l = 0. It means that the two values of j are numerically identical and hence it is a singlet level.

$$j_1 = l + \frac{1}{2} \text{ and } j_2 = l - \frac{1}{2}$$

$$j_1 = 0 + \frac{1}{2} \text{ and } j_2 = 0 - \frac{1}{2}$$

$$j_1 = +\frac{1}{2} \text{ and } j_2 = -\frac{1}{2}$$

(ii) In the P level, l = 1 and so the corresponding values of 'j' are 3/2 and 1/2.

$$j_1 = 1 + \frac{1}{2} \text{ and } j_2 = 1 - \frac{1}{2}$$

$$j_1 = 1 + \frac{1}{2} \text{ and } j_2 = 1 - \frac{1}{2}$$

$$j_1 = \frac{3}{2}\ j_2 = \frac{1}{2}.$$

Thus, each spectral line in 'P' level is doublet.

Similarly in the D and F levels, the values of 'j' are 5/2 and 3/2, 7/2 respectively. Again 'j' has two values and therefore each line will be a doublet. The doublet of the sharp series of spectra may be represented by the expressions

$$\bar{v} = aP_{1/2} - nS$$

and
$$\bar{v} = aP_{3/2} - nS$$

where S is a singlet and P is a doublet.

PHOTOCHEMICAL INHIBITION

There are certain substances which are able to retard the rate of a photochemical reaction when present in trace amounts. Such substances are known as *inhibitors* and the phenomenon is known as *photochemical inhibition.* Some examples are.

(i) Traces of nitric oxide and propylene lower the quantum yield of the photochemical combination of hydrogen and chlorine.

(ii) Traces of impurities like NH_3 which when present in the hydrogen and chlorine reaction lower its quantum yield from 10^6 to 10^4. In 1905, Chapman showed that the inhibitor action of NH_3 is due to the side reaction of NCl_3 which may result due to the reaction of NH_3 with Cl_2.

$$NCl_3 + 3HCl \rightleftharpoons NH_3 + 3Cl_3$$

$$NCl_3 + 3H_2O \rightleftharpoons NH_3 + 3HClO$$

At equilibrium the concentration of NH_3 is maintained and the actual formation of HCl is reduced,

(iii) SO_2 and O_2 were also found to be inhibitors in various photochemical reactions.

Explanation. It is generally accepted that photo-inhibitors interrupt the chain reactions by removing chain carrier atoms or radicals.

PERIOD OF INDUCTION

Many reactions are characterised by an initial period during which the process appears to be silent. This time is known as period of induction. Some examples are.

(i) An induction period is observed in hydrogen and chlorine reaction when certain impurities are present. With pure hydrogen and chlorine, no induction period is observed.

(ii) An induction period is also observed in the absorption of As_2O_2 by mercuric oxide in the presence of light.

Period of induction is due to presence of impurities like ammonia. These substances act as inhibitors by breaking these chains. When these impurities are completely converted into nitrogen and ammonium chloride during the reaction, then only normal photochemical reaction can take place.

PHOTOSTATIONARY STATE OR PHOTOCHEMICAL EQUILIBRIUM

A state of photochemical equilibrium is said to exist in a reaction when the rates of two opposing reactions of which at least one is light sensitive, become equal under the influence of light radiation.

Types : A number of cases of the photostationary state have been studied. These cases fall into two categories.

(a) In the First Category, only one reaction is light sensitive

$$A + B \underset{\text{dark}}{\overset{\text{light}}{\rightleftharpoons}} C + D$$

Some examples of such type are.

(i) *Dissociation of Nitrogen Dioxide*. When the vapour of nitrogen dioxide is exposed to light radiation of wavelength less than 3700Å, it dissociates to form nitric oxide and oxygen but the combination of these products is a dark reaction. Hence the equilibrium is

$$2NO_2 \underset{\text{dark}}{\overset{\text{light}}{\rightleftharpoons}} 2NO + O_2$$

When the reaction just starts, the pressure of the gas rises due to the decomposition of the nitrogen dioxide but becomes constant as soon as the stationary state it reached.

Dimerisation of anthracene. Another example of a photo-stationary state is the dimerisation of anthracene in solution in which the forward process is light sensitive but the back ward reaction is a thermal reaction, and so the equilibrium is

$$2_{14}H_{10} \underset{\text{dark}}{\overset{\text{light}}{\rightleftharpoons}} C_{28}H_{20}$$

Let us calculate the equilibrium constant of the reaction of category (a),

light $$A + B \underset{\text{dark}}{\overset{\text{light}}{\rightleftharpoons}} C + D$$

As the forward reaction is light sensitive, its rate will not depend upon the concentrations of A and B but upon the intensity of absorbed light. Hence.

Rate of the forward reaction = $k_1 I_{abs}$ where k_1 is constant for the forward reaction. As the backward reaction is a dark or thermal reaction, its rate will depend upon the concentrations of products C and D, *i.e.,*

Rate of backward reaction = k_2 [C] [D].

At photostationary state.

Rate of forward reaction = Rate of backward reaction

or $$\frac{k_1}{k_2} = \frac{[C][D]}{I_{abs}} \quad \text{or} \quad K = \frac{[C][D]}{I_{abs}}$$

where K is the equilibrium constant and is defined as the ratio of velocity constants of two opposing reactions.

The equilibrium constants for photochemical equilibria are constant only for a given light intensity, and vary as the latter is charged. Photochemical equilibrium constant is generally independent of temperature changes if the intensity of light is kept constant.

(b) In the Second Category, both the reactions are light-sensitive.

$$A + B \underset{\text{light}}{\overset{\text{light}}{\rightleftharpoons}} C + D$$

Few examples of this category are

(i) *Formation of sulphur trioxide.* It is an example of a photostationary state in which both the forward and back ward processes are light sensitive.

$$2SO_2 + O_2 \underset{\text{light}}{\overset{\text{light}}{\rightleftharpoons}} 2SO_3 .$$

The value of photochemical equilibrium constant of this reaction is almost independent of temperature from 500° to 800°C.

(ii) *Isomerisation of maleic acid into fumaric acid.* Another example is the isomerisation of maleic acid into fumaric acid in which both forward and backward reactions are light sensitive.

```
H—C—COOH     light          H—C—COOH
   ||        <====>            ||
H—C—COOH     light      HOOC—C—H
 Maleic acid               Fumaric acid
```

The value of equilibrium constant can be calculated in the similar manner as given in category (a).

Application. The concept of photostationary state has been used to explain the phenomenon of *vision.* When the light sensitive substance in eye called visual purple exposed to light, the latter gets bleached to form visual yellow.

Thus, an equilibrium state is established between visual purple and visual yellow. In dark the visual purple accumulates and eye becomes sensitive. When exposed to light further, it results in the phenomenon of *dazzling.*

BIOLUMINESCENCE

Certain living organisms emit light and show the phenomenon of chemiluminescence. It is known as bioluminescence. The phenomenon of bioluminescence was investigated by *E. N. Harvey* (1915). Some interesting examples are.

(i) The cold light produced by certain living organisms like fire-fly or glow worm is probably due to the oxidation of protein, luciferin, by the atmospheric oxygen in the presence of enzyme luciferase. The firefly emits light having a maximum intensity of wavelength of 5700Å. It can be easily judged by naked eye.

(ii) Certain marine animals also emit light. The protozoa Noctiluca of the sea produces phosphorescence. It has also been observed that many deep sea fishes have powerful organs which emit dazzling light.

PHOTOSYNTHESIS

The energy which supports the activities of most living organisms on the earth is derived directly or indirectly form the energy of sunlight through photosynthesis *Photo* = light, *synthesis* = to build up). *It is a process in which simple carbohydrates are synthesised from water and carbon dioxide in the chlorophyll containing tissues of plants in the presence of sunlight. Oxygen being a by-product is given out*

Kamen (1963) has defined photosynthesis as a *series of processes in which electromagnetic energy is converted into chemical free energy which can be used for biosynthesis.* By a single overall equation, it may be shown as below.

$$CO_2 + H_2O \xrightarrow[\text{chlorophyll}]{hv} 1/6\,(C_6H_{12}O_6) + O_2$$

In more general way, it is

$$CO_2 + H_2O \xrightarrow[\text{chlorophyll}]{hv} 1/n\,[CH_2O]_n + O_2$$

Scarce of Oxygen Liberated in Photosynthesis : The released oxygen (O_2) in photosynthesis comes from water [H_2O^{18}]. When green plants were supplied with water containing [H_2O^{18}]. The released oxygen was entirely of the O^{18} type, indicating that water is only source to release oxygen in photosynthesis. This can be represented by the following summary reaction.

$$6CO_2 + 12H_2O^{18} \rightarrow C_6H_{12}O_6 + 6H_2O + 6O_2{}^{18}.$$

Photosynthesis and Pigments. The photosynthetic products are energy-rich organic compounds. The potential chemical energy of these compounds comes from the light energy. The light' energy to be effective in photosynthesis must be absorbed by a suitable pigment. The vital role is performed by the green pigment chlorophyll, in plants.

Chlorophyll pigment. There are at least seven types of chlorophylls known : chlorophylls *a, b, c, d* and *e.* bacteriochlorophyll and bacterioviridin. All these chlorophyll molecules contain a *tetrapyrrole* skelton formed into a ring with an atom of magnesium in the centre of the ring. A so-called pyrrole molecule contains a skeleton of five atoms, four carbon and one nitrogen and five are arranged in a ring.

Four such pyrroles arranged in a ring form the 'head' of a chlorophyll molecule. Attached to this *porhpyrin* ring at one point is an alcohol (phytol) "tail", a long chain of linked carbons. Relatively minor variations in the kinds and groupings of other atoms joined to this head and tail skeleton account for the differences among different kinds of chlorophylls.

Chlorophylls *a* and *b* are the two most abundant ones found in all the autotrophic plants except the pigment containing bacteria. Chlorophyll *b* is however, absent in blue-green, brown, and red algae. The other chlorophylls (c, d, e) are found only in algae in combination with chlorophyll a. Chlorophyll *a* possesses a – CH_3, a methyl group which is replaced by a – CHO, an aldehyde group, in chlorophyll *b*.

Path of Carbon Dioxide in Photosynthesis : The air contains 0.03% carbon dioxide. This gas diffuses into the air spaces between the cells of the mesophyll of the leaf through the open stomata. Here it combines with water to form carbonic acid. Transport of carbon dioxide in the form of carbonic acid takes place from the spongy cells to the palisade cells until the main site of photosynthesis is reached. It is now proved that in photosynthesis carbon dioxide is not reduced as such but it first forms a complex with cellular compounds and is then reduced by some 'activated' compounds produced as a result of photochemical reaction.

Since isolated choloroplasts are unable to reduce carbon dioxide in the light, it is concluded that either carbon dioxide uptake must take place outside the chloroplast or some mechanism has been destroyed during extraction of the chloroplasts. The first alternative is more probable. *Frenkel* (1941) prepared CO_2 from radioactive (C^{14}) carbon. He fed *Nitella* cells with this carbon dioxide ($C^{14}O_2$) in the dark and found that all the radioactive carbon was located in the cytoplasm. When exposed to light, 4/5 of radioactive carbon was found in the chloroplasts. The simplest interpretation of Frenkel's result is that carbon dioxide uptake occurs in the cytoplasm whereas reduction of carbon dioxide is associated with the chloroplasts.

Mechanism of Photosynthesis : Various theories have been propounded by various workers. But the earlier theories failed and still not theory is available which describes the mechanism satisfactorily. The uncertainty in the mechanism is die to.

(a) No products of the earlier stages of the process could be detected experimentally.

(b) When chlorophyll is extracted from plant, it fails to reproduce its photosynthetic property. So the photosynthesis cannot be studied under controlled experimental conditions.

(c) The complete structure of the chlorophyll was not known fully.

The mechanism of photosynthesis as currently understood is usually divided into three phases :

(1) The absorption of the light energy by the pigment system,

(2) Conversion of light energy into chemical energy by photo-phosphorylation, and

(3) Synthesis of organic compounds by the products of the photo-chemical reactions.

Enough evidence has accumulated in the recent past to show that photosynthesis consists of two stages, (a) light phase and a dark phase. The reaction of light phase is light sensitive and 'therefore' called a photochemical reaction. The reactions of the dark phase do not require light and are temperature sensitive. Such reactions are after called *Blackman reactions.*

$$\xrightarrow[\text{Photochemical reaction}]{\text{Step I}} \xrightarrow[\text{chemical reaction}]{\text{Step II}} \text{Photosynthetic Product}$$

(The component reactions of photosynthesis).

Evidences for the Existence of Light and Dark Phases : There are following evidences.

1. *Experiments with intermittent light.* When Chlorella cells were exposed to continuous illumination, *Warbarg* found that the rate of photosynthesis was less when compared to those exposed alternately to light and darkness. The increase in the rate of photosynthesis was enhanced greatly if the span of the light period was reduced. This is because light period followed by darkness allows the cells to utilize all the products of the light reaction for the reduction of CO_2 in the dark reaction. When the light was continuous or interrupted, the products of light phase accumulate because the pace of light reaction is far greater than that of dark.

2. *Temperature Experiments.* When the photosynthesis is carried out in the excess of carbon dioxide but in the presence of low light intensity, the rate of photosynthesis does not increase with

the increase in temperature. This indicates that some reaction requires light and is not affected by rise in temperature. When the concentration of carbon dioxide is low and the intensity of light is high, the rate of photosynthesis becomes double for every 10°C rise in temperature. This shows that there must be a reaction which is not affected by light. This is the dark reaction of photosynthesis.

Let us discuss these two stages separately.

LIGHT PHASE

Evidence in support of light phase. Hill and *Scrisbrick* found that isolated chloroplasts when illuminated produced oxygen from water. If certain hydrogen acceptors (oxidants) were present in the medium, they were reduced by the chloroplasts. The oxidants are called *Hill reagents* and the reaction as *Hill reaction.* Hill attributed the production of oxygen and hydrogen from water by light. *Arnonin* 1954, working on isolated chloroplasts reported that in addition to carrying out the Hill reaction, the chloroplasts could also synthesis ATP in the light from ADP and inorganic phosphate (Pt). This *photosynthetic phosphorylation* may be represented as.

$$ADP + P_t \xrightarrow{\text{Light energy}} ATP.$$

Reaction occurring in Light phase.

When chlorophyll molecule absorbs one quantum of light, it gets excited. The excited chlorophyll molecule brings about the following changes.

(i) Excited chlorophyll molecule interacts chemically with water to liberate oxygen.

$$4H_2O + \text{chlorophyll}^* \rightarrow 4[H] + 4[OH] + \text{chlorophyll}$$

$$4[OH] \rightarrow 2H_2O + O_2.$$

The nascent hydrogen reacts with NADP to form $NADPH_2$ which acts as a reducing agent.

$$NADP + 2H \rightarrow NADPH_2$$

(ii) Excited chlorophyll also brings about the reaction between ADP and inorganic phosphate Pt to form an energetic compound ATP.

$$ADP + P_i + \text{chlorophyll}^* \rightarrow ATP + \text{chlorophyll}.$$

The products obtained from light reaction enter the dark reaction of photosynthesis.

Mechanism of light phase reactions. When sun light falls on the leaves, the chlorophyll present in the leaves absorbs light photon and becomes excited. This photo excitation results in the displacement of an electron from the normal orbit (of chlorophyll) into a anew orbit. The hole left by the displaced electron is soon filled up by the return of the same electron or another one.

If the same electron is returning the process is termed as *cyclic-transfer of electron.* However, if the different electron is returning, it is termed as non-cyclic transfer of electrons.

$$\text{Chlorophyll} + h\nu \rightarrow \text{Chlorophyll}^*$$

$$\text{Chlorophyll}^* \rightarrow (\text{chlorophyll})^* + e^-$$

Non-cyclic electron transfer. It involves the following steps.

(i) First of alkyl water dissociates into hydrogen and hydroxyl ions.

$$4H_2O \rightarrow 4H^+ + 4OH^-$$

(ii) When a quantum of short wavelength of light is received by a pigment II of chlorophyll molecule, it loses an electron. The loss of electrons is immediately compensated by the electrons of 4 OH^- ions, *i.e.,* converted into water and oxygen.

$$4\,OH^- \rightarrow 2H_2O + O_2 + 4e^-$$

(iii) The electron released from above process is raised from + 0.8 eV to zero potential at which it is captured by an electron carrier called *Plastoquinone.*

(iv) After this the electron travels downhill and falls back to + 0.4 eV in a dark reaction through a series of carriers (cytochrome b_6, cyt. f, and plastocyanine) to pigment system I. The energy released in the dowahill passage of electron from cytochrome b_6 to cytochrome f is utilised to convert ADP and inorganic phosphate into ATP.

$$ADP + P_i \rightarrow ATP$$

(v) The electron is pumped from + 0.04 to – 0.4 eV with energy of another quantum absorbed by the system. I. According to *Tagawa* and *Arnon* (1962) the electron from system I is captured by any iron-containing protein, called *ferredoxin* with a potential of – 4.32 eV (Fig. 2.31).

(vi) According to *Aron* (1967) the electron is finally passed on at – 0.32 e.v. to NAD which together with two H+ released from water, becomes reduced to NHDPH + H^+ (expressed as $NADPH_2$ for convenience).

$$NADP^+ + 2e^+ + 2H^+ \rightarrow NADPH + H^+$$

Energy from ATP helps to more the electron from $NADPP_2$ to phosphoglyceric acid into carbon cycle.

Cyclic electron transfer. The expelled electron from the chlorophyll molecule is first accepted by *ferredoxin* or by an other electron acceptor of the chloroplast.

The electron then traverses via cytochrome b_6 and cytochrome f, the two native cytochromes of chloroplasts, and ultimately reaches the original chlorophyll molecule from which it has been expelled. This scheme of electron transport is called *cyclic photophosphorylation.*

In the cyclic photophosphorylation the electron, that is returned to the chlorophyll, is the same as the one that was expelled. In cyclic photophosphorylation energy is released during the transfer of electrons to cytochromes and the released energy is utilized to synthesize ATP.

II. Dark Phase : (Path of carbon in photosynthesis).

The present knowledge on the path of carbon in photosynthesis comes mainly from the work of *Melvin Calvin* and his associates,. They allowed the algae supplied with carbonlabelled CO_2 to photosynthesize for very brief periods of time and killed them instantaneously with boiling alcohol. Such studies using tracer technique revealed that products of CO_2 assimilation even for periods less than a minute were mainly sugars and amino acids shown in Fig. 2.31.

Calvin's group was interested to know the first stable product of photosynthesis. For this they allowed the algae to photosynthesis for 5 second and extracted the compounds after killing the organisms. The first stable product of very brief period of photosynthesis was found to be 3 carbon compound phosphoglyceric acid (PGA). Since PGA is a 3 carbon compound, it was thought that CO_2 is accepted by a 2-carbon molecule to form 3 carbon PGA. But no such compound was detectable in plants.

Later, *Benson* working in the same laboratory found that a 5-carbon sugar Ribulose diphosphate (RuDP) is the acceptor molecule of CO_2. On accepting CO_2 it was established that an unstable carbon intermediate is found and it cleaves into 2 molecules of PGA.

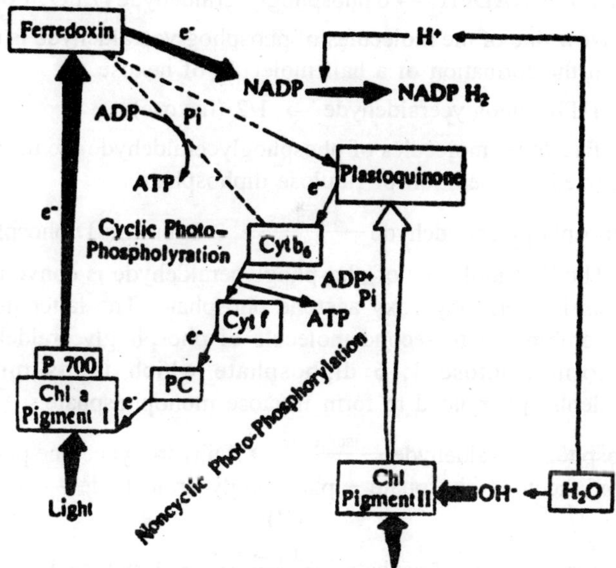

Fig. 2.31

The end products of photosynthesis, namely sugars or starches, are synthesized from PGA by reversed sequence of EMP pathways of glycolysis. It was also established that some of the PGA is used to regenerate RuDP by accepting some more CO_2 molecules. Thus, there is a regular cycle of reactions that ultimately produce the end product of photosynthesis.

The formation of hexose sugar in the dark reaction may be summarised as follows.

(a) 3 molecules of carbondioxide combine with 3 molecules of RuDP (Ribulose diphosphate) to form 6 molecules of phosphoglyceric acid (PGA).

$$3CO_2 + 3RuDP \rightarrow 6PGA$$

(b) 6 molecules of phosphoglyceric acid (PGA) combine with 6 molecules of ATP, which are formed in the light reaction, to form 6 molecules of diphosphoglyceric acid (DPGA).

$$6PGA + 6ATP \rightarrow DPGA + 5\ ADP$$

(c) 6 molecules of DPGA are reduced to 6 molecules of phosphoglyceraldehyde by 6 molecuies of $NADPH_2$ which are produced in light reaction.

6 DPGA + 6 $NADPH_2$ $\rightarrow$ 6 phosphoglyceraldehyde H_2PO_4 + 6 NADP

(d) Now one of the molecules of phosphoglyceraldehyde is utilised in the formation of a half molecule of hexose.

1 Phosphoglyceraldehyde $\rightarrow$ 1/2 (hexose).

The five other molecules of phosphoglyceraldehyde are utilised for the continued regeneration of ribulose diphosphate.

5 Phosphoglyceraldehyde $\xrightarrow{\text{Changes}}$ 3 Ribulose Diphosphate.

(i) The first molecule of phosphoglyceraldehyde is converted into its isomer, dihydroxy acetone phosphate. The latter molecule combines with second molecule of phosphoglyceraldehyde to form fructose 1, 6 diphosphate which in its turn gets dephosphorylated to form fructose monophosphate.

Phosphoglyceraldehyde $\xrightarrow{\text{Isomerises}}$ Dihydroxyacetone phosphate

Dihydroxy acetone phosphate + phosphoglyceraldehyde $\longrightarrow$
(II)

Fructose, 6 monophosphate $\xrightarrow{ADP \rightarrow ATP}$ Fructose 1, 6 diphosphate.

(ii) The fructose 6 monophosphate now combine with the third molecule of phosphoglyceraldehyde to give rise erthrose monophosphate (C_4) and xyluose monophosphate (C_5).

Fructose – 6 – monophosphate + Phosphoglyceraldehyde
(III)

$\longrightarrow$ Erythrose monophosphate (C_4)

+ Xylulose monophosphate (C_5)

(iii) The erythrose monophosphate now combines with the fourth molecule of phosphoglyceraldehyde to form sedoheptulose diphosphate which loses one phosphate to become sedoheptulose monophosphate. It then combines with the fifth molecule of phosphoglyceraldehyde to give rise to one molecule each of xylulose monophosphate and ribose monophosphate.

Erythrose monophosphate + Phosphoglyceraldehyde.

$\longrightarrow$ sedoheptulose diphosphate

$\downarrow$ ADP – ATP

Xylulose monophosphate + Ribose monophosphate $\xleftarrow[\text{(V)}]{\text{Phosphoglyceraldehyde}}$ Sedoheptulose monophosphate

The ribose monophosphate gives rise to one molecule of ribulose monophosphate. The two molecule of xylulose monophosphate, one from this step and another form step (ii), are reversibly converted into its isomer ribulose monophosphate.

In this way five molecules of phosphoglyceraldehyde are converted into three molecules of ribulose monophosphate which in turn gets phosphorylated to form ribulose diphosphate. Now by doubling the number of CO_2 molecules of six, the path of carbon in dark phase of photosynthesis.

LATENT IMAGE

The usefulness of present-day photographic materials is a result of the behaviour of the tiny individual particles of silver halide which are contained in photographic emulsion. In many cases these particles are extremely sensitive to light and store up the effect to an exceedingly small amount of light. The effect can, in turn, be multiplied many fold by the action of a developer. *This stored up effect of light is known as the latent image.*

It is so small that it cannot be developed itself. Consequently, the nature of the latent image and the nature of the development process are closely related.

Def. *When a photographic plate is exposed to light radiation for a certain time, an invisible image is formed on it. This invisible image is known as latent image.*

FACTS

(i) In the formation of latent image, no change in the emulsion has been detected.

(ii) The formation of latent image is very sensitive to traces of substances such as Ag, Ag_2 S etc.

(iii) When the photographic plate carrying latent image is exposed to mild reducing agent like pyrogallol, the latent image is developed to a negative plate.

Studies of the behaviour of the latent image under different conditions of development and exposure and its reactions with certain other chemicals have led to a fairly clear conception of its nature and the mechanism of its formation. An understanding of these processes requires knowledge of the structure of the silver halide grains themselves.

Theories of Latent Image Formation : When a film is exposed to light radiation, the absorbed energy ejects an electron from the bromide ion.

$$Br^- + hv \rightarrow Br + 1e^-.$$

The electron thus set free is captured by a silver ion to form a silver atom.

Thus, the overall reaction is the photochemical dissociation of silver bromide into silver and bromine by absorbing one quanta of light, *i.e.,*

$$AgBr + hv \rightarrow Ag + Br.$$

Thus during the formation of latent image, silver is formed and bromine atom being removed by gelatin.

Objection. The main problem is that light is absorbed all over the surface of the photographic film but silver atoms make their appearance only at a few isolated points within each grain to form a latent image. How does it happen ? How is the latent image formed ? Various theories regarding the formation of latent image are.

I. Paul's View : According to Paul there are several empty holes in the crystal. When a positive ion is missing at one point a negative ion will be missing at some other point. Thus, there exists some holes in the crystal.

When light is incident on a photographic plate, the electrons are given out and finally fall inside the holes. Thus,

$$X^- + hv \rightarrow \qquad X + e^-$$

the latent image was formed at these holes. When the exposed photographic plate is developed, negative image is formed at these holes. This theory does not explain the formation of latent image in complete manner.

\+ – + – + –
– + – + – +
\+ – hole – + –
– + – + – +
\+ – + – + –
– + – hole – +
\+ – + – + –
– + – + – +

Fig. 2.32 : Formation of latent image.

II. Webb's Theory : Recent advances in quantum mechanics and the applications of these principles to the structure and behaviour of crystals have contributed a great deal to knowledge of the problem of latent image formation. Webb (1936) first applied the principles of quantum mechanics to an explanation of the formation of the latent image.

Webb described the latent image as a concentration of electrons at a speck, or trap. The electrons are supposed to be transferred to a higher energy level by the absorption of light and in this energy level they could wander through the crystal at will as in the photo-conductance effect. The sensitivity specks were supposed to contain energy levels slightly lower than this conductances level, so that when an electron travels to such a specks, it gives up a bit of energy in going to this new level and is trapped. This gave an excellent explanation of the speck concentration of the effect of absorbed light but did not explain the formation of metallic silver as latent image substance.

III. Gurney and Mott's Theory : Gurney and Mott presented a theory for the formation of latent image. His main postulates are.

1. The energy structure of a crystal of silver halide can be described as in Fig. 2.33. The Ag and Br represent the silver and bromide ions which are present alternately in a crystal. The crosshatches labelled as P represent the energy level of the electrons connected with the bromide ions. This energy band is filled and electrons cannot move around in it. The clear energy band labelled S represents the energy level corresponding to the conductance level in metallic silver. In a crystal of silver halide all the valence electrons are associated with the bromide ions and none with silver so that this band is empty.

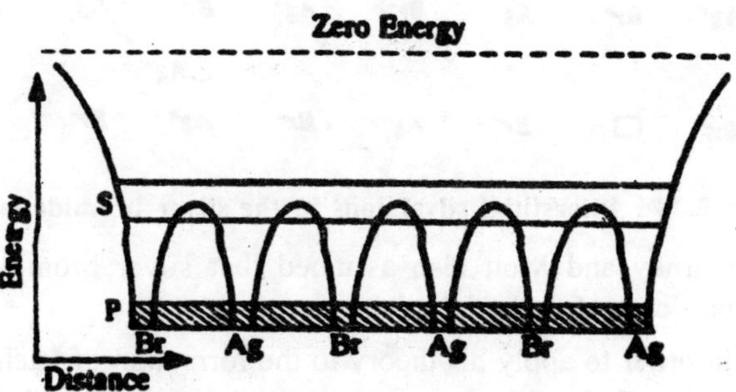

Fig. 2.33 : Quantum mechanical description of the energy level occurring in a crystal of silver bromide.

Electrons cannot have energy represented by the areas not included by these two bands. The photo-conductance phenomenon can be explained by this diagram–when light is absorbed, its energy is transferred to one of the electrons in the P and the electron is lifted to the S band where it is free of wander around just as the conductance electrons flow around in metallic silver.

2. Another concept which was applied to theory of photolysis by Gurney and Mott was the presence of interstitial ions in a crystal lattice. A perfect crystal would contain some ions which are not in their proper places but in the interstitial positions. Presumably these are caused by thermal motion of the ions in the crystals. This condition is shown in Fig. 2.34.

   These ions and their movement through the crystal, either by going from the interstitial position to another or by moving to one of the holes which has been lifted by another interstitial ion, account for the every small electrical conductivity of crystals in the dark.

   These ions might be considered as being in solution in a crystal just a salt is dissolved in water. The conductivity of such crystals follows the same laws as the conductivity of ionised salt solutions. This conductivity decreases with decreasing temperature.

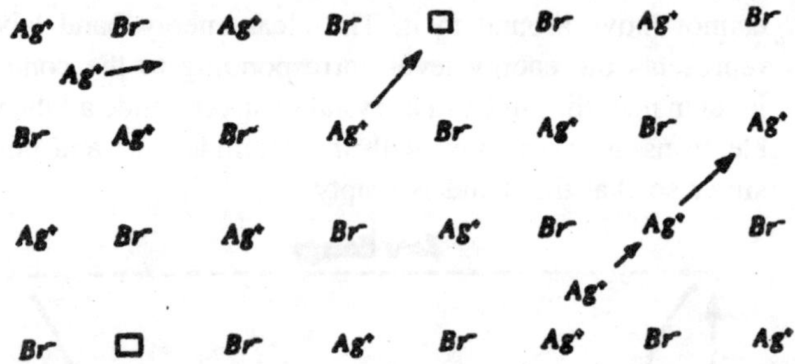

Fig. 2.34 : Interstitial silver ions in the silver bromide lattice.

3. Gurney and Mott also assumed that silver bromide contains particles of colloidal silver.

4. In order to apply his theory to the formation of latent image as well as to the photolytic formation of metallic silver he postulated that the sensitivity specks, silver sulphide or whatever their

complete composition might be, can serve the same function as the colloidal silver specks, *i.e.*, they can act as traps for electrons in the conductance level of the crystal. The theory involves the following points.

(i) When a film is exposed to light radiation, the primary absorption of light occurs in Br– ions of silver bromide crystal.

$$Br^- + h\nu \rightarrow Br^- + le$$

producing an excited electron.

(ii) The bromine atom is termed as a positive hole and is transferred from its original position in the crystal lattice to the surface of the crystal by the following process.

$$Br + Br^- \rightarrow Br^- Br$$

On the surface, the positive hole is trapped by coming in contact with an impurity centre line Ag_2S, probably by absorbing an electron from the sensitizer.

(iii) The excited electrons [from (i)] are free to move on through the crystal lattice. The moving electrons come in contact with a specks of silver and the trapped. This charges the speck negatively and this negative charge acts at some of interstitial silver ions which migrate to speck.

The positive charges on the speck due to silver ions are neutralised by the excited electrons.

$$Ag^+ + le \rightarrow Ag.$$

Thus, the above process is repeated and a large number of silver atoms collect as an invisible speck at one point in a crystal. The same process is repeated in other silver bromide crystals on the film and ultimately leads to the formation of latent image.

Objections : These are.

(i) This theory had not taken sufficient account of the possibility of the recombination of photo released electrons with the positive holes which are formed at the same time.

$$\underset{\text{positive hole}}{Br} + le \rightarrow Br^-$$

(ii) *Gurney* assumed the existence of interstitial silver ions on a photographic film. *Mitchell* showed that a very small number of interstitial silver ions is found to exist on a photographic plate. This number is not sufficient to explain Gurney-Mott's theory.

IV. Mitchell's Theory (1957) : It involves the following steps.

(i) Upon the absorption of a quantum of light, a positive hole and photo-electron are formed.

$$Br^{-}\ h\nu \rightarrow \underset{\text{Positivee hole}}{Br} + \underset{\text{Photo electron}}{le}$$

(ii) The positive hole is trapped at a sensitivity speck in the crystal and this liberates an interstitial silver ion.

(iii) The liberated interstitial silver ions absorbed at some distortion or kink in the crystal lattice.

Then, the absorbed positively charged silver ion absorbs the photo-electron.

$$Ag^{+} + le \rightarrow Ag.$$

This single silver atom is called the latent pre-image.

(iv) Another interstitial silver ion released by the trapping of a positive hole is absorbed this latent pre-image.

$$Ag^{+} + Ag \rightarrow \begin{matrix} Ag^{+} \\ | \\ Ag \end{matrix}$$

Again, the silver ion traps a photo-electron to give a two-atom specks. This is called latent sub-image.

$$\begin{matrix} Ag^{+} \\ | \\ Ag \end{matrix} + le \rightarrow \begin{matrix} Ag \\ | \\ Ag \end{matrix}$$

(v) A third interstitial silver ion released by trapping of positive hole is absorbed on the latent sub-image and again, the absorbed silver ion traps a photo-electron, leading to the formation of latent image.

General Views : The series of experiments carried out in recent years have shown that the above theories are not satisfactory and they require further investigations.

SOLVED EXAMPLES

Examples 1:

Radiation of wave-length 2540 A° was passed through a cell containing 10 ml of a solution of 0.0495 molar oxalic acid and a 0.01 molar uranyl sulphate. After the absorption of 8.81 × 10³ ergs of radiation the concentration of oxalic acid was reduced to 0.0383 molar. Calculate the quantum yield for the photo-chemical decomposition of oxalic acid at given wave-length.

Solution:

10 ml. of 0.0495 molar oxalic acid

$$= \frac{10 \times 0.0495}{1000} = 0.000495 \text{ mole}$$

10 ml of 0.0383 molar oxalic acid

$$= \frac{10 \times 0.0383}{1000} = 0.000383 \text{ mole}$$

$\therefore$ Amount of oxalic acid decomposed

$$= (0.000495 - 0.000383 = 0.000112 \text{ mole}$$

No. of molecules of oxalic acid decomposed by light.

$$= 0.000112 \times 6.023 \times 10^{23} \text{ molecules.}$$

Also, the No. of quanta absorbed.

$$= \frac{\text{Energy absorbed}}{\text{Energy of one quantum}} = \frac{8.81 \times 10^8 \text{ ergs}}{hv}$$

$$= \frac{8.81 \times 10^8 \text{ ergs}}{hc/\lambda} \qquad \left[\because v = \frac{c}{\lambda}\right]$$

$$= \frac{8.81 \times 10^8 \lambda}{hv} = \frac{8.81 \times 10^8 \times 2540 \times 10^{-8}}{6.625 \times 10^{-27} \times 3 \times 10^{10}}$$

$$= \frac{8.81 \times 2540}{6.625 \times 3 \times 10^{-17}}$$

$$\because \text{ Quantum yield} = \frac{\text{No. of molecules reacted}}{\text{No. of quanta absorbed}}$$

$$\text{Quantum yield} = \frac{0.000112 \times 6.023 \times 10^{23}}{\dfrac{8.81 \times 2540}{6.625 \times 3 \times 10^{-17}}} = 0.5937.$$

Example 2:

A substance when dissolved in water at 10^{-3} M concentration absorbs 15 per cent of an incident radiation in a path of 1 cm length. What should be the concentration of the solution in order to absorb 90 per cent of the same incident radiation ?

Solution:

First case. Concentration, c = 10–3 M, absorption = 10% = 0.1

or $\quad I/I_0 = 1 - 0.1 = 0.9, \; t = 1$ cm

According to Beer's law, $I/I_0 = 10^{-act}$

or $\quad 0.90 = 10^{-a \times 10^{-3}} \times 1$ or $a = 45.8$

Second case. Concentration c = ?. Absorption = 90% = 0.90

$\therefore \quad I/I_0 = 1 - 0.90 = 0.10, \; t = 1$ cm

$\therefore \quad I/I_0 = 10$ act or $0.10 = 10^{-45.8 \times c \times 1}$

or $\quad c = 0.0218$ M

For the lowest orbit, n = 1, it means that

$$r = \frac{n^2h^2}{4\pi^2me^2} = \frac{(6.65\times10^{-27})^2}{4\times(3.14)^2\times(9\times10^{-28})\times(4.78\times10^{-10})^2}$$

$$= 0.53 \times 10^{-8} \text{ cm.} = 0.53 \text{ Å}.$$

Example 3:

Calculate the velocity of the electron in the Bohr's orbit given that

$h = 6.625 \times 10^{-27}$ erg, sec.,

$m = 9.11 \times 10^{-28}$ gm.

$e = 4.819 \times 10^{-10}$ e s.u.

Solution:

In Bohr's hydrogen atom, centripetal force is equal to the electrostatic force. Thus,

$$\frac{mu^2}{r} = \frac{e^2}{r^2} \text{ or } u = \frac{e}{\sqrt{(mr)}}$$

Now $$r = \frac{h^2}{4\pi^2me^2} \text{ or } mr = \frac{h^2}{4\pi^2e^2}.$$

Taking square root of both sides, we get

$$\sqrt{(mr)} = \sqrt{\left(\frac{h^2}{4\pi^2 e^2}\right)} \text{ or } \sqrt{(mr)} = \frac{h}{2\pi e}.$$

Substituting this value in equation (A), we get

$$u = \frac{2\pi e^2}{h} = \frac{2 \times 3.14 \times (4.819 \times 10^{-10})^2}{6.625 \times 10^{-27}}$$

$$= 2.192 \times 10^8 \text{ cm/sec.}$$

Example 4:

Give e= 4.78 × 10^{-10} e.s.u., h = 6.65 × 10^{-27} *ergs, sec, and m* = 9 × 10^{-33} *gm., calculate the radius of the lowest orbit of the hydrogen atom.*

Solution:

According to Bohr's theory, the radius of the n th orbit of hydrogen atom is given by

$$r = \frac{n^2 h^2}{4\pi^2 me^2} \qquad [\because Z = 1]$$

Example 5:

Calculate the radius of the second Bohr orbit of hydrogen atom and velocity of electron in this orbit. $\frac{1}{4\pi\varepsilon_0} = 9.10^9 \frac{N-m^2}{C^2}$, *m* = 9.1 × 10^{-31} *kg, e* = 1.6 × 10^{-19} *C and h* = 6.63 × 10^{-34} *J-s)* .

Solution:

2.1 Å, 1.1 × 10^6 m/s.

Example 6(a):

Calculate the speed of electron in the n th orbit of the hydrogen atom. If relativ stic is necessary.

(e = 1.6 × 10^{-19} *C, h* = 6.63 × 10^{-34} *J-s, c 3.0* × *108 m* s^{-1}

and $\frac{1}{4\pi\varepsilon_0}$ = 9 × 10^9 *N* m^2/C^2*).*

Solution:

Let v be the speed of the electron (mass m, charge e) in the n th orbit whose radius is r, all in SI units. The charge on hydrogen nucleus is + e. As the centripetal force is provided by the electrostatic attraction, we have

$$\frac{mv^2}{r} = \frac{1}{4\pi\varepsilon_0}\frac{e^2}{r^2} \qquad ...(1)$$

Further, from Bohr's postulate, the angular momentum of the electron is an integral multiple of h/2π, that is,

$$mvr = \frac{nh}{2\pi}. \quad n = 1, 2, 3, ... \qquad ...(2)$$

Dividing eq. (1) by eq. we get

$$v = \frac{1}{4\pi\varepsilon_0}\frac{2\pi e^2}{nh}.$$

Substituting the given values, we get

$$v = (9 \times 109 \text{ N m}^2/\text{C}^2) \times \frac{2 \times 3.14 \times (1.6 \times 10^{-19}\text{ C})^2}{n\,(6.63 \times 10^{-34}\text{ Js})}$$

$$= \frac{1}{n}\,(2.18 \times 106) \text{ m s}^{-1}.$$

This is the required value. Now

$$\frac{v}{c} = \frac{1}{n}\,\frac{2.18 \times 10^6}{3.0 \times 10^8} = \frac{0.0073}{n}.$$

Thus, for n = 1, 2, 3, ... we have

$$\frac{v}{c} = 0.0073, 0.0036, 0.0024,...$$

Hence the relativistic correction in necessary for n = 1 orbit only.

Example 6(b):

Calculate the wavelengths of the first two lines of the Balmer series and the series limit. ($h = 6.63 \times 10^{-34}$ J s, $c = 3.0 \times 10^8$ ms^{-1}, $e = 1.6 \times 10^{-19}$ C, $m = 9.1 \times 10^{-31}$ kg, $\varepsilon_0 = 8.85 \times 10^{-12}$ C^2/N m^2).

Solution:

Let us first calculate the Rydberg constant from the given data. We have

$$R = \frac{me^4}{8\varepsilon_0{}^2ch^3}$$

$$= \frac{(9.1\times10^{-31}\,\text{kg})\times(1.6\times10^{-19}\,\text{C})^4}{8\times(8.5\times10^{-12}\,\text{C}^2/\text{N}\,\text{m}^2)^2\times(3.0\times10^8\,\text{ms}^{-1})\times(6.63\times10^{-34}\,\text{Js})^3}.$$

Now, the wavelengths of the spectral lines of Balmer series are given by

$$\frac{1}{\lambda} = R\left(\frac{1}{2^2}-\frac{1}{n^2}\right), n = 3, 4, 5, ...$$

For the first line, n = 3.

$$\therefore \quad \frac{1}{\lambda_1} = R\left(\frac{1}{2^2}-\frac{1}{3^2}\right) = \frac{5R}{36}$$

or
$$\lambda_1 = \frac{36}{5R} = \frac{36}{5\times(1.097\times10^7\,\text{m}^{-1})} = 6.563 \times 10^{-7}\ \text{m}$$

$$= 6563 \times 10^{-10}\text{m} = 6563\ \text{Å}.$$

For the second line, n = 4. We can see that l23 = 4861 Å.

For the series limit, n = ∞.

$$\therefore \quad \frac{1}{\lambda_\infty} = \frac{R}{4}$$

or
$$\lambda_\infty = \frac{4}{R} = \frac{4}{1.097\times10^7}$$

$$= 3.646 \times 10^{-7}\ \text{m} = 3646\ \text{Å}.$$

Example 7:

Given that the Rydberg constant for hydrogen is 109678 cm^{-1}, find the long and short wavelength limits of the Lyman series.

Hint : For long wavelength limit (first line) n = 2 and for short wavelength limit n = ∞.

Solution:

1216 Å, 912 Å.

Example 8:

Calculate the wavelength of the eight line of the Balmer series of the hydrogen atom. Rydberg constant for hydrogen atom is

1.097 × 10^7 m^{-1}.

Hint : For the eight line, n = 10.

Solution:

3798 Å.

Example 9(a):

The wavelength of the first line of Balmer series is 6563 Å. Calculate Rydberg constant.

Solution:

The wavelength of the spectral lines of Balmer series are given by

$$\frac{1}{\lambda} = R_H \left(\frac{1}{2^2} - \frac{1}{n^2}\right), \; n = 3, 4, 5, ...$$

For the first ($\lambda = 6563 \times 10^{-10}$ m), n = 3.

$$\therefore \quad \frac{1}{6563 \times 10^{-10}\, m} = R_H \left(\frac{1}{2^2} - \frac{1}{3^2}\right) = \frac{5}{36} = R_H$$

or

$$R_H = \frac{36}{5 \times (6563 \times 10^{-10}\, m)} = 1.097 \times 10^7\ m^{-1}.$$

Example 9(b):

Find the wavelength of the photon emitted when the hydrogen atom goes from n = 10 state to the ground state. ($R_H = 1.097 \times 10^{-3}$ Å^{-1}).

Solution:

Since the atom drops to the ground state (n = 1), the photon emitted belongs to the Lyman series. The wavelengths of the spectral lines in this series are given by

$$\frac{1}{\lambda} = R_H \left(\frac{1}{1^2} - \frac{1}{n^2}\right), \; n = 2, 3, 4, ...$$

For n = 10, we have

$$\frac{1}{\lambda} = R_H \left(1 - \frac{1}{100}\right) = \frac{99}{100} R_H$$

or

$$\lambda = \frac{100}{99\, R_H}$$

$$= \frac{100}{99 \times (1.097 \times 10^{-3}\ \text{Å}^{-1})} = 921\ \text{Å}.$$

Example 10:

The first line of the Balmer series in the spectrum of hydrogen has a wavelength of 6563 Å. Calculate the wavelength of the first line of Lyman series in the same spectrum.

Solution:

The wavelength of the spectral lines of hydrogen spectrum are given by

$$\frac{1}{\lambda} = R\left(\frac{1}{n_f^2} - \frac{1}{n_i^2}\right),$$

where R is Rydberg constant. For the first member of Balmer series $n_f = 2$ and $n_i = 3$.

$$\therefore \quad \frac{1}{\lambda_1} = R\left(\frac{1}{2^2} - \frac{1}{3^2}\right) = \frac{5R}{36}.$$

For the first member of the Lyman series, $n_f = 1$ and $n_i = 2$.

$$\therefore \quad \frac{1}{\lambda_1'} = R\left(\frac{1}{1^2} - \frac{1}{2^2}\right) = \frac{3R}{4}.$$

From (i) and (ii), we have

$$\frac{\lambda_1}{\lambda_1'} = R\,\frac{5R}{36} \times \frac{4}{3R} = \frac{5}{27}.$$

$$\therefore \quad \lambda_1' = \frac{5}{27}\,\lambda_1 = \frac{5}{27} \times 6563$$

$$= 1215 \text{ Å}.$$

Example 11:

The first member of Balmer series of hydrogen has a wavelength of 6563 Å. Calculate the wavelengths of the second and fourth members.

Solution:

The wavelengths of the spectral lines of Balmer series are given by

$$\frac{1}{\lambda} = R\left(\frac{1}{2^2} - \frac{1}{n^2}\right), \; n = 3, 4, 5, \ldots$$

For the first member, n = 3.

$$\therefore \quad \frac{1}{\lambda_1} = R\left(\frac{1}{2^2} - \frac{1}{3^2}\right) = \frac{5R}{36} \qquad \text{....(1)}$$

For the second member, n = 4.

$$\therefore \quad \frac{1}{\lambda_2} = R\left(\frac{1}{2^2} - \frac{1}{4^2}\right) = \frac{3R}{16} \qquad \text{....(2)}$$

Dividing eq. (2) by (1), we get

$$\frac{\lambda_1}{\lambda_2} = \frac{3R}{16} \times \frac{36}{5R} = \frac{27}{20}.$$

$$\lambda_2 = \lambda_1 \times \frac{20}{27}.$$

But λ_1 = 6563 Å (given).

$$\therefore \quad \lambda_2 = (6563 \text{ Å}) \times \frac{20}{27} = 4861 \text{ Å}.$$

Similarly λ_4 = 4102 Å.

Example 12:

Calculate the (i) energy required to ionise hydrogen atom in the ground state (ii) limit of Balmer series. (Rydberg constant is 1.097 × 10^7 m^{-1}, h = 6.63 × 10^{-34} J s, c 3.0 × 10^8 m s^{-1}, 1 eV = 1.6 × 10^{-10} J).

Solution:

(i) The energy required to ionise, that is, to remove an electron from the hydrogen atom in the ground state (n = 1) to infinity (where the energy is zero) is numerically equal to the energy of the electron in the n = 1 orbit. Now, the energy of the electron in n th orbit of hydrogen is given by

$$E_n = -\frac{me^4}{8\varepsilon_0^2 h^2}\left(\frac{1}{n^2}\right)$$

$$= -\frac{Rhc}{n^2}. \qquad \left[\because R = \frac{me^4}{8\varepsilon_0^2 ch^3}\right]$$

For the ground (first) orbit, n = 1.

$$\therefore \quad E_1 = -\,Rhc$$

$$= -(1.097 \times 10^7 \text{ m}^{-1})(6.63 \times 10^{-34}\text{J})(3 \times 10^8 \text{m s}^{-1})$$

$$= -21.8 \times 10^{-19}\text{ J}$$

$$= -\frac{21.8\times10^{-19}}{1.6\times10^{-19}} = -13.6\text{ eV}.$$

Hence the energy required to remove the electron from n = 1 orbit to infinity is 13.6 eV.

(ii) The Balmer series is represented by

$$\frac{1}{\lambda} = R\left(\frac{1}{2^2}-\frac{1}{n^2}\right),\ n = 3, 4, 5, \ldots$$

For the series limit n = ∞ so that

$$\frac{1}{\lambda} = \frac{R}{4}.$$

$$\therefore \quad \lambda = \frac{4}{R} = \frac{4}{1.097\times10^{7}\text{ m}^{-1}}$$

$$= 3.646 \times 10^{-7}\text{ m} = 3646\text{ Å}.$$

Example 13:

The ground-state energy of an electron in the hydrogen atom is – 13.6 eV. Compute the energy of:

(i) n = 3 state in He^{+},

(ii) n = 2 state in Li^{++}.

Solution:

The energy of an electron in the nth state of a hydrogen like atom is given by

$$E_n = -\frac{mZ^2e^4}{8\varepsilon_0^{\,2}h^2}\left(\frac{1}{n^2}\right)$$

$$= -\frac{RZ^2hc}{h^2}. \qquad \left[\because R = \frac{me^4}{8\varepsilon_0^{\,2}ch^3}\right].$$

For H, Z, = 1, so that

$$(E_n)_H = -\frac{Rhc}{n^2}.$$

When n = 1 (ground state) $(E_1)_H$ = – 13.6 eV (given).

$\therefore$ Rhc = 13.6 eV.

Hence $(E_2)_H = -\frac{Rhc}{4} = -3.4$ eV

and $(E_3)_H = -\frac{Rhc}{9} = -\frac{13.6}{9} = -1.5$ eV.

(i) For He^+, Z = 2, so that

$$(E_3)_{He} = -\frac{4Rhc}{9} = 4\,(E_2)_H = -6.0 \text{ eV.}$$

(ii) For Li^{++}, Z = 3, so that

$$(E_2)\,L_i = -\frac{9Rhc}{4} = 9\,(E_2)_H = -30.6 \text{ eV.}$$

Example 14(a):

(i) How much energy is required to remove an electron from hydrogen atom in the ground state and also in a state with n = 8?

(ii) Calculate the corresponding energies of the singly-ionised helium atom (m = 9.11 × 10–31 kg, e = 1.6 × 10^{-19} C, h = 6.63 × 10^{-34} J s, e_0 = 8.85 × 10^{-12} $C^2/N\,m^2$ and 1 eV 1.6 × 10^{-19} J).

Solution:

(i) The energy required to remove an electron from the hydrogen atom in the ground state (n = 1) to infinity (where the energy is zero) is numerically equal to the energy of the electron in the n = 1 orbit. Now the energy of the electron in n th orbit of hydrogen (Z = 1) is given by

$$E_n = -\frac{me^4}{8\varepsilon_0^2 h^2}\left(\frac{1}{n^2}\right).$$

For the ground orbit, n = 1.

$$\therefore \quad E_1 = \frac{me^4}{8\varepsilon_0{}^2 h^2}$$

$$= -\frac{(9.11\times10^{-31})\times(1.6\times10^{-19})^4}{8\times(8.85\times10^{-12})^2\times(6.63\times10^{34})^2}$$

$$= -2.17 \times 10^{-18} \text{ joule}$$

$$= -\frac{2.17\times10^{-18}}{1.6\times10^{-19}} = -13.6 \text{ eV.}$$

Hence the energy required to remove the electron from n = 1 orbit to infinity is 13.6 eV. This is known as the binding energy of the hydrogen atom.

Again, for n = 8 orbit,

$$E_8 = -\frac{me^4}{8\varepsilon_0{}^2 h^2}\left(\frac{1}{8^2}\right) = \frac{E_1}{64}$$

$$= -\frac{13.6}{64} = -0.213 \text{ eV.}$$

Hence the energy required to remove an electron from n = 8 orbit to infinity is 0.213 eV.

(ii) The energy expression for

$$He^+ = -\frac{m Z^2 e^4}{8\varepsilon_0{}^2 h^2}\left(\frac{1}{n^2}\right)$$

$$= -\frac{4 m e^4}{8\varepsilon_0{}^2 h^2}\left(\frac{1}{n^2}\right) = -4\,(E_n)_H.$$

This is, the energies for He^+ are 4 times the corresponding energies for H. Thus the results are

$$4 \times 13.6 = 54.4 \text{ eV}$$

$$4 \times 0.213 = 0.85 \text{ eV.}$$

Example 14(b):

A photon of energy 12.1 eV is completely absorbed by a hydrogen atom, originally in its ground state, so that the atom is excited. What is the quantum number of this state ? The ground state (negative) energy of hydrogen atoms is 13.6 eV.

Solution:

Let the required quantum number be n. Then

$$E_1 - E_n = -12.1 \text{ eV}$$

or $$E_1 - \frac{E_1}{n^2} = -12.1 \text{ eV.} \qquad \left[\because E_n = \frac{E_1}{n^2}\right]$$

But $$E_1 = -13.6 \text{ eV.}$$

$$1 - \frac{1}{n^2} = \frac{-12.1}{-13.6} = 0.89$$

or $\frac{1}{n^2} = 1 - 0.89 = 0.11$

or $n^2 = \frac{1}{0.11} = 9.$

$\therefore$ $n = 3.$

Example 14(c):

A beam of electron bombards a sample of hydrogen. Through what minimum potential difference must be electrons have been accelerated if the first line of the Balmer series is to be emitted ? The ionisation potential of hydrogen atom is 13.6 volts. Explain how many possible spectral lines can be expected if the atom finally attains the normal state.

Solution:

Normally the hydrogen atoms are in the ground state (n = 1). The first Balmer line is emitted when the atom returns from n = 3 to n = 2 state. Hence in order to emit this line the atom must be first raised to the n = 3 state by electron bombardment. Therefore, the energy of the bombarding electrons must be equal to the difference of energy between n = 1 and n = 3 states ($E_1 \sim E_3$).

The energy of hydrogen atom in the n th state is

$$E_n = -\frac{Rhc}{n^2}.$$

If E_1 is the energy in the ground (n = 1) state, then $E_1 = -Rhc$. Thus

$$E_n = \frac{E_1}{n^2}$$

For n = 3, we have $E_3 = \frac{E_1}{9}$

$$\therefore\ E_1 - E_3 = E_1 - \frac{E_1}{9} = \frac{8}{9} E_1.$$

But $E_1 = -13.6$ eV.

$$\therefore\ E_1 - E_3 = \frac{8}{9} \times (-13.6) = -12.1 \text{ eV}.$$

Hence the bombarding electrons must be accelerated by 12.1 volts.

If the atom finally attains the normal state, the possible number of spectral lines is 3 ($3 \to 2$, $3, \to 1$, $2 \to 1$).

Example 15:

The wavelength of the first line of Balmer series of hydrogen is 6562.8 Å. Calculate (i) the ionisation potential and the first excitation potential of the hydrogen atom. ($h = 6.63 \times 10^{-34}$ J-s, $c = 3 \times 10^8 m\ s^{-1}$).

Solution:

(i) The wavelengths of the Balmer lines are given by

$$\frac{1}{\lambda} = R\left(\frac{1}{2^2} - \frac{1}{n^2}\right), \qquad n = 3, 4, 5, ...$$

For the first line n = 3 and l = 6562.8 Å = 6562.8×10^{-10} m.

$$\therefore \quad \frac{1}{6562.8 \times 10^{-10}} = R\left(\frac{1}{2^2} - \frac{1}{3^2}\right) = \frac{5R}{36}$$

or

$$R = \frac{36}{5 \times 6562.8 \times 10^{-10}\ \text{m}} = 1.097 \times 10^7\ \text{m}^{-1}.$$

This ionisation potential of an atom is numerically equal to the (ionisation) energy required to remove an electron completely from the atom in the normal state. In hydrogen atom, the electron stays in the first orbit (n = 1). Hence the energy required to remove this electron to infinity (where the energy is considered to be zero) is numerically equal to the energy of the electron in the first orbit. We know that electron energy in hydrogen atom is given by

$$E_n = -\frac{Rhc}{n^2}.$$

In the ground state, n = 1.

$$\therefore \quad E_1 = -Rhc$$

$$= -13.6 \text{ eV.} \qquad \text{(as obtained in the leas problem).}$$

Here the ionisation potential of the hydrogen atom is 13.6 volts.

(ii) The first excitation potential of the atom is the energy required to shift the electron from n = 1 to n = 2 orbit, that is, $E_1 \sim E_2$. Now,

$$E_1 = -13.6 \text{ eV.}$$

and

$$E_2 = \frac{E_1}{4} = -3.4 \text{ eV.} \qquad \left[\because E_n \propto \frac{1}{n^2}\right]$$

$\therefore \quad E_1 \sim E_2 = 13.6 - 3.4 = 10.2$ eV.

Hence the first excitation potential is 10.2 volts.

Example 16:

The wavelength of a yellow line of sodium is 5896 Å. Calculate its wave number and frequency.

Solution:

The wave number $\bar{v}$ is the reciprocal of wave-length λ, that is,

$$\bar{v} = \frac{1}{\lambda}.$$

Here λ = 5896 Å = 5896 × 10^{-8} cm.

$$\therefore \quad \bar{v} = \frac{1}{5896 \times 10^{-8}\ \text{cm}} = 16960 \text{ per cm.}$$

The frequency v is related to the wavelength l by

$$c = v\lambda,$$

where c is the speed of light. Thus

$$v = \frac{c}{\lambda} = \frac{3 \times 10^{8}\ \text{cm/s}}{5896 \times 10^{-8}\ \text{cm}} = 5.1 \times 10^{12}\ \text{s}^{-1}$$

Example 17:

The ionisation potential of an atom is 14.2 eV. Calculate the series limit in its absorption spectrum. Dates as in last problem.

Solution:

The series limit in the absorption spectrum of an atom corresponds to the energy which when absorbed by a ground-state atom, ionises it. Thus, if V be the ionisation potential of an atom, the wavelength at the series limit is given by

$$\lambda = \frac{hc}{eV}$$

$$= \frac{(6.62 \times 10^{-34}\ \text{Js}) \times (3 \times 10^{8}\ \text{ms}^{-1})}{(14.2 \times 1.6 \times 10^{-19}\ \text{J})}$$

$$= 0.874 \times 10^{-7}\ \text{m}$$

$$= 874\ \text{Å}$$

Example 18:

In a Franck-Hertz experiment the first dip in the current-voltage graph for hydrogen was observed at 10.2 volts. Calculate the wavelength of light emitted by hydrogen atom when existed to the first excitation level. ($h = 6.62 \times 10^{-34}$ J s, $c = 3.0 \times 10^{8}$ m/s, $e = 1.6 \times 10^{-19}$ C.)

Hint : $\lambda = \dfrac{hc}{eV}$.

Solution:

1217 Å.

Example 19:

With Franck-Hertz type of experiment on sodium, the first spectral line to appear is the D-line, $\lambda = 5.89 \times 10^{-7}$ m. What is the first excitation potential of sodium ? Given : $h = 6.63 \times 10^{-34}$ J, s, c 3 × 108 m/s and 1 eV = 1.6×10^{-19} J.

Solution:

Let V volt be the excitation potential. Then the (excitation) energy imparted to the electron will be eV joule, where e coulomb is the charge on the electron. This energy is re-emitted as photon (radiation) when the electron returns to the normal state. If v be frequency of the emitted radiation, the photon energy will be hv. Thus

$$eV = hv$$

or
$$V = \frac{hv}{e}.$$

If l be the wavelength of the emitted radiation, then $v = c/\lambda$.

$$\therefore \quad V = \frac{hv}{e\lambda}$$

$$= \frac{(6.63\times10^{-34}\,\text{Js})(3\times10^{8}\,\text{ms}^{-1})}{(1.6\times10^{-19}\,\text{C})(5.89\times10^{-7}\,\text{m})}$$

$$= 2.1\ (\text{J/C}) = 2.1\ \text{V}.$$

Example 20(a):

The first two excitation potentials of atomic hydrogen in Franck-Hertz experiment are 10.2 and 12.09 volts. Draw an energy level diagram and show all possible transitions for emission and absorption along their wavelengths.

(h = 6.63 × 10^{-34} J s, c = 3.0 × 108 m/s, 1 eV = 1.6 × 10^{-19} J.)

Solution:

The energy level diagram, and the *three* possible transitions (a), (b) and (c) for emission are shown in the figure.

The frequency of radiation resulting from the transitions (a) is given by

$$\nu_a = \frac{E_2 - E_1}{h},$$

and the corresponding wavelength is

$$\lambda_a = \frac{hc}{E_2 - E_1}. \qquad [\because c = \nu\lambda]$$

Now, from the Fig., $E_2 - E_1$ = 10.2 eV = 10.2 × (1.6 × 10^{-19}) J. Therefore

$$\lambda_a = \frac{(6.63 \times 10^{-34}\,\text{J s}) \times (3.0 \times 10^8\,\text{m s}^{-1})}{(10.2 \times 1.6 \times 10^{-19}\,\text{J})}$$

$$= 1.216 \times 10^{-7}\ \text{m} = 1216 \times 10^{-10}\ \text{m} = 1216\ \text{Å}.$$

Similarly, $$\lambda_b = \frac{hc}{(E_3 - E_1)}$$

$$= \frac{(6.63 \times 10^{-34}) \times (3.0 \times 10^8)}{(12.09 \times 1.6 \times 10^{-19})}$$

$$= 1.026 \times 10^{-7}\ \text{m} = 1026 \times 10^{-10}\ \text{m} = 1026\ \text{Å}.$$

Also, $$\lambda_c = \frac{hc}{(E_3 - E_2)}$$

$$= \frac{hc}{(E_3 - E_1) - (E_2 - E_1)}$$

$$= \frac{hc}{(12.09 - 10.2)\,\text{eV}} = \frac{hc}{1.89\,\text{eV}}$$

$$= \frac{(6.63 \times 10^{-34}) \times (3.0 \times 10^8)}{(1.89 \times 1.6 \times 10^{-19})}$$

$$= 6.567 \times 10^{-7}\ \text{m} = 6567 \times 10^{-10}\ \text{m} = 6567\ \text{Å}.$$

In absorption, only the transitions starting from n = 1 shall be observed which correspond to 1216 Å and 1026 Å.

Example 20(b):

(a) *Compare the assumptions made by Planck in discussing cavity radiation, those by Einstein in connection with the photoelectric effect and those by Bohr in connection with the hydrogen spectrum.*

(b) *Explain the implication of the fact that- the energies of the hydrogen atom orbits are negative.*

(c) *Can a hydrogen atom absorb a photon of energy greater than the binding energy of the atom ?*

Solution:

(a) *Assumptions of Planck, Einstein and Bohr* : Planck had assumed that the atoms of a hot body behave as oscillators and have discrete (quantised) energies. They do not emit radiant energy continuously, but only in 'jumps' or 'quanta'. Planck, however, still maintained that radiation propagates continuously through space as electromagnetic waves.

Einstein, in order to explain the photoelectric effect, went a step ahead. He proposed that the radiation not only is emitted as quantum at a time but also propagates as individual quanta (photons). He thus treated the propagation of radiation as particle propagation rather than wave propagation.

Bohr's in order to explain the hydrogen spectrum, adopted the Planck's quantum hypothesis that the radiation is emitted discontinuously from the atom. He started with the quantisation of the angular momentum of the electron in the orbit which ultimately results in the quantisation of energy of the atom.

(b) *Negative Energy of Hydrogen Orbits* : An electron revolving in an hydrogen orbit has a negative potential energy by virtue of its attraction towards, the nucleus, and also kinetic energy (which is positive) by virtue of its motion. The potential energy is greater in magnitude .than the kinetic energy so that the net energy is negative.

The negative energy signifies that the electron cannot escape from The atom. A positive energy for a nucleus electron combination would mean that electron is not bound to the nucleus. Such a combination cannot constitute an atom.

Example 20(c):

A μ^- meson (charge – e, mass = 207 m, where m is mass of electron) can be captured by a proton to form a hydrogen- like "mesic" atom. Calculate the radius of the first Bohr orbit, the binding energy, and the wavelength of the first line in the Layman series for such an atom. The mass of the proton is 1836 times the mass of electron. The radius of first Bohr orbit and the binding energy of hydrogen are 0.53 Å and 13.6 eV respectively. $R_¥$ = 109737 cm^{-1}.

Solution:

The reduced mass of the system is

$$m = \frac{(207\,m)(1836\,m)}{207\,m + 1836\,m} = 186\ m.$$

From Bohr theory, the radius of the first orbit (n = 1) of a hydrogen -like atom for Z = 1, is given by (taking finite mass of nucleus in consideration).

$$r_1 = 4\pi\varepsilon_0 \frac{h^2}{4\pi^2 \mu e^2} = 4\ \pi\varepsilon_0 \frac{h^2}{4\pi^2 (186m) e^2}$$

$$= \frac{1}{186}\left(4\pi\varepsilon_0 \frac{h^2}{4\pi^2 me^2}\right).$$

The quantity in the bracket is the first Bohr orbit of hydrogen atom which is 0.53 Å. Therefore

$$r_1 = \frac{1}{186} \times 0.54\ \text{Å} = 2.85 \times 10^{-3}\ \text{Å}.$$

Again, from Bohr's theory, the ground-state energy for a hydrogen-like atom with Z = 1 is given by

$$E_1 = -\frac{\mu e^4}{8\varepsilon_0{}^2 h^2} = -\ 186\ \frac{me^4}{8\varepsilon_0{}^2 h^2}$$

$$= -\ 186 \times 13.6 = -\ 2530\ eV.$$

Hence the binding energy is 2530 eV.

The wavelength of the Lyman lines are given by

$$\frac{1}{\lambda} = R_M\left(\frac{1}{1^2} - \frac{1}{n^2}\right),\ n = 2, 3, 4,...$$

where R_M is the Rydberg constant for the music atom. For the first line, n = 2 so that

$$\lambda = \frac{4}{3R_M}.$$

Now, $R_M = \frac{\mu e^4}{8\varepsilon_0^2 ch^3}$ and $R_¥ = \frac{me^4}{8\varepsilon_0^2 ch^3}$, so that

$$R_M = \frac{\mu}{m} R_¥ = 186\ R_¥ = 186 \times 109737\ \text{cm}^{-1}.$$

Here $\lambda = \frac{4}{3R_M} = \frac{4}{3 \times 186 \times 109737\ \text{cm}^{-1}}$

Example 21:

A positronium atom is a system consisting of a position and an electron. Calculate the reduced mass, the Rydberg constant and the wavelength of the first Balmer line for positronium. (Give m = 9.1 × 10^{-31} kg, R_H = 1.09737 × 10^{-3} $Å^{-1}$ and H_α = 6563 Å).

Solution:

The positron has the same mass m as the electron and has equal but positive charge. The reduced mass of the electron-positron atom is therefore

$$m = \frac{(m)(m)}{m+m} = \frac{1}{2}\ m = 4.55 \times 10^{-31}\ \text{kg},$$

while the reduced mass of electron in hydrogen is very nearly m.

The Rydberg constant $\left(\frac{\mu e^4}{8\varepsilon_0^2 ch^3}\right)$ for positronium is therefore half that for hydrogen (with infinitely heavy nucleus). Thus

$$R_P = \frac{1}{2} R_H = 0.54868 \times 10^{-3}\ Å^{-1}.$$

The wavelength of first Balmer line (H_α) for hydrogen atom is given

$$\frac{1}{\lambda_H} = R_H \left(\frac{1}{2^2} - \frac{1}{3^2}\right),$$

while that for positronium atom is $\frac{1}{\lambda_P} = R_P \left(\frac{1}{2^2} - \frac{1}{3^2}\right)$.

Thus $$\frac{\lambda_P}{\lambda_H} = \frac{R_H}{R_P} = 2.$$

$$\therefore \quad \lambda_P = 2\lambda_H = 2 \times 6563 = 13126\ Å.$$

Example 22:

Find the recoil speed of hydrogen atom after it emits a photon in going from n = 3 to n = 1 state. Electron mass is 9.11×10^{-31} kg and $h = 6.626 \times 10^{-34}$ J s.

Solution:

The energy of the hydrogen atom in the n th state is given by

$$E_n = -\frac{Rhc}{n^2},$$

where R is Rydberg constant. From this, we get

$$E_1 - E_3 = -\frac{Rhc}{1^2} + \frac{Rhc}{3^2} = -\frac{8}{9} \text{ R h c.}$$

The energy of the emitted photon is

$$\Delta E = E_1 \sim E_3 = \frac{8}{9} \text{ R h c.}$$

The momentum of the photon is

$$p = \frac{\Delta E}{c} = \frac{8}{9} \text{ R h.}$$

By conservation of momentum, the recoil momentum of the hydrogen atom will be equal (and opposite) to the momentum of the emitted photon. The recoil speed of the atom is

$$v = \frac{\text{momentum}}{\text{mass}} = \frac{8}{9}\frac{Rh}{m_H}.$$

But $m_H = 1836$ m, where m is electron mass.

$$\therefore \quad v = \frac{8}{9}\frac{Rh}{(1836\,m)}.$$

Putting the given values of h, m and using $R = 1.097 \times 10^7 \text{ m}^{-1}$, we get

$$v = \frac{8}{9}\frac{(1.097 \times 10^7 \text{ m}^{-1})(6.626 \times 10^{-34} \text{ Js})}{1836\,(9.11 \times 10^{-31} \text{ kg})}$$

$$= 3.86 \text{ m/s.}$$

Example 23(a):

The ionisation potential of hydrogen tom is 13.6 volt. Find the wavelength of the Lyman series limit.

Solution:

When an atom absorbs energy so that its two stationary states becomes existed simultaneously, two superimposed sets of radiations are produced to give a 'beat' variation. If v¢ and v¢¢ represent the frequencies of two superimposed vibrations, the frequency v of the emitted beat is

$$v = v' - v''.$$

The wave mechanical vibrations are related to the energies of the corresponding state by the usual quantum expressions E′ = hv′ and E′ – E″ = hv″ so that

$$E' - E'' = h\,(v' - v'')$$

$$\Delta E = 13.61\left(\frac{1}{(1)^2} - \frac{1}{\infty}\right) = 13.61 \text{ eV}.$$

Therefore, ionisations potential = 13.61 eV.

Example 23(b):

A beam of monochromatic photons of energy 9 eV is incident on hydrogen gas all of whose atoms are in the ground state. It is found that the beam is fully transmitted without absorption. Why ? The ground state energy of an electron in the hydrogen atom is $E_1 = -13.6$ eV.

Solution:

The minimum energy that can be absorbed by ground-state hydrogen atom is $E_1 \sim E_2$, which would excite it to the next state (n = 2). Now,

$$E_1 \sim E_2 = E_1 \sim \frac{E_1}{4} \qquad \left[\because E_n = \frac{E_1}{n^2}\right]$$

$$= \frac{3}{4} E_1$$

$$= \frac{3}{4} \times 13.6 = 10.2 \text{ eV}.$$

Hence photons of energy 9 eV cannot be absorbed by hydrogen atoms.

Example 24:

Calculate the ionisation potential of hydrogen from the following date.

Solution:

According to Bohr's theory, the energy of an electron in the nth orbit of hydrogen atom is

$$E = -\frac{2\pi^2 e^4 m}{n^2 h^2}$$

When the atom is in the normal state, the only electron in the hydrogen atom in the first orbit, *i.e.*, n = 1.

$$\therefore \quad E = -\frac{2\pi^2 m e^4}{h^2}$$

or $$E = \frac{2\times(3.14)^2 \times 9\times 10^{-28} \times (4.8\times 10)^4}{(6.6\times 20^{27})^2} = 2.165 \times 10^{-11} \text{ eg.}$$

To remove the electron from first orbit to infinity 2.165×10^{-11} erg of energy must be supplied. The amount of energy is called the ionisation potential of hydrogen atom.

Ionisation potential of hydrogen atom

$$= 2.165 \times 10^{-11} \text{ erg}$$

$$= \frac{2.165\times 10^{-11}}{1.6\times 10^{-12}} \text{ electron volts} = 13.53 \text{ eV.}$$

Example 25:

Energy in a Bohr's orbit is given to be equal to – B/n^2 *where* B = 2.179×10^{-11} *erg. Calculate the frequency of radiation and also the wave number when the electron jumps from the third orbit to the second* (h = 6.62×10^{-27} *erg sec) orbit.*

Solution:

Energy of a Bohr's orbit is given by

$$E = \frac{B}{n^2} = \frac{2.179\times 10^{-11}}{n^2} \text{ erg.}$$

For the third orbit, n = 3

$$E_2 = \frac{2.179\times 10^{-11}}{9} = -0.2421 \times 10^{-11} \text{ erg.}$$

For the second orbit, n = 2

$$E_2 = \frac{2.179\times 10^{-11}}{4} = -0.54475 \times 10^{-11} \text{ erg.}$$

$$E_3 - E_2 = (0.54475 \times 10^{-11}) - (0.2421 \times 10^{-11})$$
$$= 0.30265 \times 10^{-11} \text{ erg.}$$

When electron jumps from the third to the second orbit, a photon is emitted, the energy of which is given by

$$hv = E_3 - E_2 \text{ where v is the frequency}$$

$$v = \frac{E_3 - E_2}{h} = \frac{0.30265 \times 10^{-11}}{6.62 \times 10^{-27}} = 4.572 \times 10^{14} \text{ sec}^{-1}.$$

The corresponding wave number is given by

$$\bar{v} = \frac{1}{\lambda} = \frac{v}{c} = \frac{4.572 \times 10^{14}}{3 \times 10^{-10}} = 15240 \text{ cm}^{-1}$$

Example 26(a):

Give using spectral notation for the following states of the atom.

(i) n = 4, L = 2, S = 0

(ii) n = 4, L = 1, S = 1, J = 0 and

(iii) n = 3, L = 2, multiplicity 2.

Solution:

(i) Multiplicity (2S + 1) = 1, J = L + S = 2

∴ State will be 4 $^1D_2$.

(ii) Multiplicity (2S + 1) = 3 ∴ State will be 4 $^3P_0$.

(iii) Multiplicity will be (2S + 1) = 2 or S = 1/2

As L = 2, J = 5/2 or 3/2.

The two states which are positive are 3 $^2D_{5/2}$ or $^2D_{3/2}$. *Give using spectral notation for the following states of the atom.*

(i) n = 4, L = 2, S = 0

(ii) n = 4, L = 1, S = 1, J = 0 and

(iii) n = 3, L = 2, multiplicity 2.

Solution:

(i) Multiplicity (2S + 1) = 1, J = L + S = 2

∴ State will be 4 $^1D_2$.

(ii) Multiplicity (2S + 1) = 3 ∴ State will be 4 $^3P_0$.

(iii) Multiplicity will be (2S + 1) = 2 or S = 1/2

As L = 2, J = 5/2 or 3/2.

The two states which are positive are 3 $^2D_{5/2}$ or $^2D_{3/2}$.

Example 26(b):

Calculate the energy in calories per mole or per Einstein for radiations of wavelength 100 Å..

Solution:

Energy per quantum = hv

But h = 6.62 × 10^{-27} erg-sec.

$$\therefore\ V = \frac{c}{\lambda} = \frac{3\times10^{10}}{1000\,\text{Å}} = \frac{3\times10^{10}\,\text{cm.sec}^{-1}}{1000\times10^{-8}\,\text{cm}} = 3\times10^{15}\ \text{sec}^{-1}$$

Energy per quantum = hv

$= (6.62 \times 10^{-27}$ erg. sec) $(3 \times 10^{15}\ \text{sec}^{-1})$

$= 19.86 \times 10^{-12}$ ergs.

But N = 6.02 × 10^{28} molecules/mole

and hv = 19.86 × 10^{-12} erg.

Therefore, the energy per mole or per Einstein

= N hv

$= (6.02 \times 10^{23}$ molecules mole^{-1}) 19.86 × 10^{-12} ergs/mole

$= 11.94 \times 10^{12}$ ergs/mole

$= \dfrac{11.94\times10^{12}}{10^7}$ Jule/mole [∵ 1 Joule = 10^7 ergs]

$= \dfrac{11.94\times10^{12}}{4.184\times10^7}$ cal/mole [∵ 1 cal = 4.184 Joules]

$= \dfrac{2.8590\times10^5}{10^3}$ K cal/mole [∵ 1 K cal = 10^3 cal]

= 285.90 K cal/mole.

$$= \frac{285.90}{23.06} \text{ electron volt } [\because 1 \text{ eV} = 23.06 \text{ K cal/mole}]$$

$$= 12.390 \text{ electron-volts.}$$

Example 27:

$$B \rightarrow C$$

1.0×10^{-5} mole of B was formed on absorption of 6.62×10^7 ergs at 3600 Å. Calculate the quantum yield or efficiency.

Solution:

No. of moles reacting

$$= 1.0 \times 10^{-5} \times 6.02 \times 10^{23} \text{ molecules}$$

$$= 6.02 \times 10^{18} \text{ molecules}$$

No. of quanta absorbed

$$= \frac{\text{Total energy absorbed}}{\text{Energy of one quantum}} = \frac{6.62 \times 10^7 \text{ ergs}}{hv}$$

$$= \frac{6.62 \times 10^7}{hc/\lambda}$$

$$= \frac{6.62 \times 10^7 \times \lambda}{hc} \qquad \left[\because v = \frac{c}{\lambda}\right]$$

But $\quad l = 3600 \text{ Å} = 3600 \times 10^{-8}$ cm.

$c = 3 \times 10^{10}$ cm/sec, $h = 6.62 \times 10^{-27}$ erg/sec.

$$\text{No. of quanta absorbed} = \frac{6.62 \times 10^7 \times 3600 \times 10^{-8}}{6.62 \times 10^{-27} \times 3 \times 10^{10}} = 1.2 \times 10^{19}$$

$$\text{Therefore, equation yield} = \frac{\text{No. of molecules reacting}}{\text{NO. of quanta absorbed}}$$

$$= \frac{6.02 \times 10^{18}}{1.2 \times 10^{19}} = 0.506.$$

Example 28:

The bond energy in a molecule is 142.95 cal/mole. What is the longest wave-length of light capable of dissociating this molecule ?

Solution:

Energy per mole = N hv

where N= 6.02×10^{23} molecules, h = 6.62×10^{-27} erg-sec

$$v = \text{frequency} = \frac{c}{\lambda} = \frac{3.0 \times 10^{10}}{\lambda}$$

$$\therefore \text{ Energy per mole } = \frac{6.02 \times 10^{23} \times 6.62 \times 10^{-27} \times 3 \times 10^{10}}{\lambda} \quad ...(A)$$

The energy per mole = 142.95 cal/mole

$= 142.95 \times 10^3$ cal/mole

$= 142.95 \times 10^3 \times 4.184 \times 10^7$ ergs/mole ...(B)

Equating eqs. (A) and (B), we get

$$\frac{6.02 \times 10^{23} \times 6.62 \times 10^{-27} \times 3 \times 10^{10}}{142.95 \times 10^3 \times 4.184 \times 10^7}$$

$$= 142.95 \times 103 \times 4.184 \times 107$$

or
$$\lambda = \frac{6.02 \times 10^{23} \times 6.62 \times 10^{-27} \times 3 \times 10^{10}}{142.95 \times 10^3 \times 4.184 \times 10^7} \text{ cm}$$

$$= 2000 \times 10^{-8} \text{ cm} = 2000 \text{ Å}.$$

Example 29:

The dissociation energy of hydrogen is 102900 cal/mole. If H_2 is dissociated by illumination with radiation of wave-length 2537 Å. What fraction of the radiant energy will be converted in to kinetic energy ?

Solution:

Dissociation energy = 102900 cal/mole

$$= \frac{102900}{6.023 \times 10^{23}} = 1.708 \times 10^{-19} \text{ cal/molecule.}$$

Also, one quantum of light is able to dissociate one molecule of hydrogen. Therefore, Energy of one quantum = $hv = \frac{hc}{\lambda}$.

But $h = 6.625 \times 10^{-27}$ erg sec,

$c = 3 \times 10^{10}$ cm/sec

$\lambda = 2537$ Å $= 2537 \times 10^{-8}$ cm

$$\therefore \text{ Energy of one quantum } = \frac{6.625 \times 10^{-27} \times 3 \times 10^{10}}{2537 \times 10^{-8}} \text{ ergs}$$

$$= \frac{3 \times 6.625 \times 10^{-9}}{2537} \text{ ergs}$$

$$= \frac{3 \times 6.625 \times 10^{-9}}{2537 \times 4.18 \times 10^{7}} \text{ cal}$$

[$\because$ 1 cal 4.18 × 10^7 ergs]

But the dissociation energy per mole = 1.708 × 10^{-19} cal

$\therefore$ Energy converted into kinetic energy

$$= (1.874 \times 10^{-19} - 1.708 \times 10^{-19}) \text{ cal}$$

$$= 0.166 \times 10^{-1?} \text{ cal}$$

Hence, fraction converted into K.E. = $\frac{0.166 \times 10^{-19}}{1.874 \times 10^{-19}}$.

Example 30:

Calculate the time taken by the electron to traverse the first orbit in the hydrogen atom. Electron mass and charge are 9.1 × 10^{-31} kg and 1.6 × 10^{-19} C and h = 6.63 × 10^{-34} J-s.

Solution:

The radius of the n the Bohr orbit is

$r_n = 4\pi\varepsilon_0 \frac{n^2 h^2}{4\pi^2 m e^2}$ and the velocity of electron in the n th orbit is

$$v_n = \frac{1}{4\pi\varepsilon_0} \frac{2\pi e^2}{nh}.$$

The time taken by the electron to traverse the n the orbit is therefore

$$T_n = \frac{2\pi r_n}{v_n}$$

$$= 2\pi (4\pi\varepsilon_0)^2 \frac{n^2 h^2}{4\pi^2 m e^2} \frac{nh}{2\pi e^2}$$

$$= \frac{4\varepsilon_0^2 h^3 n^3}{me^4}.$$

For the first orbit, n = substituting the given values of h, m, e and ε_0 = 8.85 × 10^{-12} $C^2/N\text{-}m^2$, we get

$$T_1 = \frac{4.(8.85 \times 10^{-12})^2 (6.63 \times 10^{-34})^3}{(9.1 \times 10^{-31})(1.6 \times 10^{-19})^4} = 1.5 \times 10^{-16} \text{ s.}$$

Example 31:

Find an expression for the radius of the electron orbit in the hydrogen atom in its n th state. What will be the approximate quantum number n for an electron in an orbit of radius 1 mm ? (Take required values from problem 1).

Solution:

The basic equation in the Bohr's theory of hydrogen atom are.

$$\frac{mv^2}{r} = \frac{1}{4\pi\varepsilon_0}\frac{e^2}{r^2} \qquad ...(1)$$

and

$$mvr = \frac{nh}{2\pi}. \qquad ...(2)$$

Squaring (2) and dividing by (1), we get

$$r = 4\,\pi\,\varepsilon_0\,\frac{n^2h^2}{4\pi^2me^2}.$$

Substituting the given values, we get

$$r = \frac{1}{9\times10^9\ Nm^2/C^2}\times \frac{n^2\,(6.63\times10^{-34}\ Js)^2}{4\times(3.14)^2\times(9.1\times10^{-31}\ kg)\times(1.6\times10^{-19}\ C)^2}$$

$= 0.53 \times 10^{-10}\ n^2$ meter

$= 0.53\ n^2$ Å.

For r = 1 mm = 10^7 Å, we have

$$10^7 = 0.53\ n^2.$$

$$\therefore \qquad n^2 = \frac{10^7}{0.53} = 18.87 \times 10^6 \text{ or n; } 4350$$

Example 32:

The average life-time of an electron in an excited state of hydrogen atom is about 10^{-8} s. How many revolutions does an electron in the n = 2 state make before dropping to the n=1 state? (R = 1.097 × 107 m^{-1}).

Solution:

Let v be the velocity of electron (mass m, charge e) in an orbit of radius r. This basic equations are.

$$\frac{mv^2}{r} = \frac{1}{4\pi\varepsilon_0}\frac{e^2}{r^2}$$

and $$mvr = n\frac{h}{2\pi}.$$

These equations give

$$v = \frac{nh}{2\pi mr}, \quad r = \frac{n^2h^2\varepsilon_0}{\pi me^2}.$$

The number of revolutions of the electron in the orbit per second is

$$f = \frac{v}{2\pi r} = \frac{2Re}{n^3} \qquad \text{(as proved in Q. 8)}$$

For n = 2 state, we have

$$f = \frac{2\times(1.097\times10^7\ m^{-1})\times(3\times10^8\ ms^{-1})}{8}$$

$$= 8.2 \times 10^{14}\ s^{-1}.$$

Hence the number of revolutions of the electron in its life-time of 10^{-8} second is $(8.2 \times 10^{14}) \times 10^{-8} = 8.2 \times 10^6$.

Example 33:

An orange photon of wavelength 600 nm is emitted from an atom. Find the difference in energy in the two atomic states involved. Find the same result for the red 6563 Å line of hydrogen. ($h = 6.63 \times 10^{-34}$ j s, $c = 3 \times 10^8\ m\ s^{-1}$, $1\ eV = 1.6 \times 10^{-19}$ J).

Solution:

By Bohr's postulate, the emitted frequency is given by

$$v = \frac{\Delta E}{h},$$

where ΔE is the difference in energy. But $v = c/\lambda$. Therefore

$$\frac{c}{\lambda} = \frac{\Delta E}{h} \text{ or } \Delta E = \frac{hc}{\lambda}.$$

Here $\lambda = 600\ nm = 600 \times 10^{-9}$ m.

$$\therefore \quad \Delta E = \frac{(6.63\times10^{-34}\ Js)\times(3.0\times10^8\ ms^{-1})}{600\times10^{-9}\ m}$$

$$= 3.31 \times 10^{-19}\ J$$

$$= \frac{3.31 \times 10^{-19}}{1.6 \times 10^{-19}} = 2.07 \text{ eV}.$$

For $\lambda = 6563 \text{ Å} \times 10^{-10}$ m, we can show that

$$\Delta E = 3.03 \times 10^{-19} \text{ J} = 1.9 \text{ eV}.$$

Example 34(a):

What is the smallest wavelength in the spectral lines of Paschen series in the hydrogen atom ?

Solution:

The formula for Paschan series is

$$\frac{1}{\lambda} = R\left(\frac{1}{3^2} - \frac{1}{n^2}\right), \; n = 4, 5, 6, \ldots$$

For the smallest wave length, $n = \infty$.

$$\therefore \quad \lambda = \frac{9}{R} = \frac{9}{1.097 \times 10^7 \text{ m}^{-1}}$$

$$= 8.204 \times 10^{-7} \text{ m} = 8204 \text{ Å}.$$

Example 34(b):

The series limit of Balmer series is at 3646 Å. Calculate the wavelength of the first member Ha of this series.

Solution:

The formula for Balmer series is

$$\frac{1}{\lambda} = R\left(\frac{1}{2^2} - \frac{1}{n^2}\right), \; n = 3, 4, 5,$$

For series limit $n = \infty$ ad $\lambda = \lambda_\infty = 3646$ Å.

$$\therefore \quad \frac{1}{3646 \text{ Å}} = \frac{R}{4} \text{ or } R = \frac{4}{3646 \text{ Å}}.$$

Again, for the first member $n = 3$.

$$\therefore \quad \frac{1}{\lambda_1} = R\left(\frac{1}{2^2} - \frac{1}{3^2}\right) = \frac{5R}{36}$$

$$\text{or} \quad \lambda_1 = \frac{36}{5R} = \frac{36}{5 \times \frac{4}{3646 \text{ Å}}} = 6563 \text{ Å}.$$

3

Wave Particles

EMISSION OF α-PARTICLES FROM A RADIOACTIVE ELEMENT

On the basis of the tunnel effect we give an explanation of the emission of a-particles from a ratio-active element.

Elements with atomic nt aber $Z \geq 81$ are usually unstable and consequently undergo spontaneous disintegration with the emission of α-particles, β-particles and γ-rays (An α-particle consists of two neutrons and two protons and hence it is the nucleus of the helium atom). This phenomenon is called natural radio activity.

There is a narrow range of kinetic energies of α-particles emitted by radioactive elements. An α-particle emitted by thorium, ${}_{90}Th^{281}$, has lowest energy (4.007) Mev) and that emitted by thorium C', ${}_{54}Po^{212}$ has highest energy (8.95 Mev). Since an α-particle carries positive charge 2e, when it has teen emitted from a nucleus of atomic number Z, its potential energy, at a large distance from the residual nucleus of atomic number Z – 2 is given by

$$V(r) = \frac{(Z-2)e}{4\pi\varepsilon_0 r}(2e) = \frac{2(Z-2)e^2}{4\pi\varepsilon_0 r}$$

where r is the distance of the α-particle from the centre of the nucleus. If distance r is decreased the potential energy increases till it becomes maximum equal to V_0 at a position $r = r_0$ just outside the nucleus. The maximum potential energy is equal to the work which must be done against the repulsive electrostatic force to move the a-particle from $r = \infty$ to $r = r_0$. When it is inside the nucleus, it is acted upon by short range and extremely powerful attractive force due to the nucleus.

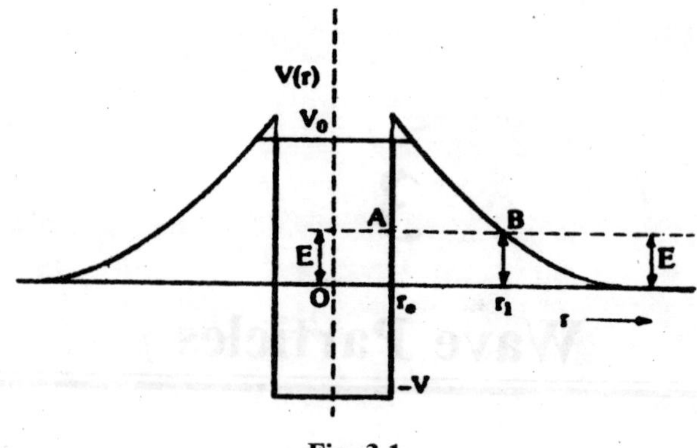

Fig. 3.1

These forces are effective over a short distance r_0 from the centre of the nucleus. This distance is called the *effective radius of the nucleus*. Due to the attractive forces the potential energy of the a-particle inside the nucleus is negative. Therefore, the graph of the potential energy V (r) of an a-particle against its distance r from the centre of the nucleus is in the form of a potential well as shown in Fig. 3.1. The maximum positive potential energy of an a-particle just outside the nucleus is given by

$$V_0 = \frac{2(Z-2)e}{4\pi\varepsilon_0 r_0}.$$

For $_{84}Po^{210}$ which emits an a-particle with kinetic energy 5.3 MeV, if we substitute

$$Z = 84,\ \varepsilon = 1.6 \times 10^{-10} \text{ coul,}$$

$$\varepsilon_0 = 8.85 \times 10^{-12} \text{ coul}^2/\text{Nm}^2 \text{ and}$$

$$r_0 = 1.5\times10^{-15} A^{1/2} = 1.5 \times 10^{-15} \times 210^{1/3} \text{ m}$$

We get, $V_0 \approx 27\,\text{MeV}$

Thus, the kinetic energy of an α-particle emitted from a radioactive nucleus is much less than the maximum positive potential energy just outside the nucleus.

Further of Classical Mechanics to explain the Emission of α-Particles

When an a-particle is emitted from a radio active nucleus with kinetic energy E, this energy . This energy is shown by the dashed line AB in

Fig. 3.1. The energy E is equal to the potential energy of the a-positive at $r = r_1$.

$$\therefore \qquad E = \frac{2(Z-2)e^2}{4\pi\varepsilon_0 r_1}$$

In the region from $r = r_0$ to $r = r_1$, the kinetic energy K of the α-particle is given by

$$K + \frac{2(Z-2)e^2}{4\pi\varepsilon_0 r} = E \frac{2(Z-2)e^2}{4\pi\varepsilon_0 r_1}$$

where $r_0 < r_1$

$$K = -\frac{2(Z-2)e^2}{4\pi\varepsilon_0}\left(\frac{1}{r} - \frac{1}{r_1}\right).$$

Since $r < r_1$, this region is the region of negative kinetic energy for the α-particle. This region of the electrostatic field around the nucleus is called potential barrier region for the α-particle. The distance AB = $r_1 - r_0$ is the width of the barrier for an α-particle of kinetic nucleus is the height of the potential barrier. Since energy V_0 of the a-particle just outside the according to the laws of classical mechanics, if an a-particle is originally inside the nucleus it can never come out.

Quantum mechanical Explanation of the Emission of Particles

In quantum mechanics, it is assumed that an a-particle in the nucleus is in constant to and fro motion. The motion is confined to the region surrounded by the potential barrier. Therefore, it is associated with a standing wave. The wave is propagated through the potential barrier and it emerges from the barrier as a progressive wave with a small amplitude.

The wave function Ψ (r) of an α-particle in the three regions is shown qualitatively in Fig. 3.2. In the nucleus, the wave function has an oscillating character in the barrier-region the wave function is exponentially attenuated, and outside the barrier region the wave function has again an oscillation character with a small amplitude. Since $|\Psi(r)|^2$ gives the probability of finding a particle at distance r, we conclude that there is a finite probability of transmission of an a particle through the potential barrier.

The quantum mechanical theory of the emission of α-particle was first given by G. Gamow, and independently by R. W. Gurney E. U. Condon in 1928.

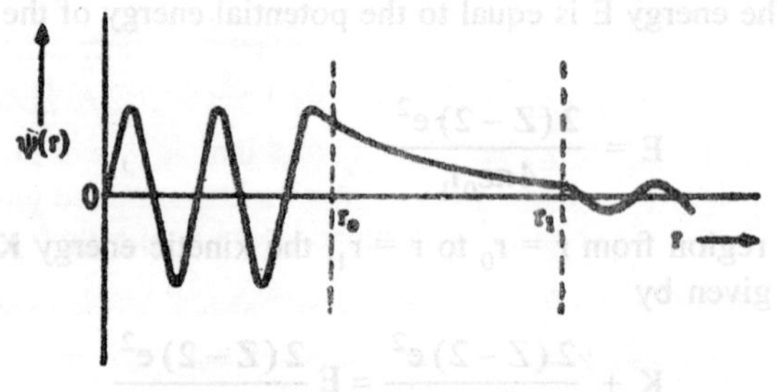

Fig. 3.2

PARTICLE IN A ONE DIMENSIONAL BOX

One of the simplest application of wave mechanics is found in the treatment of a particle confined to move within a box. Suppose the particle travels only along the x-axis and is confined between x = 0 and x = a by two infinitely hard walls so that the particle has no chance of penetrating them. Suppose that the particle does not lose energy when it collides with such walls, so that its total energy remains constant. This box can, then be represented by a potential well or box of width 'a' with potential walls of infinite height at x = 0 and x = a, so that the potential energy V of the particle is infinitely high, *i.e.*, infinitely on I and III sides of the box, while the inside of II, throughout the length a, V is uniform. For sake of convenience, we assume V to be zero inside the box (Fig. 3.3).

Since the potential outside the box is infinity high , the probability of finding the electron outside must be zero and hence Y must be zero is regions I and III when 0 > x > a.

Wave function in region II (in one-dimension) is

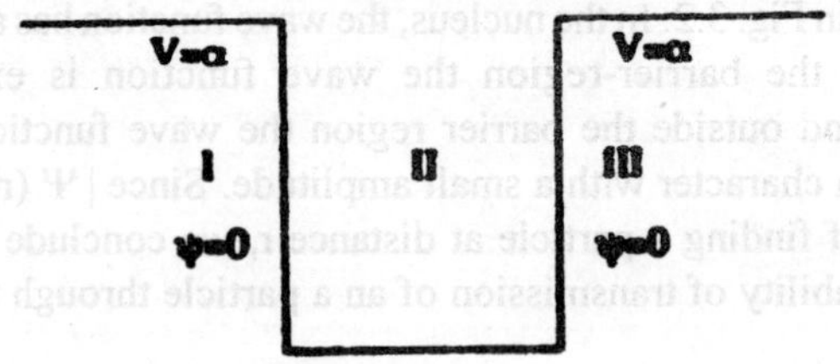

Fig. 3.3

$$\frac{d^2\Psi}{dx^2} + \frac{8\pi^2 mE}{h^2}\Psi = 0 \qquad ...(1)$$

or $$\frac{d^2\Psi}{dx^2}+\alpha^2\Psi = 0 \left[\text{where } \alpha^2 = \frac{8\pi^2 mE}{h^2}\right] \quad ...(2)$$

Solution of this is $\Psi = A^{i\alpha x} + Be^{-i\alpha x}$

where A and B are arbitrary constants. Let us take the solution,

$$\Psi = A^{i\alpha x} = A(\cos \alpha x + i \sin \alpha x) \quad ...(3)$$

$$= A \cos \alpha x + Ai \sin \alpha x = A \cos \alpha x + A' \sin \alpha x \quad ...(4)$$

At $x = 0, \Psi = 0$, hence by Eq. (4), $A = 0$

So the solution of the equation is $\Psi = A' \sin \alpha x$

But at $x = a, \Psi = 0$, hence $0 = \sin n\pi$, where n is an integer,

$$\sin \alpha a = \sin n\pi$$

$$\alpha = n\pi/a$$

$$\alpha^2 = \frac{n^2h^2}{a^2} = \frac{8\pi^2 mB}{h^2} \quad \text{[from Eq. (2)]}$$

$$E = \frac{n^2h^2}{8ma^2} \text{ where } n = 1, 2, 3 \quad ...(5)$$

Some significant conclusions from the above derivation are.

(a) The energy of the electron is now discrete when it is moving in a closed space. The continuity conditions breaks down because of appearance of n, which makes the energy quantised. For smaller values of a, the energy increases because a^2 is in the denominator. For value of n to keep E constant, *i.e.*, the quantum states will be more in number , if 'a' is infinity, n, is also infinity *i.e.*, the electron is not quantised. Thus, a *free electron con have energy but when it is confined within a certain range of space, the energy levels become quantised. The greater the localisation, the higher is the energy.*

(b) In any atom which is regarded as a type of potential box there are several energy levels corresponding to $n = 1, 2, 3,...$

where $$n = 1, E_1 = \frac{h^2}{8ma^2}$$

$$n = 2, E_2 = \frac{4h^2}{8ma^2} \quad E_2 - E_1 = \frac{3h^2}{8ma^2}$$

$$n = 3,\ E_3 = \frac{9h^2}{8ma^2} \quad E_3 - E_2 = \frac{5h^2}{8ma^2}$$

This suggests that difference between consecutive energy levels is not constant.

(c) The minimum value of n is 1; it cannot be zero as it must always have a significant value. Hence minimum energy is given by.

$$Em/n = \frac{h^2}{8ma^2}$$

From this we conclude that when an electron is enclosed in a box, it has a minimum energy at the lowest state which is the 'zero point energy'

(d) The complete wave function. We have

$$\alpha = \frac{n\pi}{a} \text{ and } \Psi = A' \sin \alpha x = A' \sin \frac{n\pi x}{a}$$

To ascertain A', the normalising condition is

$$\int_{-\infty}^{+\infty} \Psi^2 \, dx = 1$$

$$A'^2 \int_{-\infty}^{+\infty} \sin^2 \frac{n\pi x}{a} dx = 1$$

Now the wave function $\Psi = \infty$ at and beyond x = 0 and x = a, hence the limits $-\infty$ to ∞ may be replaced by 0 to a.

$$A'^2 \int_0^a \sin^2 \frac{n\pi x}{a} dx = 1$$

$$\Rightarrow \quad A'^2 \frac{a}{2} = 1 \text{ or } A' = \pm \sqrt{\frac{2}{a}}$$

Hence $\quad \Psi = \sqrt{\frac{2}{a}} \sin \frac{nx\pi}{a}$

This is absolutely a real wave function. Two informations can be derived.

(i) For n = 1, 2, 3 ... the wave functions would be like those shown in Fig. 3.4.

(ii) 'n' is a typical quantum number, which represents the nodes in the electron waves. As n increases, the number of nodes varies in the following

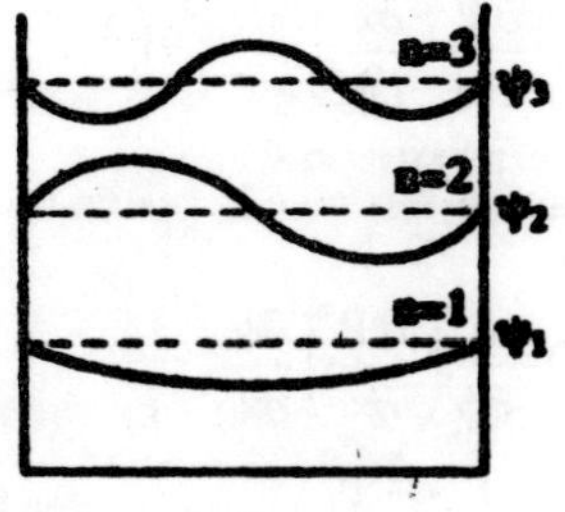

Fig. 3.4

n = 1 no mode

n = 2 1 node

n = 3 2 nodes

n = 4 3 nodes

and so on

ELECTRON IN RING

Suppose an electron of mass m is restricted to move along a circular track of radius r on which potential energy is constant (Fig. 3.5).

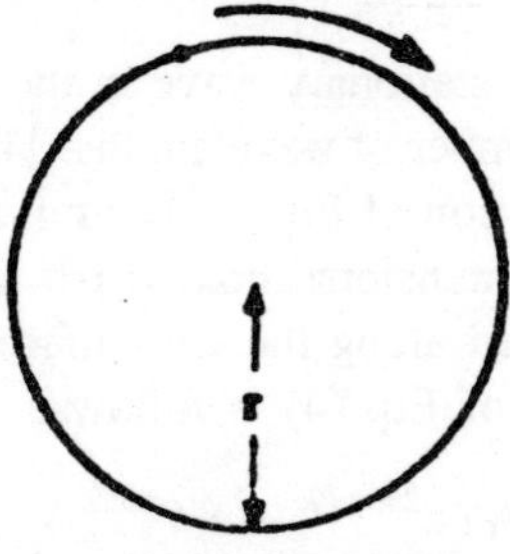

Fig. 3.5

In order to solve this problem, the polar coordinates will be used. The one-dimensional Schrodinger's equation is

$$\frac{\partial^2 \Psi}{\partial x^2} + \frac{8\pi^2 m}{h^2}(E - V)\Psi = 0 \qquad ...(1)$$

The transformation form Cartesian to polar coordinates may be carried out as follows.

Let $x = r\phi$

Then, since $\frac{\partial \Psi}{\partial x} = \frac{\partial \Psi}{\partial x} \cdot \frac{\partial x}{\partial \phi}$

$$\Rightarrow \quad \frac{\partial \Psi}{\partial x} = \frac{1}{r}\frac{\partial \Psi}{\partial \phi}$$

$$\Rightarrow \quad \frac{\partial^2 \Psi}{\partial x^2} = \frac{\partial}{\partial \phi}\left(\frac{\partial \Psi}{\partial x}\right)\frac{\partial \phi}{\partial x}$$

$$= \frac{1}{r^2}\frac{\partial^2 \Psi}{\partial \phi^2} \qquad \text{...(2)}$$

On substituting Eq. (2) in Eq. (1), we get

$$\frac{1}{r^2}\frac{\partial^2 \Psi}{\partial \phi^2} + \frac{8\pi^2 m}{h^2} \cdot (E - V)\Psi = 0 \qquad \text{...(3)}$$

Since the energy is all kinetic, it means that the potential energy is zero *i.e.*, V = 0. Thus Eq. (3) becomes as follows.

$$\Rightarrow \quad \frac{1}{r^2}\frac{\partial^2 \Psi}{\partial \phi^2} + \frac{8\pi^2 m}{h^2} \cdot E\Psi = 0$$

$$\frac{\partial^2 \Psi}{\partial \phi^2} + \frac{8\pi^2 mr^2 E}{h^2}\Psi \qquad \text{...(4)}$$

In order to set up a stationary wave in the ring, the circumference must contain a whole number of wavelengths. Therefore, $2\pi r = n\pi$ where n is an integer. The solution of Eq. (4) is similar to that of the problem of a particle in a one dimensional box. If $p = 2\pi/l$, and $a/r = \phi$, where a is the distance travelled along the circumference and Ψ is measured in radians, the solution of Eq. (4) is follows.

$$\Psi = A\,(e^{2\pi ia/\lambda} + e^{-2\pi ia/\lambda})$$

$$\Psi = A\,(e^{4pr\phi} + e^{-1pr\phi}) \qquad \text{...(5)}$$

On differentiating Eq. (5), we get

$$\frac{\partial \Psi}{\partial \phi} = A\,(ipr\, e^{ipr\,\phi} - ipr\, e^{-ipr\,\phi})$$

$$\Rightarrow \quad \frac{\partial^2 \Psi}{\partial \phi^2} = A\,(i^2p^2r^2\, e^{ipr\,\phi} + i^2p^2r^2\, e^{-ipr\,\phi})$$

$$= i^2p^2r^2\, A\,(e^{ipr\,\phi} + e^{-ipr\,\phi})$$

$$= i^2p^2r^2\Psi \text{ [using Eq. 5]}$$

$$= -p^2r^2\Psi \; [\because i^2 = -1]$$

$$= -\frac{4\pi^2r^2}{\lambda^2}\Psi\left[\because p = \frac{2\pi}{\lambda}\right] \qquad ...(6)$$

On substituting Eq. (6) in (4), we get

$$-\frac{4\pi^2r^2}{\lambda^2}\Psi + \frac{8\pi^2mr^2E}{h^2}\Psi = 0$$

$$\Rightarrow \qquad E = \frac{h^2}{2m\lambda^2}$$

$$\text{But} \qquad \lambda = \frac{2\pi r}{n}$$

$$\therefore \qquad E = \frac{\pi^2h^2}{8\pi^2mr^2}$$

where n has the values 1, 2, 3,... . Therefore, the energy can take one of a series of eigen values.

THE PARTICLE A THREE-DIMENSIONAL BOX

Now we shall consider the case of a particle enclosed inside a rectangular box, with edges a, b, c in length. The potential function (x, y, z) has a constant value of zero in following regions.

$$V(x, y, z) = 0 < x > a,$$

$$V(x, y, z) = 0 < y > b,$$

and

$$V = (x, y, z) = 0 < z > c.$$

The potential outside the box at the walls is infinite. The Schrodinger's time independent wave equation inside the box is given by

$$\frac{\partial^2\Psi}{\partial x^2} + \frac{\partial^2\Psi}{\partial y^2} + \frac{\partial^2\Psi}{\partial z^2} + \frac{8\pi^2m}{h^2}E\Psi = 0 \qquad ...(1)$$

The equation (1) is separated by substituting

$$Y(x, y, z) = X(x)\,Y(y)\,Z(z) = XYZ \qquad ...(2)$$

Differentiating equation (2) with respect to x, y and z separately by keeping the remaining two factors as constant, we get

$$\frac{\partial^2\Psi}{\partial x^2} = \frac{\partial^2 X}{\partial x^2}\cdot YZ \qquad ...(3)$$

$$\frac{\partial^2\Psi}{\Psi y^2} = \frac{\partial^2 Y}{\partial y^2}\cdot XZ \qquad ...(4)$$

$$\frac{\partial^2\Psi}{\partial z^2} = \frac{\partial^2 Z}{\partial z^2}\cdot XY \qquad ...(5)$$

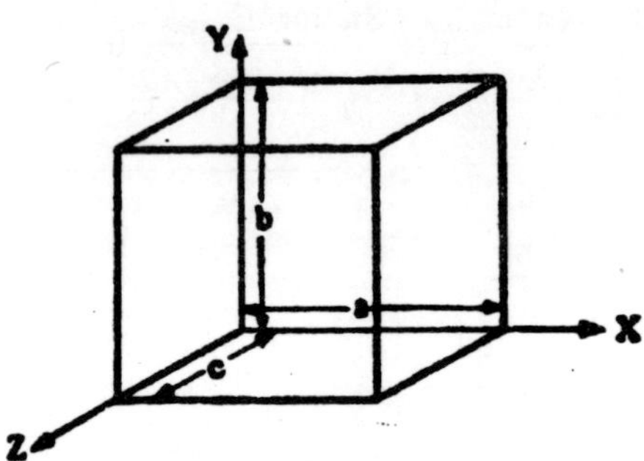

Fig. 3.6

Substituting equations (3), (4) and (5) in equation (2), we get

$$\frac{\partial^2 X}{\partial x^2}YZ + \frac{\partial^2 Y}{\partial y^2}XZ + \frac{\partial^2 Z}{\partial z^2}XY + \frac{8\pi^2 m}{h^2}E\Psi = 0 \qquad ...(6)$$

Also substituting equation (2) in (6), we get

$$\frac{\partial^2 X}{\partial x^2}YZ + \frac{\partial^2 Y}{\partial y^2}XZ + \frac{\partial^2 Z}{\partial z^2}XY + \frac{8\pi^2 m}{h^2}EXYZ \qquad ...(7)$$

Dividing equation (7) by XYZ, we get,

$$\frac{1}{X}\frac{\partial^2 X}{\partial x^2} + \frac{1}{Y}\frac{\partial^2 Y}{\partial y^2} + \frac{1}{Z}\frac{\partial^2 Z}{\partial z^2} = -\frac{8\pi^2 m}{h^2}E \qquad ...(9)$$

For the given energy of the particle, the term ($-\ 8\pi^2 mE/h^2$) is constant and each term on the left side is a function of one variable only. If we allow only one of these. (x or y or z) to vary at a time and keep the other two a constants (say we vary x keeping y and z constant), the sum of the three terms is still equal to the constant on the right hand side.

This means that each of the three terms on the left is itself a constant and is independent of the other variable, present in it. Let is represent the constants for the three terms as $-\alpha_1^2$, $-\alpha_2^2$ and $-\alpha_3^3$. These have a minus sign because the term on the right side of equation has minus sign. This gives three differential equations.

$$\frac{1}{X}\frac{\partial^2 X}{\partial x^2} = -\alpha_1{}^2 \quad ...(9)$$

$$\frac{1}{Y}\frac{\partial^2 Y}{\partial y^2} = -\alpha_2{}^2 \quad ...(10)$$

$$\frac{1}{Z}\frac{\partial^2 Z}{\partial z^2} = -\alpha_3{}^3 \quad ...(11)$$

Substituting equations (9), (10) and (11) in (8), we get

$$(-\alpha_1^2)+(-\alpha_2^2)+(-\alpha_3^2) = -\frac{8\pi^2 m}{h^2}E \quad ...(12)$$

$$\Rightarrow \qquad \alpha_1^2+\alpha_2^2+\alpha_3^2 = \frac{8\pi^2 m}{h^2}E \quad ...(13)$$

From solutions of equations (9), (10) and (11), we get

$$\left.\begin{aligned} X &= A_1 \cos\alpha_1 x + B_1 \sin\alpha_1 x \\ Y &= A_2 \cos\alpha_2 y + B_2 \sin\alpha_2 y \\ Z &= A_3 \cos\alpha_3 z + B_3 \sin\alpha_3 z \end{aligned}\right\} \quad ...(14)$$

where A_1, A_2 and A_3 and B_1, B_2 and B_3 are constants. The values of these constants can be obtained by applying the boundary conditions. We know that Ψ vanishes at the surface of infinite potential, hence

$$\left.\begin{aligned}\Psi = 0 \text{ when } x &= a\\ x &= 0\end{aligned}\right\} \qquad \left.\begin{aligned}y &= 0\\ y &= b\end{aligned}\right\} \qquad \left.\begin{aligned}z &= 0\\ z &= c\end{aligned}\right\}$$

Applying these boundary conditions, we get

$$A_1 = A_2 = A_3 = 0$$

Further

$\because$ $B_1 \neq 0$ $\therefore$ $\sin\alpha_1 a = 0$ *i.e.*, $\alpha_1 a = n_x\pi$ or $\alpha_1 = \dfrac{n_x\pi}{a}$

$\because$ $B_2 \neq 0$ $\therefore$ $\sin\alpha_2 b = 0$, *i.e.*, $\alpha_2 b = n_r\pi$ $\alpha_2 = \dfrac{n_r\pi}{b}$

$\because$ $B_3 \neq 0$ $\therefore$ $\sin\alpha_3 c = 0$, *i.e.*, $\alpha_3 c = n_z\pi$ or $\alpha_3 = \dfrac{n_z\pi}{c}$

Hence $\qquad X = B_1 \sin\dfrac{n_x\pi x}{a}$ $\qquad$ [From eq. (13)] $\qquad$ ...(14)

where n_x is any integer, nx = 1, 2, 3...

Similarly $\Psi = B_2 \sin \dfrac{n_r \pi y}{b}$, $n_r = 1, 2, 3, \ldots$ ...(15)

and $Z = B_3 \sin \dfrac{n_z \pi z}{c}$, $n_z = 1, 2, 3, \ldots$ ...(16)

$$\therefore \quad \Psi (n_x, n_y, n_z, x, y, z) = B_1 B_2 B_3 \sin \frac{n_x \pi x}{a} \sin \frac{n_y \pi y}{b} \sin \frac{n_z \pi z}{c}$$

$$\Rightarrow \quad \Psi (n_z, n_y, n_z, x, y, z,) = k \sin \frac{n_x \pi x}{a} \sin \frac{n_y \pi y}{b} \sin \frac{n_z \pi z}{c} \quad ...(17)$$

where k is normalization constant. The value of k can be obtained by applying the normalised condition, *i.e.*,

$$\int_0^a \int_0^b \int_0^c \Psi \Psi^* \, dx \, dy \, dz = 1 \quad ...(18)$$

$$\Rightarrow \quad k^2 \int_0^a \int_0^b \int_0^c \sin^2 \frac{n_x \pi x}{a} \sin^2 \frac{n_r \pi r}{b} \sin^2 \frac{n_z \pi z}{c} \, dx \, dy \, dz = 1 \quad ...(19)$$

But $\int_0^a \sin^2 \dfrac{n_x \pi x}{a} dx = \dfrac{a}{2}$

Similarly $\int_0^a \sin^2 \dfrac{n_y \pi y}{b} dy = \dfrac{b}{2}$

and $\int_0^a \sin^2 \dfrac{n_z \pi z}{c} dz = \dfrac{c}{2}$

$$\Rightarrow \quad k^2 \frac{a}{2} \cdot \frac{b}{2} \cdot \frac{c}{2} = 1 \text{ or } k = \frac{2\sqrt{2}}{\sqrt{(abc)}}$$

$$\Rightarrow \quad \Psi (n_x, n_y, x_z, x, y, z) = \frac{2\sqrt{2}}{\sqrt{(abc)}} \sin \frac{n_x \pi x}{a} \sin \frac{n_y \pi y}{b} \sin \frac{n_z \pi z}{c}$$

Again from equation (12), $\alpha_1^2 + \alpha_2^2 + \alpha_3^3 = \dfrac{8\pi^2 mE}{h^2}$

$$\therefore \quad \frac{n_x^2 \pi^2}{a^2} + \frac{n_y^2 \pi^2}{b^2} + \frac{n_z^2 \pi^2}{c^2} = \frac{8\pi^2 mE}{h^2}$$

$$\Rightarrow \quad E = \frac{h^2 \pi^2}{8m\pi^2} \left[\frac{n_x^2}{a^2} + \frac{n_r^2}{b^2} + \frac{n_z^2}{c^2} \right]$$

$$\Rightarrow \qquad E = \frac{h^2}{8m}\left[\frac{n_x^{\ 2}}{a^2}+\frac{n_y^{\ 2}}{b^2}+\frac{n_z^{\ 2}}{c^2}\right] \qquad ...(20)$$

p = 6	123,	132,	213,	231,	312,	321
p = 1	222,					
p = 3	113,	131,	311			
p = 3	122,	212,	221			
p = 3	112,	121,	211			
p = 1	111.					

But if we consider a box that is cubical in shape such that a = c = c, energy can be ex pressed by

$$E = \frac{h^2}{8ma^2}(n_x^{\ 2}+n_y^{\ 2}+n_z^{\ 2}).$$

For the lowest quantum state (111), in which n_x, n_r, and n_z, respectively, are equal to unity, it is seen that $E = 3h^2/8ma^2$. There is only one set of quantum numbers that gives this energy state, and this level is said to be non-degenerate.

THE FREE PARTICLE

A particle is said to be free when no force is acting upon it and has constant potential energy which may be assumed to be zero. Hence the total energy of the particle is all the kinetic. The free particle is similar to a plane progressive wave moving along a straight line, we shall consider the motion of particle in one dimension and then in three dimensions.

Suppose m is the mass of particle which is moving along x-axis. Since there is no field, the potential energy V is zero. Therefore, the Schrodinger's wave equation in one-dimension is as follows.

$$\frac{\partial^2\Psi}{\partial x^2}+\frac{S\pi^2 mE}{h^2}\Psi = 0 \qquad ...(1)$$

$$\frac{\partial^2\Psi}{\partial x^2}+x^2\Psi = 0$$

where $\qquad \alpha^2 = \dfrac{8\pi^2 mE}{h^2} \qquad ...(2)$

A solution satisfying the above equation is

$$\Psi = A \sin \alpha x$$

$$\Psi = A \sin \frac{\sqrt{(8\pi^2 mE)}}{h} \cdot x \qquad \text{[Use Eq. (2)]}$$

$$\Psi = A \sin \frac{2\pi}{h} \sqrt{(2mE)}$$

where A is an arbitrary constant. For every positive value of E, the right hand side of Eq. (3) yields single-valued, finite and continuous quantities. Hence this is a solution of the wave equation as can be seen by direct substitution. In this case, for all +ve values of E, the free particle would show a continuity of energy states. It means that all possible values of E from zero to infinity are possible. In other words we have a continuous spectrum of energy values. *Thus, energy is not quantitised in the case of free particle.*

We now pass on to wave equation three dimensions.

$$\frac{\partial^2 \Psi}{\partial x^2} + \frac{\partial^2 \Psi}{\partial y^2} + \frac{\partial^2 \Psi}{\partial z^2} + \frac{8\pi^2 m}{h^2} E\Psi = 0 \qquad ...(3)$$

Eq. (1) is separated by substituting

$$\Psi (x, y, z) = X_{(x)} \ Y_{(y)} \ Z_{(z)} = XYZ \qquad ...(4)$$

Differentiating Eq. (4) with respect of x, y, and z separately by keeping the remaining two factors as constant, we get

$$\frac{\partial^2 \Psi}{\partial x^2} = \frac{\partial^2 X}{\partial x^2} \cdot YZ \qquad ...(5)$$

$$\frac{\partial^2 \Psi}{\partial y^2} = \frac{\partial^2 Y}{\partial y^2} \cdot XZ \qquad ...(6)$$

$$\frac{\partial^2 \Psi}{\partial z^2} = \frac{\partial^2 Z}{\partial z^2} \cdot XY \qquad ...(7)$$

Substituting Eqs. (5), (6) and (7) in Eq. (2) we get

$$\frac{\partial^2 X}{\partial x^2} + YZ + \frac{\partial^2 X}{\partial y^2} \cdot XZ + \frac{\partial^2 Z}{\partial z^2} XY + \frac{8\pi^2 m}{h^2} E\Psi = 0 \qquad ...(8)$$

Also substituting Eq. (4) in Eq. (8), we get

$$\frac{\partial^2 X}{\partial x^2} + YZ + \frac{\partial^2 Y}{\partial y^2} XZ + \frac{\partial^2 Z}{\partial z^2} XY = -\frac{8\pi^2 m}{h^2} EXYZ \qquad ...(9)$$

Dividing Eq. (9) by XYZ, we get

$$\frac{1}{X}\frac{\partial^2 X}{\partial x^2}+\frac{1}{Y}\frac{\partial^2 Y}{\partial y^2}+\frac{1}{Z}\frac{\partial^2 Z}{\partial z^2}=-\frac{8\pi^2 m}{H^2} \quad ...(10)$$

Also X is a function of x only, the first term does not change its values t when y and z change. Same is true for the second and third terms. Nevertheless the sum of these terms should yield a constant quantity – $8\pi^2 mE/h^2$ for any choice of x, y and z. Let us assume y and x constant and covert Eq. (10) as an equation in x only. Then, $\frac{1}{X}\frac{\partial^2 x}{\partial x^2}$ is independent of x, y and z and each equals a constant.

Suppose $$\frac{1}{X}\frac{\partial^2 x}{\partial x^2}=k_x \quad ...(11)$$

Similarly $\frac{1}{Y}\frac{\partial^2 Y}{\partial y^2}$ = kr and $\frac{1}{Z}\frac{\partial^2 Z}{\partial z^2}=k_x$ with the condition that

$$k_x+k_y+k_z=-\frac{8\pi^2 m}{h^2}E.$$

It is convenient to put kx $=\frac{8\pi^2 m}{h^2}E_x$ which gives the Eq. (11) in the form

$$\frac{\partial^2 X}{\partial x^2}+\frac{8\pi^2 m}{h^2}E_x X=0 \quad ...(12)$$

The solution of the above equation is

$$X_{(x)}=A_x \sin\left\{\frac{2\pi}{h}\sqrt{2Em_x}\,(x-x_0)\right\} \quad ...(13)$$

where A_x and x_0 are two arbitrary constants. The latter x_0 defined the locations of the zeros of the since function. Eq. (13) is, therefore, a general solution. The equation for y and z are exactly analogous to this one and are given as

$$Y_{(y)}=A_y \sin\left\{\frac{2\pi}{h}\sqrt{2mE_y}\,(y-y_0)\right\} \quad ...(14)$$

and $$Z_{(z)}\ A_z \sin\left\{\frac{2\pi}{h}\sqrt{2mE_z}\,(z-z_0)\right\} \quad ...(15)$$

So the overall eigen function may be obtained as the product of the Eqs. (13), (14) and (15), *i.e.*,

$$\Psi\,(z, y, z)=A \sin\left\{\frac{2\pi}{h}\sqrt{2mE_x}\,(x-x_0)\right\}\times$$

$$\sin\left\{\frac{2\pi}{h}\sqrt{2mE_y}\,(y-y_0)\right\}\times\sin\left\{\frac{2\pi}{h}\sqrt{2mE_z}\,(z-z_0)\right\}$$

where $A = A_x A_y A_z$ ...(16)

If it is assumed that the wave has all its energy in the x-direction and there is none in the y- and z- directions, then

$$\Psi(x, y, z, t) = A_x\left[\sin\frac{2\pi}{h}\sqrt{2mE_x}\,(x-x_0)\right]e^{-2\pi i f(t)}$$

On comparing this with $\Psi = C_x\, e^{2\pi i(x/\lambda - ct)} = C_x\, e^{2\pi/x/\lambda} \times e^{-2\pi/vt}$, we get

$$\lambda = \frac{h}{\sqrt{(2mE_x)}} \qquad ...(17)$$

In classical mechanics the energy E_x of the free particle of mass m moving with velocity V_x is given by

$$E_x = \frac{1}{2}mv_x^2 \qquad ...(18)$$

On substituting Eq. (18) in Eq. (1.7), we get

$$\lambda = \frac{h}{\sqrt{(2m,\frac{1}{2}mv_x^2)}} = \frac{h}{mv_x} \qquad ...(19)$$

The above equation is same as the de-Broglie's equation,

PARTICLES IN POTENTIAL BARRIERS

In the case of free particles, the particles is not quantised. Let us now consider the case of imposing constraints upon the particle by assuming that the motion is confined within fixed boundaries. This condition can be visualised by considering potential barriers.

I. Single Potential Barrier Problem

Let us first consider the collision of a particle with a square potential barrier of constant magnitude V. In this problem, the particle is approaching from region ZI of negative x and strikes the potential barrier V = OP (Fig. 3.7) Classically, if the energy of the electron is less than that of the barrier top, it should always be reflected and, therefore cannot be presented in the region. If but if the energy is greater, then the electron is always transmitted to region II. Here we shall prove from quantum mechanical considerations that the electron has a definite probability of being both reflected and transmitted even if the energy becomes less than the top of the barrier.

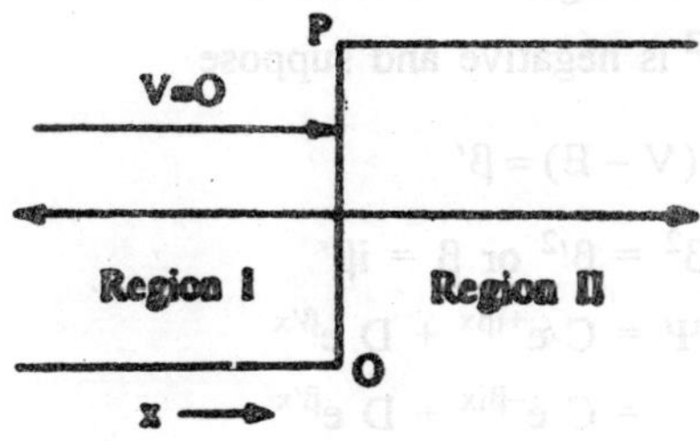

Fig. 3.7

Schrodinger's equation for the electron in region I (since V = 0) is given in one dimension by

$$\frac{\partial^2\Psi}{\partial x^2}+\frac{8x^2mE}{h^2}\Psi \quad ...(1)$$

or

$$\frac{\partial^2\Psi}{\partial x^2}+\alpha^2\Psi \quad ...(2)$$

where

$$\alpha^2=\frac{8\pi^2mE}{h^2} \quad ...(3)$$

The solution of Eq. (1) is

$$\Psi = A\,e^{I\alpha x}+B\,e^{-I\alpha x} \quad ...(4)$$

where A and B are arbitrary constants.

In Eq. (4), the positive exponent reveals that the electron is moving along a since (or cosine) wave and the negative exponent reveals that it is reflected back along a since (or cosine) wave.

Schrodinger's equation for the electron in region It is given in one dimensions by

$$\frac{\partial^2\Psi}{\partial x^2}+\frac{8\pi^2m}{h^2}(E-V)\Psi=0 \quad ...(5)$$

$$\frac{\partial^2\Psi}{\partial x^2}+\beta^2\Psi=0 \quad ...(6)$$

where

$$\beta^2=\frac{8\pi^2m(E-V)}{h^2} \quad ...(6A)$$

The solution of Eq. (6) is

$$\Psi = C\,e^{i\beta x} + D\,e^{-i\beta x}$$

where C and D are integration constants.

But E < V, β^2 is negative and suppose

$$\frac{8\pi^2 m}{h^2}(V - E) = \beta'$$

or $$\beta^2 = \beta'^2 \text{ or } \beta = i\beta'$$

Therefore $$\Psi = C\,e^{+i\beta x} + D\,e^{\beta' x} \quad ...(7)$$

$$= C\,e^{-\beta i x} + D\,e^{\beta' x} \quad ...(8)$$

Now Ψ has all real quantities because β' is real. Therefore, Ψ ceases to be a sine or cosine function in region II. De Moivres theory is no longer applicable and Ψ is only an exponential function. On applying the boundary condition to Eq. (8) it can be seen that at $x = \infty$, $\Psi = D\,e^{\beta' x} = \infty$ while Ψ should be zero at infinity. Hence D = 0.

Therefore, $$\Psi = C\,e^{-\beta' x} \quad ...(9)$$

Eq. (9) is the equation for electron wave in region II. But how can Ψ have a definite value in region II unless the electron is present there ? Hence although the energy is less than the potential barrier, yet the election has a definite probability f being present is region II from the quantum mechanical point of view.

The graphical interpretations of these wave functions give rise to some intersecting facts Graphically, the wave functions in regions I and II possess several alternatives.

In region I, the wave is sinusoidal whereas in region II it of the exponential form. In Fig, 3.8(a), the waves on the two sides are not continuous while in Fig. 3.8(b), the waves on the two sides are continuous. Which is correct?

At x = 0, $\Psi_t = A + B$ in region I, and at x = 0, $\Psi_{tt} = C$ in region II. But the wave function for the same electron must be continuous at the same point x = 0. The curve Fig (3.8a) has two values of amplitude at x = 0 which is wrong.

Therefore A + B = C must hold at x = 0 which means that the wave must be continuous. Therefore, Fig 3.8(b) is the correct graphical representation.

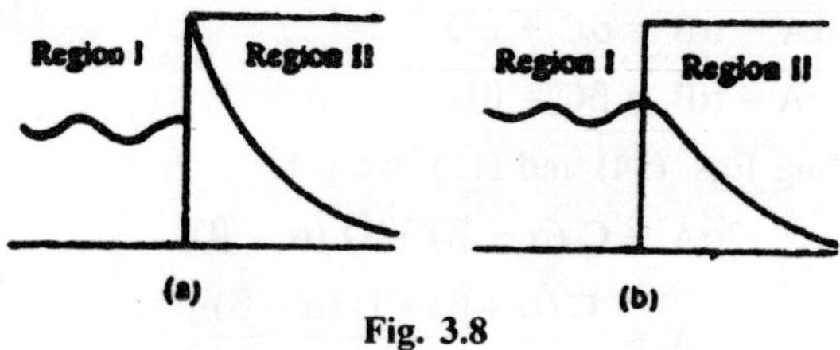

Fig. 3.8

The Barrier Whose V < E

Let the energy of the electron be higher than the height of the potential barrier. From classical considerations it is evident that the electron can go into region II. The wave function in region II is given by Eq. (7) *i.e.*,

$$\Psi = C\,e^{i\beta x} + D\,e^{-i\beta x} \qquad ...(10)$$

In the above solution, β^2 is positive and b is no more imaginary. Therefore, the solution is allright.

The wave functional in region I is given by Eq. (4). *i.e.*,

$$\Psi_t = A\,e^{i\alpha x} + B\,e^{-i\alpha x} \qquad ...(11)$$

At x = 0, the wave function from I to II should be continuous. It means that at x = 0,

$$A + B = C + D \qquad ...(12)$$

Further it follows that the first derivative, $\partial Y/\partial x$, should also be continuous at the same point (x = 0) so that

$$i\,\alpha\,A\,e^{i\alpha x} - i\,\alpha\,\beta\,e^{-i\alpha x}$$

$$= i\,\beta\,C\,e^{i\beta x} - i\,\beta\,D\,e^{-\alpha x}$$

At $\quad x = 0,\ i\,x\,A - i\,\alpha\,\beta = i\,C - i\,\beta\,D$

$$\alpha\,A - a\,\beta = \beta\,C - \beta\,D \qquad ...(13)$$

In region I, A is representing the movement of the wave from I to in while B is representing movement in the opposite direction.

In region II, C is representing the movement in the forward direction (left to right) while D is representing the movement in the backward direction. According to classical theory, the electron may move from region I to region II but can be reflected back into I, because E > V and, therefore, b must be zero. But quantum mechanically, $\beta \neq 0$. From Eqs. (12) and (13), therefore,

$$\alpha A + \alpha B = \alpha C + \alpha D \quad ...(14)$$

and
$$\alpha A - \alpha B = \beta C - \beta D \quad ...(15)$$

On adding Eqs. (14) and (15), we get

$$2\alpha A = C\,(\alpha + \beta) + D\,(\alpha - \beta)$$

$$A = \frac{C\,(\alpha+\beta) + D\,(\alpha-\beta)}{2\alpha} \quad ...(16)$$

On subtracting Eq. (15) from Eq. (14), we get

$$2\alpha B = (\alpha - \beta)\,C + (\alpha + \beta)\,D$$

$$B = \frac{C\,(\alpha-\beta) + D\,(\alpha-\beta)}{2\alpha} \quad ...(17)$$

On dividing Eq. (17), by Eq. (16) we get

$$\frac{B}{A} = \frac{(\alpha-\beta) + C + (\alpha+\beta)\,D}{(\alpha+\beta)\,C + (\alpha-\beta)\,D} \quad ...(18)$$

In region II, there is no more and potential barrier. Therefore, the wave cannot be reflected. Hence D = 0. Therefore. Eq. (18) becomes as follows.

$$\frac{B}{A} = \frac{\alpha-\beta}{\alpha+\beta} \quad ...(19)$$

This, ration represents the coefficient of the reflected wave to that of the moving wave. Therefore, the amplitude may be obtained as follows.

$$\left|\frac{B^2}{A^2}\right| = \left(\frac{\alpha-\beta}{\alpha+\beta}\right)^2 \quad ...(20)$$

This is know as reflectancy.

But
$$\alpha = \sqrt{\left(\frac{8\pi^2 mE}{h^2}\right)} \quad \text{[See Eq. (3)]} \quad ...(21)$$

and
$$\beta = \sqrt{\left(\frac{8\pi^2 m(E-V)}{h^2}\right)} \quad ...(22)$$

Therefore
$$\left|\frac{B^2}{A^2}\right| = \left(\frac{\alpha-\beta}{\alpha+\beta}\right)^2 = \left|\frac{\sqrt{E} - \sqrt{(E-V)}}{\sqrt{E} + \sqrt{(E+V)}}\right|^2 \quad ...(23)$$

Suppose the energy of the electron be 2V, *i.e.*, E = 2V. Therefore, Eq. (23) becomes as follows.

$$\left|\frac{B^2}{A^2}\right| = \left|\frac{\sqrt{(2V)} - \sqrt{V}}{\sqrt{(2V)} + \sqrt{V}}\right|^2 = \left|\frac{\sqrt{2}-1}{\sqrt{2}+1}\right|^2$$

$$\left|\frac{\sqrt{(2-1)} \times (\sqrt{2}-1)}{\sqrt{(2+1)} \times (\sqrt{2}-1)}\right|^2 = \left|\frac{(\sqrt{2}-1)^2}{2-1}\right|^2 = 0.025.$$

Thus, the percentage of reflectance is 2.5%. From this it follows from quantum mechanics that when the energy of the electron is twice the height of potential barrier, then the electron cannot move absolutely from region I to region II and there is a definite probability of at least 2.5% that the waves should be reflected back in the region from which they have started.

As Eq. (23) does not contain Planck's constants, it is applicable to particles of all masses.

Case III. Barrier of Definite Thickness (Tunnelling Effect) : Suppose OD is the thickness of the potential barrier shown in Fig. 3.9(a). The particle with E < V is starting from region I and may be, according to previous discussions, both in II and III.

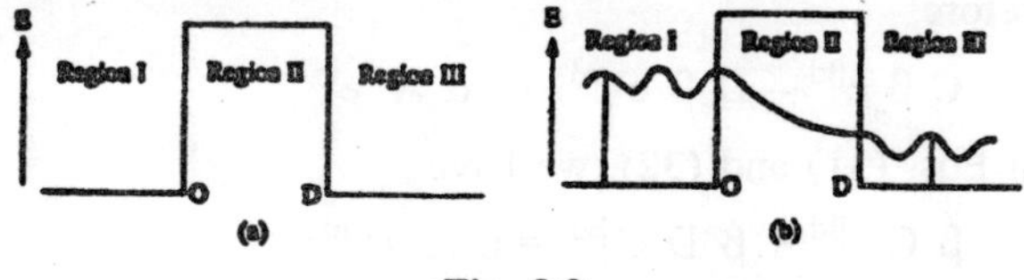

Fig. 3.9

Hence the corresponding wave functions in three regions would be

$$\Psi_I = A\,e^{i\alpha x} + B\,e^{-i\alpha x} \qquad ...(25)$$

$$\Psi_{II} = C\,e^{\beta x} = +\,D\,e{-}bx \qquad ...(26)$$

$$\Psi_{III} = A'\,e^{i\alpha x} + B'\,e^{-i\alpha x} \qquad ...(27)$$

Since in region III, there is no more any potential barrier, the wave cannot be reflected.

Thus $B'\,e^{-i\alpha x}$ which is the reflected part must be zero. Hence Eq. (27) becomes as follows.

$$\Psi_{III} = A'\,e^{i\alpha x} \qquad ...(28)$$

From the condition of continuity, Ψ_1, Ψ_2, and Ψ_3, as well as their derivatives with respect to x must be continuous. Hence, the graphical representation will be as shown in Fig. 3.9(b).

As the amplitude of Ψi is much greater than that of Ψ_{iii}, it means that the probability of finding the electron in region I is much greater than that in III. From condition of continuity, at x = 0

$$A + B = C + D$$

Also, $\frac{\partial \Psi}{\partial x}$ would also be continuous at x = 0 which gives

$$i\alpha A e^{i\alpha x} = i\alpha\ B\ e^{-i\alpha x} = C\ \beta\ e^{\beta x} - D\ \beta\ e_{-\beta x}$$

On putting x = 0 in the above equation, we get

$$\Rightarrow \quad i\alpha\ A - i\alpha B = C\beta - D\beta$$

$$\Rightarrow \quad A - B = \frac{(C - D)\beta}{i\alpha} \text{ nat } x = 0 \qquad \text{...(29)}$$

Let us now consider the wave functions in regions III and III. At x = d,

$$Y_{II} = C\ e^{\alpha d}\ D'\ e^{\beta d} \qquad \text{...(30)}$$

and $$Y_{III} = A'\ e^{i\alpha d} = \Psi_n \qquad \text{...(31)}$$

But $\partial\Psi/\partial x$ is also continuous at x = d.

Therefore,

$$C\ \beta\ e^{\beta d} - D\ \beta\ e^{-\alpha d} = i\ \alpha\ A'\ e^{i\alpha d} \qquad \text{...(32)}$$

From Eqs. (31) and (32), we have

$$\beta\ C\ e^{\beta d} + \beta\ D\ e^{-\beta d} = \beta\ A'\ e^{i\alpha d} \qquad \text{...(33)}$$

On adding Eqs, (32) and (33), we get

$$2\ \beta\ C\ e^{\beta d} = A'\ e^{i\alpha d}\ (\beta + i\alpha)$$

$$\Rightarrow \quad C = \frac{A' e^{i\alpha d}\ (\beta + i\alpha)}{2\beta\ e^{\beta d}}$$

$$= \frac{A'(\beta + i\alpha)\, e^{(i\alpha - \beta)d}}{2\beta} \qquad \text{...(34)}$$

On subtracting Eq. (33) from (32), we get

$$D = \frac{A'(\beta - i\alpha)\, e^{(i\alpha - \beta)d}}{2\beta} \qquad \text{...(35)}$$

But $\quad A + B = C + D$ (from continuity)

and $\quad A - B = \frac{(C - D)\beta}{i\alpha}$ [from Eq. 29]

$$\therefore \qquad A = \frac{i\alpha C + i\alpha D + C\beta - D\beta}{2i\alpha}$$

or
$$A = \frac{C(i\alpha + \beta) + D(i\alpha - \beta)}{2i\alpha} \qquad ...(36)$$

On substituting Eqs. (34) and (35) in (36), we get

$$A = \frac{A'(i\alpha+\beta)(i\alpha+\beta)e^{(i\alpha-\beta)d}}{4i\alpha\beta} - \frac{A'(i\alpha-\beta)^2 e^{(i\alpha+\beta)d}}{4i\alpha\beta}$$

$$= \frac{A'e^{i\alpha d}}{4i\alpha\beta}\left[(\beta^2 + 2i\alpha\beta - \alpha^2)e^{-\beta d} + (\alpha^2 + 2i\alpha\beta - \beta^2)e^{\beta d}\right]$$

$$= \frac{A'e^{i\alpha d}}{4i\alpha\beta}[\beta^2 e^{-\beta d} + 2i\alpha b\, e^{-\beta d} - \alpha^2.e^{\beta d} + 2i\alpha\beta\, e^{\beta d} - \beta^2 e^{\beta d}]$$

$$= \frac{A'e^{i\alpha d}}{4i\alpha\beta}\alpha^2 (e^{\beta d} - e^{-\beta d}) - \beta^2 (e^{\beta d} - e^{-\beta d})$$
$$+ 2\, i\, \alpha\, \beta\, (e^{\beta d} + e^{\beta d} + e^{-\beta d})]$$

$$A = \frac{A'e^{i\alpha d}}{4i\alpha\beta}[2(\alpha^2 - \beta^2)\sinh\beta d + 4i\beta\cosh\beta d] \qquad ...(37)$$

We know that A represents the coefficient of wave in region I moving forward whereas A′ represents the coefficient of wave in region III also moving forward. Therefore, $\left|\frac{A'}{A}\right|$ represents the relative probability of the electron in these two regions.

But the electron can be in region III only if it may leak through the potential barrier.

Let us now discuss in what way does the probability depend upon the length, breadth and height of the potential barrier. We know.

$$P = \left|\frac{A'}{A}\right|^2$$

But
$$A^2 = A'^2 (e^{i\alpha d} e^{-i\alpha d}\left[\frac{2(\alpha^2 - \beta^2)\sinh\beta d}{4i\alpha\beta} + \cosh\beta d\right]$$
$$+ \left[\cosh\beta d\, \frac{2(\alpha^2 - \beta^2)\sinh\beta d}{4i\alpha\beta}\right]$$

$$= A'^2 \left[\cosh^2 \beta d + \frac{(\alpha^2 - \beta^2)^2 \sinh^2 \beta d}{4\alpha^2\beta^2} \right]$$

(from the complex conjugate of A)

$$\text{Hence } P = \left| \frac{A'}{A} \right| = \frac{1}{\left[\cosh^2 \beta d + \frac{(\alpha^2 - \beta^2)}{4\alpha^2\beta^2} \sinh^2 \beta d \right]} \quad ...(38)$$

From the above equation it is evident that the value of P is a definite quantity. This proves that the particle has leaked through the potential barrier.

Let us now find the conditions when P = 0. Under what conditions the particle cannot really leak through the barrier ? When P = 0, Eq. (38) becomes as follows.

$$\cosh^2 \beta d + \frac{(\alpha^2 - \beta^2)}{4\alpha^2\beta^2} \sinh^2 \beta d = \infty$$

or $\beta d = \infty$ (because the functions are hyperbolic)

But $\beta d = \frac{2\pi}{h} \sqrt{(2m(V-E)d}$

and $\beta d = \infty$ indicates either (i) $\beta = \infty$,

i.e., $V = \infty$ or (ii)

$d = \infty$.

As m many be of macroscopic order and h is always of the order of 10^{-27} very small) and is negligible, it follows that

$$\beta = \frac{2\pi\sqrt{[2m(V-E)]d}}{(\text{small quantity})} \approx \infty$$

in the macroscopic domain. It means that in classical domain, if the height of the barrier is infinity, there is no probability of a particle being in II. Hence in the macroscopic domain where β is very large, there is no probability of small particles leaking through potential barriers if $V = \infty$.

In the microscopic domain, h cannot be neglected as compared to electronic mass, thus β cannot be ∞ for P to zero. In order to make P = 0 in this case, d should be ∞. It means that there is no probability of finding the particle in region III when the thickness of barrier is large.

The leakage through potential barriers, which is expected from the classical point of view, goes by the name of quantum i mechanical tunnelling effect. Enough evidence is available which demonstrates that the tunnelling phenomena occur in nature. The most direct evidence is found in the radioactive disintegration of atomic nuclei by alpha decay. The emission of electrons from metals at low temperatures and high electric field strengths is also controlled by tunnelling effect.

SOME POSTULATES

Postulate I. The functional f is continuous. Thus, the function Ψ can be subjected to differentiation.

Postulate II. The probability of finding the particle in the space with in r and r + dr is given by $\Psi^* \Psi$ dr.

where Ψ (r) is the function and Ψ^* (r) is its complex conjugate.

The probability of finding the particle in a finite interval between r_1 and r_2 is given by.

$$\text{Probability} = \int_{r_1}^{r_2} \Psi^* \Psi \, dr$$

where there is only one particle which is present some where, the total probability of the particle at some place is given by,

$$\text{Probability} = \int_{-\infty}^{\infty} \Psi^* \Psi \, dr = 1$$

Postulate III. The only possible values which a measurement of the observable whose operator is P, can give are the eigen values P_n of the equation. $P\Psi = P_n \Psi_n$.

This is true provided $\int \Psi_n * \Psi_n \, dV < \infty$ and that Ψ_n is single valued.

As an example, consider the measurable values of linear momentum of the particle. If the particle is some where on the X-axis (*i.e.*, only one–dimensional) we have $Px \rightarrow i \dfrac{h}{2\pi} \dfrac{\partial}{\partial x}$ as the operator used. So, the equation becomes, $-i \dfrac{h}{2\pi} \dfrac{\partial \Psi}{\partial x} = Px_n \Psi_n$.

Here Px_n is the measurable value of the momentum. Similarly, for other components Py_n, and Pz_n are obtained using $\dfrac{\partial \Psi_n}{\partial y}$ and $\dfrac{\partial \Psi_n}{\partial z}$.

If we want to find the total momentum, the equation is

$$-i\frac{h}{2\pi}\nabla\Psi_n\ P_n\ \Psi_n$$

Ψ_n is called an eigen function and P_n is an eigen value of the linear momentum.

The total momentum of the particle will be P_n. So, P_n^2

$$= P_{x_n}^2 + P_{y_n}^2 + P_{z_n}^2$$

$$= Px_n + Px_n + Py_n + Py_n + Pz_n + Pz_n$$

Postulate IV. The classical mechanical concepts such as position, velocity, linear momentum, angular momentum, energy and time occupy a central position in quantum mechanics. As in classical mechanics, they are regarded as the measurable properties of any system.

However, the behaviour of these properties is quite different from that predicated by classical mechanics. In addition, centring new properties such as spin and symmetry with respect to the inter change d of identical particles also appear.

These properties do not appear in classical mechanics. So, all these variables are called as the dynamical variables.

Every dynamical variable can be assigned a linear operator. Operators will be considered as latent observables leading to measurable quantities only when they operate on the state function Ψ, so that, we have the following table.

Since all the physically measurable quantities are real, only linear and Hamiltonian operators are permissible in quantum mechanics. The operators in quantum mechanics do not necessarily commute.

When they do commute, both the variables associated with them can be measured simultaneously and accurately. There will be no uncertainty in either of them.

If the two operators do not commute, the variable associated with them cannot be measured simultaneously and accurately. If one of them can be measured accurately, the other one cannot be measured simultaneously with accuracy.

	Dynamical variable	*Quantum mechanical operator*
1. Position co-ordinate	x	Multiplication by x
	y	Multiplication by y
	z	Multiplication by $\frac{h}{1}\frac{\partial}{\partial x}$
2. Linear momentum	P_x	Differential operators $\frac{h}{i}\frac{\partial}{\partial y}$
	P_y	Differential operators $\frac{h}{i}\frac{\partial}{\partial z}$
	P_z	Differential operators
3. Angular momentum,	M_x	$\frac{h}{1}\left[y\cdot\frac{\partial}{\partial z}-z\frac{\partial}{\partial y}\right]$
where i – 1	M_y	$\frac{h}{i}\left[z\cdot\frac{\partial}{\partial x}-x\frac{\partial}{\partial z}\right]$
	Mz	$\frac{h}{i}\left[x\cdot\frac{\partial}{\partial y}-y\frac{\partial}{\partial x}\right]$
4. Energy operator (Hamiltonian operator). $\hat{H}=\frac{p^2}{2m}+v-\frac{\hbar^2}{2m}\nabla^2+v$		

The possible state function, Y of a system is given by the solution of the differential equation, $\hat{H}\Psi=-\frac{h}{i}\cdot\frac{\partial}{\partial t}\Psi$, where $\hat{H}$ is the energy operator, the Hamiltonian operator for the system and $h=\frac{h}{2\pi}$. This is the time dependent Schrodinger equation. One can expand this as.

$$\left[-\frac{h^2}{2m}\nabla^2+v(x,y,z,t)\right]Y=\frac{h}{i}\frac{\partial}{\partial t}\cdot\Psi$$

Postulate V. If A is the operator equivalent of an observable quantity, this will be the total energy and a is the precise value of that property in a given state of the system, than it is postulate that operation by A on the function Ψ which is supposed to be suitable for this operator, is equivalent to multiplying Y by the quantity a. Thus.

$$A\Psi=a\Psi.$$

For the given state, known as the eigen state of the system Ψ is the eigen function for the operator A, and a is the corresponding eigen value it may be mentioned that in certain cases two or more eigen functions correspond to the same eigen value and belong to the same eigen state.

The state is then said to be degenerate. The operator A is also known as Hermitian operator.

Postulate VI. When a great many measurements of any dynamical variable are made on an assembly of system whose state function is Ψ, the average value obtained is given by $< a >$.

$$\langle a\rangle = \frac{\int_{-\infty}^{+\infty}\Psi^* A\Psi\, dt}{\int_{-\infty}^{+\infty}\Psi^*\Psi\, dt} \quad \text{where } \Psi \text{ is not normalized.}$$

where d_τ represents and element in the configuration space. We have

$$\langle a\rangle = \int_{-\infty}^{+\infty}\Psi^* A\Psi\, dt\text{, when } \Psi \text{ is normalized.}$$

Here A is the operator corresponding to the dynamical variable and integration is carried out over all the space.

Postulate VII. If Ψ_i and Ψ_j represent two different eigen functions both of which are satisfactory solutions of the wave equation for a given system, then these functions will be normalized if.

$$\int \Psi i\, \Psi i^*\, dt = 1 \text{ and} \int \Psi j\, \Psi j^*\, dt = 1.$$

Further, if the two eigen functions have the property that

$$\int \Psi i\, \Psi j^* = 0 \text{ or} \int \Psi i^*\, \Psi j^* = 0$$

then they are said to be mutually orthogonal.

Postulate VIII. If the two operators commute then there exists a set of functions which are simultaneously eigen functions of both the operators.

Under such conditions, the operators yield real observables simultaneously. For example, $\hat{H}$ and Mz commute and hence they have the same eigen functions. Therefore the observables or the eigen values associated with them can be measured simultaneously.

$$\therefore \quad [\hat{H}, Mz] = 0$$

$$\hat{H}\Psi = E\Psi \text{ and } Mz\Psi = m\, h\Psi$$

$$\left[\frac{h^2}{2m}\nabla^2 + v\right]\Psi = E\Psi$$

In spherical polar co-ordinates $\therefore$ $Mz \rightarrow \frac{h}{i}\frac{\partial}{\partial \phi}$.

Any two non-commuting operators cannot yield discrete or real observables simultaneously as they do not have the same eigen functions. Fox example, x and Px, are non-commuting operators, so that

$$[X, Px] = x\,Px - Pxx = \frac{h}{i} \neq 0$$

$$\nabla^2 = \frac{\partial^2}{\partial x^2} + \frac{\partial^2}{\partial y^2} + \frac{\partial^2}{\partial z^2} \quad ...(1)$$

Thus, Schrodinger wave equation *viz.*,

$$\frac{\partial^2 \Psi}{\partial x^2} + \frac{\partial^2 \Psi}{\partial y^2} + \frac{\partial^2 \Psi}{\partial z^2} \frac{8\pi^2 m(E-V)}{h^2}\Psi = 2$$

may be written in terms of Laplacian operator as

Hamiltonian Operator

Schrodinger wave equation as written above may be rewritten in the from

$$\nabla^2 \Psi + \frac{S\pi^2 m(E-V)}{h^2}\Psi = 0 \quad ...(2)$$

Hamiltonian Operator

Schrodinger wave equation as written above may be rewritten in the form

$$\nabla^2 \Psi = -\frac{8x^2 m(E-V)}{h^2}\Psi$$

or $$\nabla^2 \Psi = -\frac{8\pi^2 m}{h^2}(E\Psi - V\Psi)$$

or $$-\frac{h^2}{8\pi^2 m}\nabla^2 \Psi + V\Psi = E\Psi$$

or $$\left(-\frac{h^2}{8\pi^2 m}\nabla^2 + V\right)\Psi = E\Psi \quad ...(3)$$

This equation implies that the operation $\left(-\frac{h^2}{8\pi^2 m}\nabla^2 + V\right)$ carried on the function Ψ, is equal to the total energy multiplied with the function

Ψ. The operator $\left(-\frac{h^2}{8\pi^2 m}\nabla^2 + V\right)$ is called *Hailtonian operator* and is represented by $\hat{H}$. Thus

$$\hat{H} = \frac{h^2}{8\pi^2 m}\nabla^2 + V \qquad \text{...(4)}$$

Eqn. (3) may, therefore, be written as

$$\hat{H}\Psi = E\Psi \qquad \text{...(5)}$$

This is another short-hand from of writing the Schrodinger wave equation.

As mentioned earlier, Ψ is called the eigen function and E is called the eigen value.

Hermitian Operators

If u_1 and u_2 are some suitable functions of x and A is an operator such that the following relations hold good, then the operator A is a Hermitian operator.

$$\int u_1 * A * u_2 \, d\tau = \int u_2 \, A * u_1 * d\tau$$

In the above equation, the quantities with asterisks are complex conjugates of the corresponding quantities without asterisks. It is also to be remembered that the linear combination of Hermitian operators are also Hermitian.

Postulates of Quantum Mechanics

We have seen that on the basis of the wave nature of a moving particle, a equation for the particle can be given. It is called Schrődinger's wave equation. But just giving a wave equation is not sufficient. It has to be developed so as to obtain a quantitative description of matter on the atomic scale, and even in the nuclear scale.

There are different methods of using the wave equation so as to obtain the quantitative results like the values of momentum energy, angular momentum, etc. of a particle. In the new quantum mechanics, where the problem starts with the formulation of Schrodinger equation, the results are obtained using certain postulates. The advantage of the method will be that the values of an observable can be obtained quickly without fully solving the wave equation.

To start with, we take the problems of a single particle non-relativistically.

Postulate I. *The wave function Ψ (r, t) gives the complete knowledge of the behaviour of the particle. Similarly Ψ (r) gives the stationary state which is independent of time.*

The function Ψ which is a solution of the differential equation must be well behaved. In other words, it must be finite, single valued and continuous over the complete range of variable which must extend upto infinity. Further the squares of the values of the functions must also be finite when integrated over the complete range of variables. If Ψ is such a function and Ψ^* its complex conjugate, this conditions requires that

$$\int \Psi \Psi * d\tau$$

(where $d\tau$ represents element in the configuration space) should be finite.

We have seen that because of the uncertainty principle the knowledge of any property of a particle is not completely definite. By a 'complete knowledge' we mean that we get a description that contains as much information as possible which is consistent with the uncertainty relation.

SOLVED EXAMPLES

Example 1:

Find the height of the potential barrier for a-particle emitted from Random ($_{96}Rn^{222}$) assuming that the effective nuclear radius is given by

$$r_0 = 1.5 \times 10^{-15} A^{2/3} m$$

Solution:

$$V = \frac{2(Z-2)e^2}{4\pi\varepsilon_0 r_0}$$

$$= \frac{2(86-2)(1.6\times10^{-10})^2}{4\pi\times8.85\times10^{-12}\times1.5\times10^{-15}\times(222)^{1/3}} J$$

$$= \frac{2\times84\times1.6^2\times10^{-11}}{4\pi\times8.85\times1.5\times6.055} = \frac{430.08\times10^{-11}}{1009.57}$$

$$= 0.426 \times 10^{-11} J$$

$$V_0 = \frac{0.426\times10^{-11}}{1.6\times10^{-13}} = \frac{42.6}{1.6} = 26.62 \text{ MeV.}$$

Example 2(a):

Calculate the transition energy when the electron in the above example moves from n= 2 to n = 1. What is the wavelength of photon emitted ?

Solution:

We have

$$E = \frac{n^2h^2}{8ma^2} = n^2\ (6.6 \times 10^{-12})\ \text{erg/electron}$$

$$\Delta E = E_2 - E_1 = 6.6 \times 10^{-12}\ (n_2^2 - n_1^2)\ \text{per atom}$$

$$= 2.0 \times 10^{-11}\ \text{erg./atom}$$

According to Plank, $\Delta B = hv = \frac{hc}{\lambda}$

$$\lambda = \frac{hc}{\lambda E} = \frac{(6.6 \times 10^{-27}\ \text{erg .sec.})\ (3 \times 10^{10}\ \text{cm sec.})}{2.0 \times 10^{-11}\ \text{erg}}$$

$$= 9.9 \times 10^{-6}\ \text{cm} = 9.9 \times 10^{2}\ \text{Å}.$$

Example 2(b):

Calculate the kinetic energy of an electron in the ground state confined to a box 3Å in width and moving in one dimension (x-axis) only.

Solution:

$$E = \frac{n^2h^2}{8ma^2} = \frac{1^2 \times (6.6 \times 10^{-27}\ \text{erg. sec}^2)^2}{8 \times (9.1 \times 10^{-28}\ \text{g})\ (3 \times 10^{-8}\ \text{cm})^2}$$

$$= 6.6 \times 10^{-12}\ \text{erg./electron.}$$

4

De-Broglie's Concept of Wave Matter

INTRODUCTION

Upto 1924 Physicists were of the view that only light or electromagnetic radiation possesses dual character, sometimes they behave as a wave and at the other as a particle. The matter was considered exclusively corpuscular in nature. This ideology was broken by Louis de-Broglie in 1924 by proposing that De-Broglie's Concept of Matter may not be a monopoly of radiation, but a universal characteristic of nature. In his doctoral dissertation, he suggested a new overwhelming idea of the existence of matter waves purely on theoretical ground, fie enunciated that matter must also possess dual character as light. Like electromagnetic radiation, material particles such as electrons, protons and neutrons also exhibit wave-like character.

According to de-Broglie's concept, a moving particle always has a wave associated with it and the motion of the particle is guided by that wave to a similar manner as photon is controlled by a wave. Like electromagnetic waves matter waves can also be propagated in vacuum. Hence, they are not mechanical waves as the sound waves. And unlike electromagnetic waves they are also associated with the motion of electrically neutral bodies. Hence, they are not electromagnetic waves but they are new kind of waves.

For a photon, $\upsilon = E/h$ and $\lambda = h/p$. It is obvious that the left hand side of these equations involve the wave aspects of photons [frequency (υ), wavelength, (λ)], while the right hand side shows the particle aspects (energy, momentum). The Planck's constant (h) acts as a bridge between particle and wave aspects.

According to de-Broglie, the wavelength of a wave associated with moving particle depends upon the mass of the particle and its velocity. The wavelength λ of matter wave or de-Broglie wave is given by

$$\lambda = \frac{h}{mv} = \frac{h}{p} \qquad ...(1)$$

where h is the Planck's constant and p the momentum of material particle.

The expression for the wavelength of the wave associated with material particle can be derived by using the general equation of a standing wave and Lorentz transformation equations in relativity. Let us suppose a material particle, like electron or proton, is equivalent to a standing wave system in the regions of space occupied by the particle. The value of ψ at any instant, t at a point (x, y, z) may be expressed as

$$\psi = \psi_0 \sin \omega t \qquad \text{or } \psi = \psi_0 \sin (2\pi \upsilon t) \qquad ...(2)$$

where υ is the frequency of the particle observed by an observer at rest relative to the particle under consideration and % the amplitude of wave at a point (x, y. z).

If the particle is moving with a velocity v along positive direction of X-axis, then

$$\psi = \psi_0 \sin \left[\frac{2\pi\upsilon\left(t' + \frac{vx'}{c^2}\right)}{1 - v^2/c^2} \right] \qquad ...(3)$$

[$\because$ according Inverse Lorentz transformation, $t = \frac{t' + vx'/c^2}{(1 - v^2/c^2)}$]

The well known standard equation of wave motion is given by

$$\psi = \psi_0 \sin \left[2p\upsilon' \left(t' + \frac{x'}{u'} \right) \right] \qquad ...(4)$$

where ψ_0 is the amplitude, υ' the frequency and u' the phase velocity of the wave along X-axis.

Comparing equations (3) and (4), we get

$$u' = \frac{c^2}{v} \quad \text{and } \upsilon' \frac{\upsilon}{\sqrt{1 - v^2/c^2}} \qquad ...(5)$$

According to Einstein's mass-energy equation

$$E = mc^2 = h\nu' \quad \text{or } \upsilon' = \frac{mc^2}{h}$$

Therefore, the wavelength of moving particle is given by

$$\lambda = \frac{\text{velocity}}{\text{frequency}} = \frac{u'}{\upsilon'} = \frac{c^2/\nu}{mc^2/h} \quad \text{or } \lambda = \frac{h}{mv}$$

This shows that the waves of wavelength l are associated with a material particle of mass m moving with a velocity v.

WAVE-PARTICLE DUALTITY

Our earlier knowledge on the Planck's theory of thermal radiation: Einstein's explanation of photoelectric effect; Compton effect; emission and absorption of radiation by substance: black body radiation etc,. undoubtedly established that the electromagnetic radiation consists of discrete indivisible packets of energy (hυ) called photons which manifest particle character of radiation.

On the other hand, macroscopic optical phenomena like interference, diffraction and polarisation reveal and firmly confirm the wave character of electromagnetic radiation. Therefore, we concluded that the electromagnetic radiation has dual character, in certain situation it exhibits wave properties and in other it acts like a particle;

Before dealing with De-Broglie's Concept of Matter, we must know about the actual distinction between waves and particles, because these are only the modes of energy transmission. A particle means an object with a definite position in space and is identifiable by their distinct properties such as mass, momentum, kinetic energy, spin and electric charge. On the other hand, a wave means a periodically repeated pattern in space which is specified by its wavelength, frequency, amplitude of the disturbances, intensity, energy and momentum.

The particle and wave properties of radiation can never be observed simultaneously. To study the path of a beam of monochromatic radiation, we use the wave theory, while to calculate the amount of energy transactions of the same beam, we have to recourse to the photon or particle 'theory. In fact it has been found impossible to separate the particle and wave aspects of electromagnetic radiation. This strange and overwhelming ability of electromagnetic radiation to manifest itself as wave or as particle is now familiar as *wave-particle dualism*.

WAVE MECHANICS

Wave mechanics is a mathematical treatment of the behaviour of small particles and involves the application of a fundamental concept in physics, namely the concept of wave. Two important principles laid the foundation of wave mechanics, the dual nature of matter and the Heisenberg uncertainty principles. Later, the mathematics of wave mechanics was developed by various workers, the most popular of which, however, is the one initiated by Shrodingerr in 1926, and subsequently developed by others.

According to the recent work of Dirac, the validity of wave mechanics doesn't extend to quantum field theory but it remains perfectly adequate in the domain of atomic and molecular physics.

The development of wave mechanics is based on the following concepts.

1. Heisenberg's uncertainty principle.
2. Shrodinger's wave equation.
3. de-Broglie's idea of dual nature of matter.

The heart of the wave mechanics is an equation called *Schrodinger wave equation.* It can be derived directly or on the basis of certain postulates of quantum mechanics. Thus our main aim in the present chapter will be to derive Shrodinger wave equation directly as well as on the basis of postulates of quantum mechanics, to study certain results that follows from the Shrodinger wave equation and some applications of the Shrodinger wave equation.

FAILURE OF CLASSICAL MECHANICS

The phenomena associated with large size objects moving at speeds much below the speed of light *e.g.*, falling stones planetary motions etc., could be explained by *classical mechanics* (or also called Newtonian mechanics as it was put forward by Newton and is based upon Newton's laws of motion). However, it fails when applied to small particles such as electrons, atoms molecules etc.

For example, according to classical mechanics, it should be possible to determine simultaneously the position and velocity (or momentum) of a moving particle but this is contradicted by the Heisnberg's uncertainty principle. Similarly, classical mechanics assumes that the energy is emitted or absorbed continuously whereas Planck's quantum theory postulates that energy is emitted or absorbed not continuously bit discontinuously in the form of packets of energy, called *quanta.* Further the concept of quantum numbers was introduced arbitrarily to explain the atomic spectra

In view of the failure of classical mechanics to explain the phenomena associated with the small particles, a new mechanics has been put forward to explain these phenomena. One of these is called the *matrix mechanics* put forward by Heisenberg in 1925. It is purely mathematical and does not assume any atomic model. The other is called the *wave mechanics* put forward by Shrodinger in 1926.

It is based upon de Broglie concept of dual character of matter and thus takes into account the particle as well as wave nature of the material particles. However it has been shown that both mechanics are essentially equivalent so far as the basic physical concepts are concerned. Wave mechanics being comparatively simpler and more useful in application to chemistry will be discussed in the present chapter.

It is also called 'particle mechanics' of 'quantum mechanics' because it deals with the problems that arise when particles such as electrons, nuclei, atoms, molecules etc., are subjected to a force.

THE UNCERTAINTY PRINCIPLE OF HEISENBERG'S

The classical mechanics, the position and momentum of a moving electron can be determined with great accuracy. When an electron is considered as a wave, it is, however, not possible to know the exact location of the electron on the wave as it (wave) is extending throughout a region of space. Thus is the question.

" If an electron is exhibiting dual nature (wave and particles), is it possible to know the exact position of the electron in space at same given instant ?"

The answer to this question is given by Heisenberg which states that *" It is impossible to determine simultaneously the position and momentum of the electron with any desired accuracy".* The above definition is known as Heisenberg's uncertainty principle.

Mathematical Relation

Heisenberg's principle can be stated mathematically as:

$$\Delta P \times \Delta x \geq h/4\pi$$

when ΔP is the uncertainty in the determination of the momentum and Δx, the uncertainty in the determination of position. Equation (1) is known as Heisenberg's equation which can be stated in words as '*the product of uncertainty in the simultaneous determination of the position and momentum of a particle is equal to or greater than the Planck's constant.*"

Alternative Statement

Sometimes instead of measuring position and momentum of the system, its energy E and the time t for which it remains in that energy state are measured. In these cases the uncertainty in measurement is represented as

$$\Delta E \times \Delta t \geq h/4\pi$$

i.e., if the time for which the system remains in a particular energy state is short, then its energy will be more defined and for longer stay in a state, the energy will not be so well defined. The uncertainty principle is now regarded as a fundamental principle of nature Equation (2) applies to complex molecules and to internal molecular energy levels as well.

Derivation of Heisenberg's

It is well established that a moving body must be considered to be a de-Broglie wave group rather than as localised entity. This means that there is always a limit to the accuracy with which one can measure its particle properties. For instance, in a de-Broglie wave group in Fig. 4.1(a), one cannot see centre-representing the position of the body very easily whereas one can measure precisely the wave length of its component waves related to the body's momentum. On the other hand, if the de-Broglie wave group is very narrow as shown in Fig. 4.1(b), one can find out the position of its centre readily while one can never measure is the wavelength.

A simple argument based upon the nature of wave groups permits us to relate the uncertainty Δx in a measurement of particle position with the inherent uncertainty Δp in a simultaneous measurement of its momentum. Let us consider two waves of angular frequencies w_1 and w_2 and propagation constants k_1 and k_2 travelling along single direction.

$$\Psi_1 = A \sin (w_1 t - k_1 x)$$

$$\Psi_2 = A \sin (w_2 t - k_2 x)$$

The result after combination is

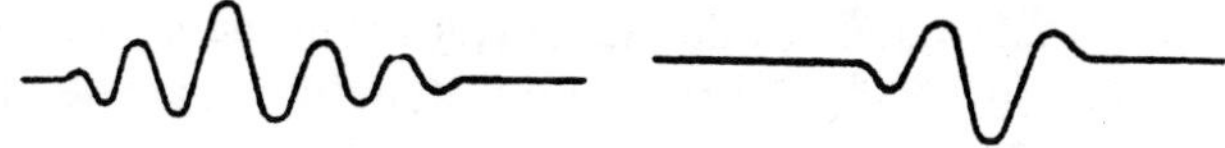

(a) The wavelength of a group at any instant can be determined but not its position.

(b) The postion of a narrow wave group at any instant can be determined but not its position.

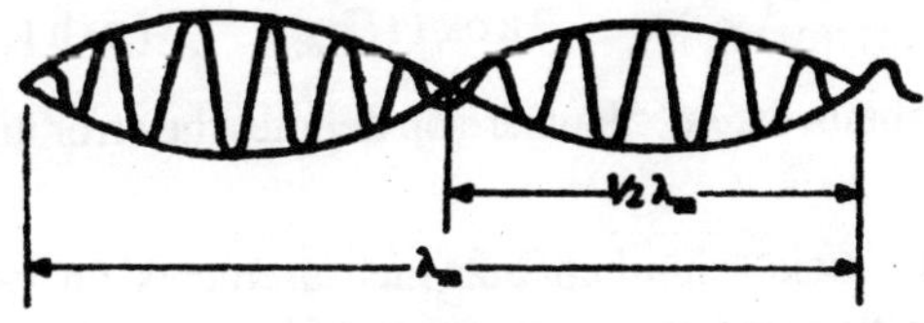

(c) Group of waves.

(c) Group of waves.

Fig. 4.1

$$\Psi = \Psi_1 + \Psi_2 = A \sin (w_1 t - k_1 x) + A \sin (w_2 t - k_2 x)$$

$$= 2A \sin (wt - kx) \cos \left(\frac{\delta w}{2} t - \frac{\delta k}{2} x \right)$$

where $w = w_1 + w_2,\ k = \dfrac{k_1 + k_2}{2}$

$$\delta w = w_1 \sim w_2,\ \delta k = k_1 \sim k_2.$$

The resultant is plotted in Fig. 4.1(c), where the loop formed will travel with group velocity V_e. Now this group velocity is equal to the particle velocity and hence the loop formed is equivalent to the position of particle. The position of the particle cannot be given with certainty, it is some where between one node and the next node. The error in the measurement of the position of the particle is therefore equal to the distance between these two nodes.

Now a node is formed when $\cos\left(\dfrac{\delta w}{2}.t - \dfrac{\delta k}{2}.x \right)$ is zero. This is possible when $\left(\dfrac{\delta w}{2}.t - \dfrac{\delta k}{2}.x \right) = \dfrac{\pi}{2}$ or $\dfrac{3\pi}{2}$ or $\dfrac{5\pi}{2}$.

Thus, if x_1 and x_2 represent the positions of two successive nodes, then at any instant t, we get

$$\frac{\delta w}{2}.t - \frac{\delta k}{2}.x_1 = (2n + 1) \frac{\pi}{2}$$

and $$\frac{\delta w}{2}.t - \frac{\delta k}{2}.x_2 = (2n + 3)\frac{\pi}{2}$$

$$\therefore \quad \frac{\delta k}{2}(x_2 - x_1)n = \pi \; x_2 - x_1 = \frac{2\pi}{\delta 2}$$

$\therefore$ Error in the measurement of the position particle is

$$\delta x = x_2 - x_1 = \frac{2\pi}{\delta k} = \frac{2\pi}{2\pi \delta k\,(1/\lambda_m)} = \frac{1}{\delta(p/h)} = \frac{h}{\delta p}$$

Thus, we obtain $\delta x, \delta p = h$ where δp denotes the error in measurement of momentum.

Now if we make simultaneous measurements on two quantities, namely the position and the momentum of the particle, the product of the two fundamental errors in approximately equal to h, the Planck's constants

If we now consider a group of large number of waves having continuously varying frequencies, we get the product of fundamental error as dp = h/4p. Thus result is used in the more accurate calculations.

Experimental Proof. Let us think of an ideal experiment by which we may try to measure both the position and the momentum of an electron. In order to do so, we get up a high powered microscope with suitable lighting arrangement (Fig. 4.2).

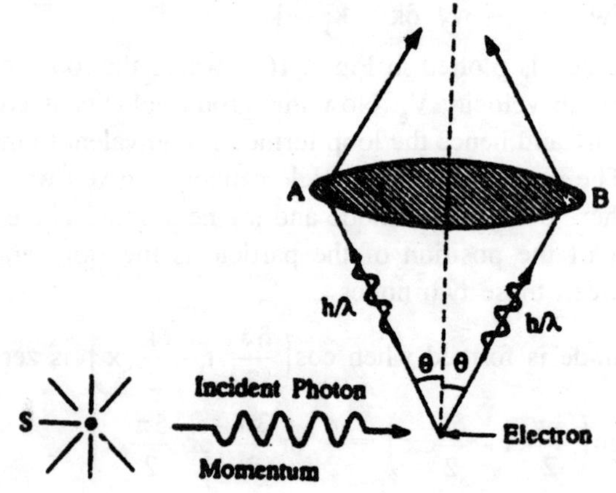

Fig. 4.2

As the photons from the source S collide with the electrons, same of these bounce into the microscope and enable the observer to see the flash of light and this to find out both the position and the momentum of the electron at some instant of time. According to classical mechanics, the observer should be able to find out their (electrons) exact position and momentum. There are, however, *two fundamental Limitations* to such an experiment.

(i) Accuracy in determining the position of electron by a microscope is limited by the laws of optics. According to optics, the resolving power of a microscope is given by

$$\Delta x = \frac{\lambda}{2\sin\theta} \qquad ...(1)$$

Δx = Minimum distance between two points in the field of view which can be distinguished as separate.

λ = Wavelength of the scattered photon, and

θ = Semi-vertical angle of the cone light coming from the illuminated electron.

If the position of the electron changes by Δx, the microscope will not be able to detect it. To make Δx very small, radiation of very short wavelength such as X-rays or gamma rays should be used. Thus, Δx will be the error or uncertainty in the measurement of the position of the electron.

(ii) In the process of determining the momentum of the electron, the interaction of electron with gamma ray photon will result in the change of momentum of the electron because of its recoil. Let a photon of wavelength, λ' and momentum h/λ' incident upon the electron collide with it and scatter into the microscope with increased wavelength λ. The scattered photon may enter the objective lens anywhere between OA and OB. In case it enters along OA. Its momentum along x-axis is given by $- h/\lambda \sin\theta$. Hence the loss of momentum or the momentum imparted to the electron in the x-direction will be

$$\frac{h}{\lambda'} + \left(\frac{-h}{\lambda}\sin\theta\right) = \frac{h}{\lambda'} - \frac{h}{\lambda}\sin\theta \qquad ...(2)$$

Similarly, the momentum imparted to the electron by the scattering of the photon along OB will be

$$\frac{h}{\lambda'} + \frac{h}{\lambda}\sin\theta \qquad ...(3)$$

The momentum imparted to the electron can, therefore, have any value between those given by equation (2) and (3). Thus, the difference between the above two moments gives the error involved in the measurements of momentum of the electron and the obtained as

$$\Delta p_x = 2h/\lambda \sin \theta. \qquad ...(4)$$

Equation (1) and (4) run contrary to each other. If we try is improve the measurement of the electron's position of decreasing l and increasing θ, we do it at the cost of accuracy in the measurement of momentum and vice-versa. Multiplying eqs. (1) and (4), we get

$$\Delta x \times \Delta p_x = h$$

which determines the limit of our knowledge about the precise measurement of the position and the momentum of the particle.

Discussion

Uncertainty principle appears to be contrary to out daily experience. We find that the calculations involving the measurement of position and momentum of both terrestrial bodies have been remarkably accurate. The reason for this is that for the bodies having appreciable mass, the uncertainties in the determination of both the position and momentum are so small that these are negligible as compared to the normal experimental error However, the situation is different in the case of small particles such as electrons and photons. Here the uncertainties are large and the classical mechanics fails to explain the behaviour of such particles. These have been explained by means of wave mechanics.

Conclusion

The consequences of the uncertainty principle are far reaching and probability takes the place of exactness in atomic theory. Thus, statement regarding the exact position and velocity (which is related to kinetic energy) of an electron must be replaced by statements concerning the probability that the electron has a given position and velocity. The Bohr concept of atom, which regards the electrons as rotating in definite orbits around the nucleus, must be abandoned and replaced by a theory which considers probability of finding the electrons in a particular region of space.

Application of Uncertainty Principle

Several phenomena in the case of particles of atomic dimensions can be understood in terms of the uncertainty principle. A few of them are given below.

(i) It easily follows from the uncertainty principle that electrons cannot exist within the nuclei of the atom.

(ii) The principle helps us to know the limit to the accuracy with which we can measure the frequency of the radiation emitted by an atom.

Shrodinger Equation of Hydrogen Atom

The hydrogen atom consists of a single electron of mass m and charge–e, moving about a nucleus of mass M and charge + e in a potential field V(r) which corresponds to coulombian attraction $\frac{e^2}{4\pi\varepsilon_0 r^2}$ between the electron and the nucleus. In general, the nucleus is not stationary.

Its motion can, however, be taken into account by treating the system as a single particle of (reduced) mass m moving in the potential field V (r), where $\mu = mM/(m + M)$. The Shrodinger equation for the motion is

$$\Delta^2\Psi + \frac{8\pi^2\mu}{h^2}[E - V(r)]\Psi = 0.$$

Converting the Laplacian operator D2 in spherical coordinates, we get

$$\frac{1}{r^2}\frac{\partial}{\partial r}\left(r^2\frac{\partial\Psi}{\partial r}\right) + \frac{1}{r^2 \sin\theta}\frac{\partial}{\partial\theta}\left(\sin\theta\frac{\partial\Psi}{\partial\theta}\right) + \frac{1}{r^2\sin^2\theta}\frac{\partial^2\Psi}{\partial\phi^2}$$

$$+\frac{8\pi^2\mu}{h^2}[E - V(r)]\Psi = 0 \quad ...(1)$$

This is the Shrodinger equation of hydrogen atom in spherical coordinates. To solve this equation, we must separate the radial (r) and angular (θ, ϕ) variables *Separation of Variables :* Let us substitute

Y (e, θ, ϕ) = R (r) Y (θ, ϕ) in eq. (i). We get

$$\frac{1}{r^2}\frac{d}{dr}\left(r^2\frac{dR}{dr}\right)Y + \frac{1}{r^2\sin\theta}\frac{\partial}{\partial\theta}\left(\sin\theta\frac{\partial Y}{\partial\theta}\right)R$$

$$+\frac{1}{r^2\sin^2\theta}\frac{\partial^2 Y}{\partial\phi^2}R + \frac{8\pi^2\mu}{h^2}[E - V(r)]RY = 0$$

Multiplying the entire equation by $\frac{r^2}{RY}$ and rearranging, we get

$$\frac{1}{R}\frac{d}{dr}\left(r^2\frac{dR}{dr}\right) + \frac{8\pi^2\mu r^2}{h^2}[E - V(r)]$$

$$= -\frac{1}{Y}\left[\frac{1}{\sin\theta}\frac{\partial}{\partial\theta}\left(\sin\theta\frac{\partial Y}{\partial\theta}\right)+\frac{1}{\sin^2\theta}\frac{\partial^2 Y}{\partial\phi^2}\right].$$

The left-hand side of this equation is a function of R only, while the right-hand side is a function of Y only. Hence each side must be equal to the same constant. Let this constant be l (l + 1). Thus, we get a radial equation

$$\frac{1}{R}\frac{d}{dR}\left(r^2\frac{dR}{dr}\right)+\frac{8\pi^2\mu r^2}{h^2}[E-V(r)] = l\,(l+1)$$

or $$\frac{1}{r^2}\frac{d}{dR}\left(r^2\frac{dR}{dr}\right)+\left[\frac{8\pi^2\mu}{h^2}\{E-V(r)\}-\frac{l(l+1)}{r^2}\right]R=0 \quad ...(2)$$

and an angular equation

$$-\frac{1}{Y}\left[\frac{1}{\sin\theta}\frac{\partial}{\partial\theta}\left(\sin\theta\frac{\partial Y}{\partial\theta}\right)+\frac{1}{\sin^2\theta}\frac{\partial^2 Y}{\partial\phi^2}\right] = l\,(l+1). \quad ...(3)$$

The later equation can be further splitted by substituting

$$Y(\theta,\phi) = \Theta(\theta)\,\Phi(\phi).$$

Thus, we get

$$-\frac{1}{\Theta\Phi}\left[\frac{1}{\sin\theta}\frac{d}{d\theta}\left(\sin\theta\frac{d\Theta}{d\theta}\right)\Phi+\frac{1}{\sin^2\theta}\frac{d^2\Phi}{d\phi^2}\Theta\right] = l\,(l+1).$$

Multiplying the entire equation by $\sin^2\theta$ and rearranging, we get

$$\frac{1}{\Theta}\sin\theta\frac{d}{d\theta}\left(\sin\theta\frac{d\Theta}{d\theta}\right)+l(l+1)\sin^2\theta = -\frac{1}{\Phi}\frac{d^2\Phi}{d\phi^2}.$$

Again, the two sides of this equation are functions of different variables and so each must be equal to the same constant m_l^2 (say). Thus, we get

$$\frac{1}{\Theta}\sin\theta\frac{d}{d\theta}\left(\sin\theta\frac{d\Theta}{d\theta}\right)+l(l+1)\sin^2\theta = m_l^2$$

or $$\frac{1}{\sin\theta}\frac{d}{d\theta}\left(\sin\theta\frac{d\Theta}{d\theta}\right)+\left[l(l+1)-\frac{m_l^2}{\sin^2\theta}\right]\Theta=0 \quad ...(4)$$

and $$-\frac{1}{\Phi}\frac{d^2\Phi}{d\phi^2}=m_l^2$$

or $$\frac{d^2\Phi}{d\phi^2} + m_l^2\ \Phi = 0\,. \qquad ...(5)$$

Thus, we have broken the Shrodinger equation into three differential equations M, (4) and (5) which are respectively in R along, in Θ alone and in Φ alone, then we try to find " acceptable" solutions of these equations, the quantum, numbers automatically come in.

Wave Functions of Hydrogen Atom

Entrance of Quantum Numbers : The solutions of eq. (5) is

$$\Phi_{m_l}(\phi) = Ae^{im_l\phi}\,, \qquad ...(6)$$

where A is integration constant. It can be shown that $\Phi_{m_l}(\phi)$ is single-valued function of position only when ml is a positive or negative integer or zero, that is

$$m_l = 0, \pm 1, \pm 2, \pm 3,...$$

m_l is known as 'magnetic quantum number'.

The solution of eq. (4) is known to be

$$\Theta_{l,m_l}(\theta) = N_{l,m_l} P_l^{|m_l|}(\cos\theta)\,, \qquad ...(7)$$

where N_{l,m_l} is constant and $P_l^{|m_l|}$ is 'associated Legendre polynomial' of degree l and order ml. The solution is found to be acceptable only when l is a positive integer, that is,

$$l = 0, 1, 2, ...$$

l is known as 'orbital quantum number'.

Further, the functions $\Theta_{l,m_l}(\theta)$ vanish if $|m_l|$ exceeds l. This means that ml can take values only upto $\pm$ l.

For solving the radial equation (2), we must specify V(r). In the present case it is $-\frac{e^2}{4\pi\varepsilon_0 r}$. Then, the equation is

$$\frac{1}{r^2}\frac{d}{dr}\left(r^2\frac{dR}{dr}\right) + \left[\frac{8\pi^2\mu}{h^2}\left\{E + \frac{e^2}{4\pi\varepsilon_0 r}\right\} - \frac{l(l+1)}{r^2}\right]R = 0$$

or $$\frac{d^2R}{dr^2} + \frac{2}{r}\frac{dR}{dr} + \left[\frac{8\pi^2\mu}{h^2}\left\{E + \frac{e^2}{4\pi\varepsilon_0 r}\right\} - \frac{l(l+1)}{r^2}\right]R = 0.$$

The solution of this equation is known to be

$$R_{n,l}(r) = N_{n,l}\left(\frac{2r}{n a_0}\right)^l e^{-r/na_0} L_{n+l}^{2l+1}\left(\frac{2r}{n a_0}\right), \quad ...(8)$$

where $N_{n,l}$ is a constant and L_{n+l}^{2l+1} is 'associated Leguerre polynomial' of degree n – l – 1 and order 2l + 1. The parameter a_0

$$a_0 = \frac{\varepsilon_0 h^2}{\pi \mu e^2}$$

and is the radius of first Bohr orbit. The solution is found to be acceptable only when n – l – 1 is a positive integer. This means that n, l and ml are restricted to the following values.

$$n = 1, 2, 3, ... \infty$$

$$l = 0, 1, 2,... (n - 1)$$

$$m_l = 0, \pm 1, \pm 2, ... \pm l.$$

n is known as 'total quantum number'.

Thus, the wave functions of hydrogen atom are given by

$$\Psi_{n,l,m_l}(r, \theta, \phi) = R_{n,l}(r)\Theta_{l,m_l}(\theta)\Phi_{m_l}(\phi),$$

where R, Θ and Φ are given by eq. (7), (7), and (6) respectively. The real form of the wave functions of hydrogen atom for the group state (n = 1, l = 0, m_l = 0) is given below.

$R^1, 0$ (r)	$\Theta_{0,0}(\theta)$	$\Phi_0(\phi)$	$\Psi_{1,00}(r, \theta, \phi)$
$\frac{2}{a_0^{3/2}} e^{-r/a_0}$	$\frac{1}{\sqrt{2}}$	$\frac{1}{\sqrt{2\pi}}$	$\frac{1}{\sqrt{\pi a_0^3}} e^{-r/a_0}$

Wave Mechanical Interpretation of n, l, ml : The radial wave equation for the hydrogen atom can be solved only when E has a continuous range of positive values, or a discrete set of *negative* values (which correspond to *bond* states),

$$E = -\frac{\mu e^4}{8\varepsilon_0^2 h^2}\left(\frac{1}{n^2}\right), \qquad n = 1, 2, 3,...$$

These are same as obtained by old quantum theory. Thus, n describes the quantisation of electron energy in the hydrogen atom.

Let us now find the interpretation of l and m_l. The wave mechanical operators for the square of the angular momentum and for the z-component of angular momentum are

$$L^2 = -\frac{h^2}{4\pi^2}\left[\frac{1}{\sin\theta}\frac{\partial}{\partial\theta}\left(\sin\theta\frac{\partial}{\partial\theta}\right)+\frac{1}{\sin^2\theta}\frac{\partial^2}{\partial\phi^2}\right] \quad ...(9)$$

$$\text{and } L_z = -\frac{ih}{2\pi}\frac{\partial}{\partial\pi} \quad ...(10)$$

The angular eq. (iii) can be written as

$$\left[\frac{1}{\sin\theta}\frac{\partial}{\partial\theta}\left(\sin\theta\frac{\partial}{\partial\theta}\right)+\frac{1}{\sin^2\theta}\frac{\partial^2}{\partial\phi^2}\right]Y_{l,m_l}(\theta,\phi) = -\,l(l+1)\,Y_{l,m_l}(\theta,\phi)$$

Multiplying both sides by $-\frac{h^2}{4\pi^2}$, we get

$$-\frac{h^2}{4\pi^2}\left[\frac{1}{\sin\theta}\frac{\partial}{\partial\theta}\left(\sin\theta\frac{\partial}{\partial\theta}\right)+\frac{1}{\sin^2\theta}+\frac{\partial^2}{\partial\phi^2}\right]Y_{l,m_l}(\theta,\phi) = l(l+1)\frac{h^2}{4\pi^2}Y_{l,m_l}(\theta,\phi).$$

Substituting eq. (9) in it, we have

$$L^2 Y_{l,m_l}(\theta,\phi) = l(l+1)\frac{h^2}{4\pi^2}Y_{l,m_l}(\theta,\phi).$$

Thus, the angular wave functions $Y_{l,m_l}(\theta,\phi)$ of hydrogen atom are also the wave functions of the square of the angular momentum operator with eigen values

(ii) $(l+1)\frac{h^2}{4\pi^2}$. This means that the electron in the hydrogen atom has an angular momentum given by

$$L = \sqrt{l(l+1}\,\frac{h}{2\pi},$$

where l = 0, 1, 2, ... (n – 1). Thus, like energy, the angular momentum is also quantised, and this quantisation is described by the orbital quantum number l. Since the magnitude of the electron angular momentum is quantised, its direction is also expected to be quantised with respect to external; magnetic field (space quantisation). Let us consider the Φ equation (6).

$$\Phi_{m_l}(\phi) = Ae^{im_l(\phi)}$$

Differentiating it, we get

$$\frac{d\Phi_{m_l}(\phi)}{d\phi} = Ae^{im_l(\phi)}(im_l) = im_l \Phi_{m_l}(\phi).$$

Multiplying both sides by $-\frac{ih}{2\pi}$, we get

$$-\frac{ih}{2\pi}\frac{d\Phi_{m_l}(\phi)}{d\phi} = \frac{m_l h}{2\pi}\Phi_{m_l}(\phi).$$

Substituting eq. (10) in it, we have

$$L_z \Phi_{m_l}(\phi) = \frac{m_l h}{2\pi}\Phi_{m_l}(\phi).$$

Thus, the wave function $\Phi_{m_l}(\phi)$ of hydrogen is also the wave function of z-component of the angular momentum operator with eigenvalues $\frac{m_l h}{2\pi}$. The means that the component of the electron angular momentum in field direction (z-axis) is given by

$$L_z = m_l \frac{h}{2\pi},$$

where $m_l = 0, \pm 1, \pm 2, \ldots \pm l$. This shows that the number of possible orientations of the angular momentum in a magnetic field (2l + 1), that is , the angular momentum is space-quantised. Thus, the magnetic quantum number n describes the space quantisation.

Electron Probability Density

As far as the exact motion of the electron within the atom is concerned, there is striking difference between old quantum theory and wave mechanics. In the old theory, the electron traces, in general elliptic orbits whose size, eccentricity and orientation are determined by the quantisation rules. In wave mechanics, on the other hand, the electron cannot pictured as being in a definite orbit. We can only find the relative probabilities of finding the electron at various locations. This is clearly a consequence of wave nature of the electron.

The electron wave function in a hydrogen atom is given by

$$\Psi_{n,l,m_l} = R_{n,l}\,\Theta_{l,m_l}\,\Phi_{m_l},$$

where $R_{n,l}$ Θ_{l,m_l} and Φ_{m_l} describe how Ψ_{n,l,m_l} varies with r, θ and respectively. The probability density $|\Psi|^2$ may therefore be written as

$$|\Psi|^2 = |R|^2\ \Theta|^2\ |\Phi|^2 .$$

Approximate pictures of $|\Psi|^2$ as functions of r, θ, and f for some simple states (for which l = 0 and hence m_l = 0).

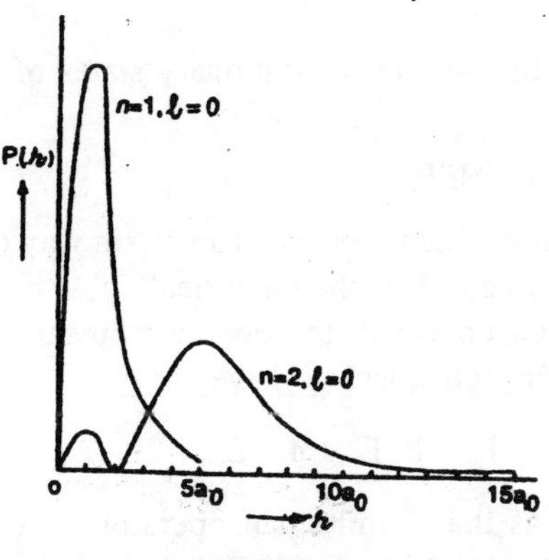

Fig. 4.3

Since we no longer have distinct electron orbits, we speak of *electron clouds* about the nucleus. In the figure, the brightness indicates roughly the density of the electron clod. (The density of the cloud at a specified point gives the probability of finding the electron at that point). It is seen that the states with l = 0 and n = 1, 2, 3, ..., are spherically symmetrical, that is the electron probability density $|\Psi|^2$ has the same value at a given r in all directions. (Electron, in other states, however, has angular preferences). For n > 1, alternative bright and dark rings appear, the dark rings correspond to nodal surfaces where the electron is ever found (zero probability).

Finally, let us see the probability of finding the electron in a hydrogen atom a distance between r and r + dr from the nucleus. This probability P (r) is given by

$$P(r)\ dr = |R_{n,l}|^2\ 4\pi r^2\ dr ,$$

where $4\pi r^2$ dr is the volume of the spherical shell whose inner radius is r and outer radius is r + dr. In Fig. 4.3, P (r) has been plotted against r for two states of hydrogen atom corresponding to n = 1 and n = 2 (l = 0 in both cases). It is seen that the principal maxima of the curves closely agree with the radii of the Bohr orbits ($r = a_0, 4a_0, ...$). Thus, whereas the electron is most likely to be found at the locations of Bohr orbits it has a finite probability of being found elsewhere. Hence, in wave mechanics the electron is, so to speak, smeared out over the whole of space.

ALTERNATIVE DERIVATION OF SHRODINGER'S WAVE EQUATION

The wave equation that applies to stationary states of a system can be written as

$$H\Psi = E\Psi \qquad ...(1)$$

where H is an operator which represents a certain way of expressing the total energy of the system. E is the numerical value of total energy and Ψ is a wave function on which the operator operates. Further for all systems that concern a chemist, we have

$$H = E = P.E. + K.E. \qquad ...(2)$$

where H is known as the Hamiltonian operator of a system after the theoretical physicists Hamilton. From Bohr's hydrogen atom,

$$V = P.E. = \frac{(+e)-(-e)}{r} = -\frac{e^2}{r} \qquad ...(3)$$

and
$$K.E. = \frac{1}{2}mu^2 \qquad ...(4)$$

As gravitational forces are about 10^{18} times smaller than the electrostatic interaction, they are therefore neglected in calculating potential energy V. Thus, the Hamiltonian of hydrogen atom in classical physics is given by the following equation [By substituting equations (3) and (4) in (2)].

$$H = \frac{1}{2}mu^2 - \frac{e^2}{r} \qquad ...(5)$$

We know, momentum = mass × velocity

or
$$p = mu \text{ or } u = \frac{p}{m} \qquad ...(6)$$

Substituting equation (6) in (5), we get

$$H = \frac{p^2}{2m} - \frac{e^2}{r} \qquad ...(7)$$

In order to convert the classical description of a system into the description of wave mechanics, the momentum p is replaced with a derivative of the form.

$$p = \frac{h}{2\pi l}\left(\frac{\partial}{\partial x}+\frac{\partial}{\partial y}+\frac{\partial}{\partial z}\right) \text{ when } i = \sqrt{(-1)} \qquad ...(8)$$

Squaring equating (8), we get

$$p^2 = \frac{h^2}{4\pi^2 l^2}\left(\frac{\partial}{\partial x}+\frac{\partial}{\partial y}+\frac{\partial}{\partial z}\right)^2$$

$$\Rightarrow \quad p^2 = \frac{h^2}{4\pi^2 l^2}\left(\frac{\partial^2}{\partial x^2}+\frac{\partial^2}{\partial y^2}+\frac{\partial^2}{\partial z^2}\right)$$

$$\Rightarrow \quad p^2 = \frac{h^2}{4\pi^2}\left(\frac{\partial^2}{\partial x^2}+\frac{\partial^2}{\partial y^2}+\frac{\partial^2}{\partial z^2}\right) \quad [\because l^2 = -1] \qquad ...(9)$$

Substituting equation (9) in (7), we get

$$H = -\frac{h^2}{8\pi^2 m}\left(\frac{\partial^2}{ax^2}+\frac{\partial^2}{ay^2}+\frac{\partial^2}{\partial z^2}\right)-\frac{e^2}{r} \qquad ...(10)$$

where H represents the wave mechanical form of the Hamiltonian operator for hydrogen atom. Substituting this form of the Hamiltonian into equation (11), we get the familiar form of the Shrodinger's equation for hydrogen atom.

$$\Rightarrow \quad H\Psi = \left[-\frac{h^2}{8\pi^2 m}\left(\frac{\partial^2}{\partial x^2}+\frac{\partial^2}{ay^2}+\frac{\partial^4}{ay^2}\right)-\frac{e^2}{r}\right]\Psi = E\Psi \qquad ...(11)$$

$$\Rightarrow \quad H\Psi = \left[-\frac{h^2}{8\pi^2 m}(\nabla^2)-\frac{e^2}{r}\right] \qquad ...(12)$$

Rearranging above equation, we get equation (11).

SHRODINGER'S EQUATION WITH RESPECT TO TIME

Let us consider a de-Broglie's wave. In order to describe the state of this wave consider two functions f (x, y, z) and g (x, y, z). The electron density at any point on the wave is proportional to $f^2 + g^2$. As this density is considered to be constant on the whole wave, it means that f and g should have a phase difference of $\pi/2$ in a plane wave, so that

$$f = \Psi_0 \cos 2\pi (ct - z/\lambda) \qquad ...(1)$$

then $g = \Psi_0 \sin 2\pi (vt - z/\lambda)$ ...(2)

Instead of two separate real functions [Eqs. (1) and (2)], it is more convenient to use a single complex function Ψ which, in the present case, is as follows.

$$\Psi = f + ig \quad ...(3)$$

Substituting equation (1) and (2) in (3), we get

$$\Rightarrow \quad \Psi = \Psi_0 \cos 2\pi (vt - z/\lambda) + I\Psi_0 \sin 2\pi (vt - z/\lambda) \quad ...(4)$$

$$\Rightarrow \quad \Psi = \Psi_0 e^{2nl} (vt - z/l) \qquad [\because e^{lx} = \cos x + i \sin x]$$

$$\Rightarrow \quad \Psi = \Psi_0 e^{2\pi lvt} e^{-2\pi tz/\lambda} \quad ...(5)$$

Differentiating Eq. (5) with respect to time t, we get

$$\frac{\partial \Psi}{\partial t} = \Psi_0 2\pi \, ive^{2\pi ivt} e^{-2\pi/\lambda} = 2\pi/v \, \Psi$$

$$\Rightarrow \quad v = \frac{\partial \Psi}{\partial t} \cdot \frac{1}{2\pi l \Psi}$$

$$[\because \Psi = \Psi_0 e^{2\pi ivt} e^{-2\pi l2/\lambda}] \quad ...(6)$$

From equation (10) of Art. 6.4, we have

$$\nabla^2 \Psi + \frac{8\pi^2 m}{h^2}(E - V)\Psi = 0$$

In above equation, replacing E by hv and multiplying throughout by $h^2/8\pi^2 m$, we get

$$\nabla^2 \Psi (h^2/8\pi^2 m) + hv - V\Psi = 0 \quad ...(7)$$

Substituting equation (6) in (7), we get

$$\nabla^2 \Psi \frac{h^2}{8\pi^2 m} + \frac{\partial \Psi}{\partial t} \frac{2}{2\pi i} - V\Psi = 0$$

$$\Rightarrow \quad -\frac{h}{2\pi l} \frac{\partial \Psi}{\partial t} = \frac{h^2}{8\pi^2 m} \nabla^2 \Psi - V\Psi$$

$$\Rightarrow \quad \frac{lh}{2\pi} \frac{\partial \Psi}{\partial t} = \frac{h^2}{8\pi^2 m} \nabla^2 \Psi - V\Psi$$

This is the Shrodinger's equation containing the time factor. It is unique among the differential equation of mathematical chemistry, as it includes the imaginary factor.

What ever be the form of the Shrodinger's equation the eigen functions contain all the information about the particle that is permitted by the uncertainty relations. This information about the particle that is permitted by the uncertainty relations. This information is in the form of probabilities. Also , it can appear as specific numbers whenever quantization is taking place.

In general practice, the solutions of the Shrodinger's equations are obtained using what are known as operators. The different types of operators are explained later in the same chapter.

EIGEN VALUES AND EIGEN FUNCTIONS

A differential equation of the second order and will have many solutions. Some of them being imaginary which have no significance. The solutions have significance only for the certain values called the *eigen-values* of the total energy E. For an atom, these correspond to the energy values associated wit different orbits in the atom. Thus, the existence of various energy levels in atom as postulated by Bohr is a direct consequency of the wave-mechanical concepts. *Solution of the wave for these definite values of E, gives the corresponding values of the wave functions (Y) known as eigen functions.*

The eigen functions must satisfy certain conditions in order to have a physical significance. They must have only one values which should be finite and continuous through the whole of space of system under consideration, *i.e.*, for all possible values of the co-ordinates (x, y, z) including infinity.

The First Interpretations of Ψ : The first attempt to this problem was made by Shrodinger himself in terms of charge-density. In any electromagnetic wave system.

$$\text{The energy density} \propto A^2 \qquad ...(1)$$

where energy density is the energy per unit volume and A is the amplitude of the wave. The number of photons per unit volume is equal to the energy density divided by hv, where hv is the energy of photon. Thus.

No of photons per unit volume,. Energy density hv

But the number of photons per unit volume is defined as the photon density. It means that

$$\text{Photon density} = \frac{\text{Energy density}}{hv} = \frac{A^2}{hv} \qquad \text{[From eq. (1)]}$$

Thus, we conclude that photon density is proportional to A^2 since hv is constant, *e.g.*,

Photon density $\propto A^2$

Similarly, if Ψ is the amplitude of the matter wave at any point in space, we may consider the particle density (number of particles per unit volume at the point) to be proportional to Ψ^2. Hence *the square of* Ψ *is a measure of the particle density.* If we multiply the particle density by electric charge (e) of the particle, we get the charge density. Therefore, *the quantity* Ψ^2 *is also the measure also the measure of charge density.*

The interpretation given above led to satisfactory results when wave mechanics was applied to the stable states of the Bohr theory. But certain difficulties arise against this interpretation of Ψ, *i.e.*,

(i) The wave packet associated with material particle must in course of time becomes dissipated to that it might not represent for long the particle concerned.

(ii) Consideration of the mutual action of the particles as a collision of the corresponding wave packet is ordinary three-dimensional space leads to very serious difficulties.

Second Interpretation of Ψ. Born (1926) in collaboration with Bohr and Heisenberg put forward the interpretation of Y. According to them. Ψ^2 does not measure the particle density or charge density at any point but it is related to the probability of finding the particle at that point at any given moment.

Accordingly, the probability 'P' of finding the electron at the point (x, y, z) is given by $P = \Psi(x, y, z)\, \Psi^*(x, y, z)$

where Ψ^* is the complex conjugate of Ψ.

As Ψ may have imaginary values one must therefore multiply it by complex conjugate in order to make P real.

The probability of finding the electron or particle at any point, may be large small of zero, but it cannot be imaginary. Of course, if Ψ is real $\Psi^* = \Psi$, and the probability P equals to the square of Ψ. Thus, we conclude that Ψ must satisfy the following conditions.

(i) it must be single values at each and every point,

(ii) it must not have the value infinite any point, and

(iii) its absolute values at all points must be such that

$$\int_{-\infty}^{+\infty}\int_{-\infty}^{+\infty}\int_{-\infty}^{+\infty}\Psi(x,y,z)\Psi^*(x,y,z)\,dx\,dz = \int\Psi\Psi^*\,dv = 1. \quad ...(2)$$

τ is a general symbol for all the co-ordinates. As there is one electron, it means that the total probability must be one.

FURTHER MATHEMATICAL CONSIDERATION OF SHRODINGER

Normalisation The Shrodinger equation is a homegeneous differential equation whose solution gives a values for Ψ. But Ψ multiplied by a constant factor would give the same differential equation. It is therefore necessary to find some conditions which will indicate which constant factor is to be used.

As the electron must be found somewhere, one can say that the probability of finding the electron in the whole of the space considered in a particular problem is unity. This may be written mathematically as

$$\int|\Psi|^2\,d\Gamma = 1$$

where Γ is the volume of the whole space. This integral determines by what constant Ψ must be multiplied for the solution of a particular problem. This constant is known as the normalisation constant. An eigen function which has been completely evaluated in this way is said to have been "normalised".

The amplitude in the wave function is determined by using this normalisation condition. For Ψ to represent a wave packet (*i.e.*, a group of waves) the above condition for normalisation should be satisfied. The numerical coefficient in Ψ, the amplitude factor should be independent of time.

Orthogonality

If we consider two wave functions Yp and Yq which correspond to two values of the energy Ep and Eq and if these wave functions are separate solutions of the Shrodinger's equation, then $\int\Psi_p\Psi_q\,d\Gamma = 0$, and the wave functions only.

Let us consider two Schordinger equations which are written in one dimension only (for the motion of a particle in a straight line).

$$\frac{d^2\Psi_p}{dx^2} + \frac{8\pi^2 m}{h^2}(E_p - V_p)\Psi_p = 0 \quad ...(1)$$

$$\frac{d^2\Psi_q}{dx^2} + \frac{8\pi^2 m}{h^2}(E_q - V_q)\Psi_q = 0 \qquad ...(2)$$

Multiplying the first equation by Ψ_q and the second by Ψ_p, and on subtracting we get

$$\frac{d}{dx}\left(\Psi_q \frac{d\Psi_p}{dx} - \Psi_p \frac{d\Psi_q}{dx}\right) + \frac{8\pi^2 m}{h^2}(E_p - V_p + V_q)\Psi_p\Psi_q = 0 \qquad ...(3)$$

Integrating with respect to x over the whole of space from ($-\infty$ to $+\infty$), since Ψ and V are zero at infinity, the first term vanishes and second term becomes as

$$\frac{8\pi^2}{h^2}(E_p - E_q)\int_{-\infty}^{+\infty} \Psi_p\Psi_q \, dx = 0 \qquad ...(4)$$

Since $E_p \neq E_q$, then

$$\int_{-\infty}^{+\infty} \Psi_p\Psi_q \, dx = 0 \qquad ...(5)$$

It is possible to make a linear combination of orthogonal functions.

Degenerate

We have already proved that for every wave function Ψ_p there must be a corresponding energy value E, *i.e.*, for every eigen function there will be one eigen value. It means that each energy state must have a wave function which will be the characteristic of the system. If for two different Y's the same value of energy be obtained, these two states (of different Ψ's) will be called *degenerate states.*

Forbidden Transitions

When certain particular eigen functions are multiplied together, they give a zero product, *i.e.*, $\Psi_p\Psi_q = 0$. (This should not be confused with orthogonality). In such a situation, the simultaneous execution of the two corresponding vibrations by the electron will not cause the emission of radiation.

These transitions are known as *forbidden transitions.* Due to these, certain lines will be missing in spectral lines. In the presence of a strong electric or magnetic field, these lines may be actually appear. In such a situation, the eigen functions are distorted by the strong fields, and so the product $\Psi_p\Psi_q^*$ is no longer zero.

Observables

On a physical system several observations and measurements can be made. These are called *observables.* May of the observables are not the properties of a system but are characteristic of the measuring system which can be carried out on it. The measurement of an observable on a system gives a number.

The state of a physical system is represented by a function of certain variables, like space coordinates, time etc. By using the rules of the quantum theory the significant information can be obtained from this function. The variables may be chosen in several ways.

The function $\Psi(r_1, r_2, r_m, t)$ is one which may represent a state, where the total number of variables is equal to the number of degrees of freedom of the system. At a given time, the state becomes a function of the space coordinates of the system only.

The quantum theory will be useful in solving the different problems in measuring an observable if two restrictions are made in selecting the function of the state.

(i) *The functions should possess an integrable square.* It should not give a value which is infinity.

$$\int \Psi^* \Psi . dV \leq \infty$$

Here Y is the function and Y* is its complex conjugate. dV is the volume element of the configuration space.

(ii) *The function Y should be single-valued.* If Ψ is a function of space coordinates, it should have a single value at a given point in space. For example, let Ψ by a function of angle θ.

If the function Ψ should be single-valued its is necessary that, the same value should be obtained if we add or subtract a multiple of 2p radians to angle q. Thus.

$$\Psi(\theta) = \Psi(\theta \pm 2n\pi)$$

where $n = 1, 2, 3, \ldots$ an integer.

Stationary State

A state is said to be in a stationary state, if the function explaining the state does not include time. Thus the function $\Psi(x, y, z)$ is a stationary state whereas function $\Psi(x, y, z, t)$ is not.

Operators

An operator is a mathematical instruction or procedure to be carried out on a function.

It is written in the form

(Operator), (function) = (Another function)

The function on which the operation is carried out is called an *operand.* The left hand side of the above equation does not mean that the function is multiplied with the operator. Evidently, and operator written alone has no significance.

A few examples are given below.

(i) $\frac{d}{dx}(x^3) = 3x^2$. Here $\frac{d}{dx}$ which stands for differention w.r.t.x is the operator, x^3 is the operand and $3x^2$ is the result of the operation.

(ii) $\int x^2\, dx = x^4/4 + C$. Here $\int()\, dx$ which stands for integration w.r.t. x is the operator, x_3 is the operand and $x^4/4 + C$ is the result of the operation.

Similarly, *taking the square or taking the square root or multiplication by a constant k* etc. are different operations which can be carried on any function (*i.e.*, the operand).

In case the symbol used for the operator is not self explanatory, a suitable letter or some symbol for the operator is used with the symbol (Δ) over it.

Algebra of Operators

The operators follow certain rules similar to those of the algebra A few of these are given below.

(1) *Addition and Subtraction of Operators :* If $\hat{A}$ and $\hat{B}$ are two different operators and f is the operand, then

$$(\hat{A} + \hat{B})f = \hat{A}f + \hat{B}f \qquad ...(6)$$

and

$$(\hat{A} - \hat{B})f = \hat{A}f - \hat{B}f \qquad ...(7)$$

(2) *Multiplication of Operators :* If $\hat{A}$ and $\hat{B}$ are two different operators and f is the operand, then the expression $\hat{A}\hat{B}$ implies that first f is operated by the operator $\hat{B}$ to get the result, say f¢, and then f¢ is operated by the operator $\hat{A}$ to get the final result (say f''), *i.e.*, $\hat{A}\hat{B}$ f implies that $\hat{B}$ f = f' and then $\hat{A}f' =$ f¢¢ so that we have

$$\hat{A} = \hat{B}f = f''$$

Thus, the order of using the operators is from *right to left* as written in the given expression.

If the same operation is to be done a number a times in succession, it is shown by power of the operator. For example.

$$\hat{A}\hat{A}f \text{ is written as } \hat{A}^2 f$$

i.e. $$\hat{A}\,\hat{A}f = \hat{A}^2 f \qquad ...(8)$$

It may be noted that usually

$$\hat{A}\,\hat{B}f \neq \hat{B}\,\hat{A}f \qquad ...(9)$$

Suppose $\hat{A} = x, \hat{B} = \frac{d}{dx}$ and $f = f(x) = x^2$. Then

$$\hat{A}\,\hat{B}f = x\frac{d}{dx}(x^2) = x\,(2x) = 2x^2$$

$$\hat{B}\,\hat{A}f = \frac{d}{dx}\,x\,(x^2) = \frac{d}{dx}\,x^3 = 3x^3$$

If, however, it is found that

$$\hat{A}\,\hat{B}f = \hat{B}\hat{A}f$$

i.e., the change of order of the operators gives the same result, the operators are said to commute.

Thus this result it unlike algebra where $x \times y = y \times x$ always.

3. *Linear operators :* An operator $\hat{A}$ is said to be linear if for the two functions f and g

$$\hat{A}\,(f+g) = \hat{A}f + \hat{A}g \qquad ...(10)$$

i.e., the operator on the sum of two functions gives the same result as the sum of two results obtained by carrying out the same operation on the two functions separately. For example, d, dx, d^2/dx^2 etc. are linear operators whereas 'taking the square', 'taking the square root' etc., are non-linear.

DE-BROGLIE'S CONCEPT OF THE DUAL NATURE OF THE ELECTRON

In Bohr's theory, *an electron is treated as a material particle of small mass moving around the nucleus along a fixed path called orbit.* Light which consists of electromagnetic radiations is known to exhibit both corpuscular (particle) and wave properties. Based on that analogy a French Physicist Louis de-Broglie (1923) suggested that.

"Matter considered to be made up of discrete particles such as atoms and molecules, may also behave like wave under proper conditions."

The wave nature of the r radiation has been well established by the phenomenon of diffraction and interference. Electrons like the light rays or X-rays has also been shown to exhibit the diffraction phenomenon. This leads to the view that *electrons like the light may also have wave properties associated with them. It means that an electron has dual nature, particle and wave.*

Where m is the mass, u is the velocity, l the wavelength and p is the momentum of an electron.

Experimental Verification of de-Broglie's Concept

(a) *Davisson and Germer's experiment :* The first experimental evidence of de-Broglie's hypothesis came from the experiments of Davisson and Germer in 1927, (Fig. 4.4) gives the arrangement used by Davisson and Germer. The various constructional parts are as under.

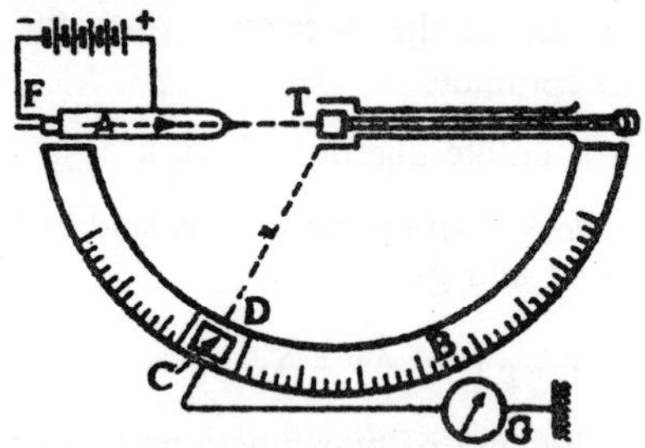

Fig. 4.4

(i) *Target (T)* : It is a single crystal of nickel. It is capable of rotating about its exist which is parallel to the axis of the electron beam. The position of the crystal can be adjusted by means of a band. A thin pencil of electrons is allowed to reflect from the crystal surface 'T' in different directions.

(ii) *Electron gun (A) :* It consists of a tungsten filament F. Electron are emitted by thermionic action. Due to the action of an electric field across the gun, a fine stream of electron emerges out of this gun.

(iii) *Outer chamber.* The whole arrangement is enclosed in an evacuated chamber, as shown in the diagram (Fig. 4.4).

(iv) *Collector (C).* It is a Faraday cylinder connected to a sensitive galvanometer G. It can be moved along a circular scale D to locate the position of maxima and minima between the angle 20° to 90°. The inner and outer walls of the cylinder are insulated from each other and a retarding potential is applied between them so that only the fastest moving electrons can enter the cylinder.

Working: Two series of experiments were performed by Davisson and Germer.

(i) In the first set, the electrons were made to fall on the crystal at a fixed angle of incidence by keeping the electron gun and collector only at one fixed position throughout the experiment. The varying the accelerating voltage, the current is measured by the galvanometer for each value of voltage. The strength of current measure by galvanometer is a measure of the intensity of diffracted beam. The intensity is then plotted against the square root of the accelerating potential, a curve with several sharp maxima is obtained as shown in Fig. 4.5. This is similar to the behaviour of X-rays reflected from a crystal. This shows that electrons like X-rays exhibit wave nature.

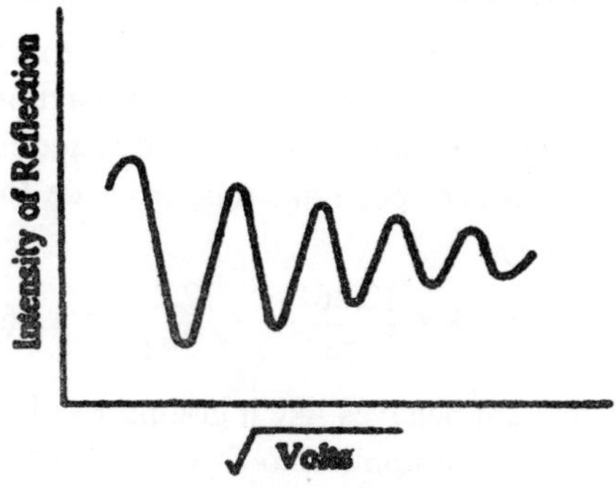

Fig. 4.5

(ii) In the second series of experiments, a beam of electrons was made to fall normally on the surface of a crystal and then the collector (C) was made to move to various positions on the scale (D) and the galvanometer current at each position is noted. The current, which was measure of the intensity of the diffracted beam of electrons was plotted against the angle between the incident beam and the beam entering the collector, known as the *collector.*

The observations were repeated for different voltages and several curves were drawn as shown in Fig. 4.6. It is observed that a bump moves upwards and attains its greatest development in the curve creasing voltage the bump moves upwards and attains its greatest development in the curve for 54 volts at a colatitude of 50°. At higher voltages, the bump gradually diminishes, there being hardly any trace of it at about 68 volts.

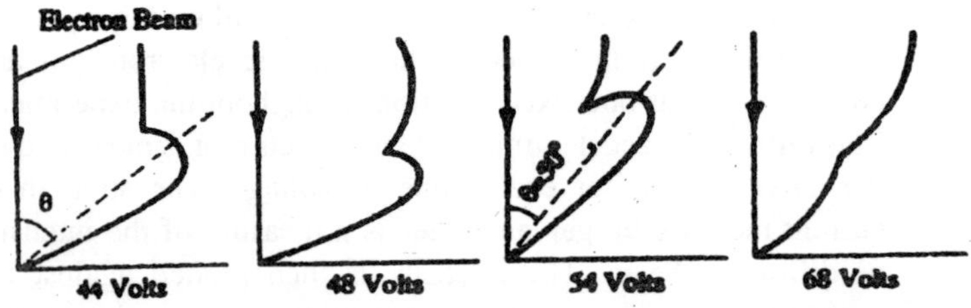

Fig. 4.6

The bump in its most important state of development offers a convincing evidence for the existence of the electron waves.

In one set of experiment, Davisson and Germer found that when the accelerating potential is raised to 54 volts, a maximum appears when the angle between the incident and the diffracted beam war 50°, *i.e.*, $\theta = 50°$ for first order spectrum, thus from de-Broglie's equation for electrons (Fig. 4.7).

$$\lambda = \frac{12.26}{\sqrt{54}} \text{Å} = \frac{12.26}{\sqrt{54}} = 1.66 \text{ Å}.$$

Since the crystal behaves like a grating for the X-rays, the standard grating formula can be applied to this case also. We have

$$n\lambda = d \sin \theta$$

for a nickel crystal of d = 2.15 Å

$$\therefore \quad 1.\lambda = 2.15 \times \sin 50° \qquad \text{or} \qquad \lambda = 1.65 \text{ Å}.$$

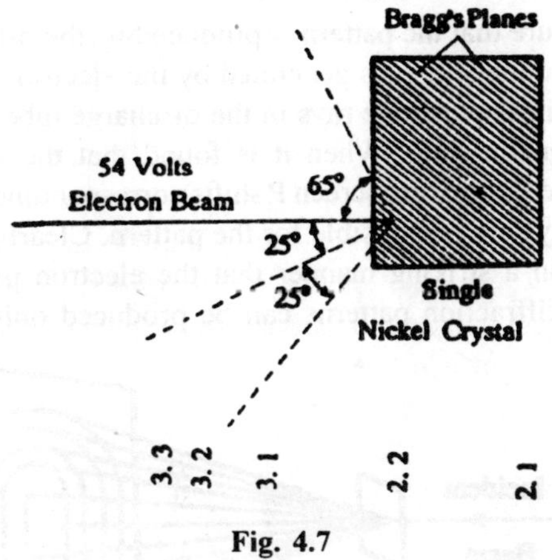

Fig. 4.7

There is a close agreement between the two results. This excellent quantitative agreement is very important, showing as it does, that a beam of electrons does really possess wave-like characteristics.

G.P. Thomson's Experiment

In 1928, G.P. Thomson extended the research on electron waves to high speed electrons ranging from 10.000 to 50,000 volts, diffracted by very thin metallic films. He used a method analogous to the Debye-Scherrer power method of X-rays analysis of crystals.

A beam of electrons obtained from filament, F accelerated by the anode A to a potential of 50,000 volts, is allowed to ass through a hole in metal block B and finally allowed to fall one a gold foil G of thickness 10^6 cm.

The electrons after diffraction were received on the photographic plate P. (Fig. 4.7). On developing it, a number of concentric rings similar to the diffraction of light were obtained confirming the wave nature of the electron (Fig. 4.9).

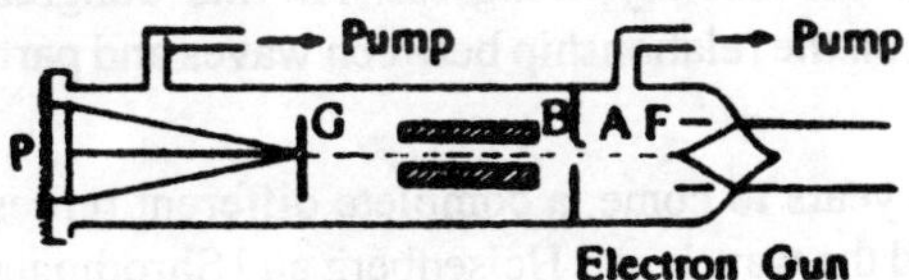

Fig. 4.8

To mark sure that the pattern is produced by the diffracted electrons and not by secondary X-rays generated by the electrons in their passage through the foil, the cathode rays in the discharge tube are deflected by means of magnetic field. When it is found that the whole pattern as observed on the fluorescent screen P shifts correspondingly, which cannot happen if X-rays are responsible for the pattern. Clearly this experiment demonstrates in a striking manner that the electron pencil behaves as waves, since diffraction patterns can be produced only by waves.

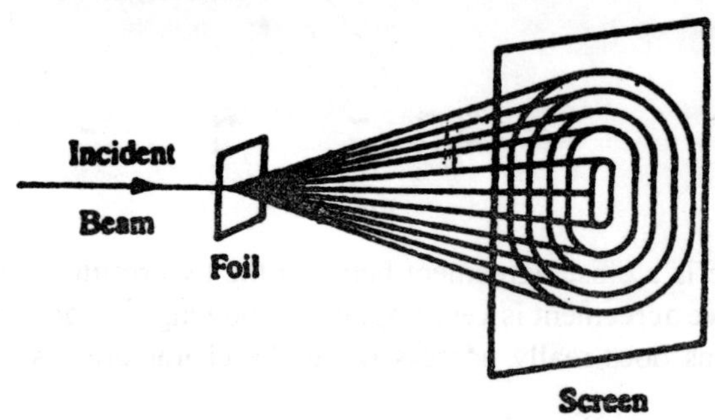

Fig. 4.9

It was also observed that the radii of the rings decreases with increases in velocity of electron, confirming the de-Broglie's relation that l decreases with increases of velocity.

Application on the de-Broglie's concept. The main applications are.

(i) The development of electron microscope is an excellent application of de-Broglie's concept.

(ii) The de-Broglie's concept has theoretically supported Bohr's second postulate.

Criticism of De-Broglie's Concept

In the summer of 1927, physicists from all over the world arrived in Brussels at the Solway Congress. At this congress, de-Broglie's representation on the relationship between waves and particles was totally rejected.

For many years to come, a complete different representation of this relationship led the way. It was Heisenberg and Shrodinger who supported and strongly represented the concepts of de Broglie at another Congress and got it accepted.

SOLVED EXAMPLES

Example 1:

Calculate the uncertainty product for a moving ball of iron weighing 500 *gm.*

Solution:

The uncertainty expression assumes the form

$$\Delta p \times \Delta x \geq \frac{h}{4\pi} \text{ nor } \Delta mu \times \Delta x \geq \frac{h}{4\pi}$$

or
$$\Delta u \times \Delta x\ ^3 \frac{h}{4m\pi} \geq \frac{6.625 \times 10^{-27}}{4 \times 3.14 \times 500} \approx 10^{-30}$$

10^{-30} is very small and is negligible. Therefore, for larger objects the uncertainty of measurements is practically nil.

Example 2(a):

Calculate the uncertainty product for a moving electron of mass 9.1091 × 10^{-28} *gm.*

Solution:

$$\Delta u \times \Delta x \geq \frac{h}{4m\pi} \geq \frac{6.625 \times 10^{-27}}{4 \times 3.14 \times 9.1091 \times 10^{-28}} \approx 0.6$$

This value is large enough in comparison with the size of the electron and is this in no way negligible.

Example 2(b):

Show that the degree of degeneracy of the n th energy level in the hydrogen atom in n^2.

Solution:

The energy values given by

$$E = -\frac{\mu e^4}{8\varepsilon_0{}^2 h^2}\left(\frac{1}{n^2}\right)$$

depend only on n and hence are degenerate with respect to both l and m_l (that the states having same n but different l's or m_l's have same energy).

For each value of n, l can take n values from 0 to n – 1, and for each value of l there are (2l + 1) values of m_l (= 0, ±, 1, + ± 2, ... ± 1). Therefore, the degeneracy of the nth energy level is

$$\sum_{l=0}^{n-1} (2l+1) = 1 + 3 + 5 \ldots + (2n - 1),$$

$$\text{a total of n terms} = \frac{n}{2}[2(1)+(n-1)2] = n^2.$$

Example 3:

Consider two hypothetical spherical shells centred on the nucleus of a hydrogen atom with radii r and r + dr. What is the probability P(r) that the electron will lie between these shells, as a function of r ? The motion of the electron in the atom is described by the wave function.

$$\Psi = \frac{1}{\sqrt{\pi a_0{}^3}} e^{-r/a_0} \cos\omega t.$$

Solution:

The volume between the shells is $dV = 4\pi r^2$ dr.

The probability of the electron being found between the shells is given by $P(r)\, dr = |\Psi|^2$ dr.

$$= \left(\frac{1}{\sqrt{\pi a_0{}^3}} e^{-r/a_0} \cos^2 \omega t\right) 4\pi r^2 dr.$$

The average probability is obtained by replacing $\cos^2 \omega\tau$ by its average value which is 1/2. Thus

$$\overline{P(r)} = \left\{\frac{1}{\pi a_0{}^3} e^{-2r/a_0} \left(\frac{1}{2}\right)\right\} 4\pi r^2$$

$$= \frac{2r^2}{a_0^3} e^{-2r/a_0}.$$

For each value of n, l can take n values from 0 to n – 1, and for each value of l there are (2l + 1) values of ml (= 0, ± 1, + ± 2, ... ± l). Therefore the degeneracy of the nth energy level is

$$\sum_{l=0}^{n-1} (2l+1) = 1 + 3 + 5 \ldots + (2n - 1), \text{ and total of n terms}$$

$$= \frac{2}{n}[2(1)+(n-1)2] = n^2.$$

5

The Applications of the Schrodinger's Equations

THE APPLICATIONS OF THE SCHRODINGER'S EQUATIONS TO THE HYDROGEN ATOM

In the field of atomic and molecular structure, the problem of structure of hydrogen atom is regarded as very important because it forms the basis for the discussion of more complex atomic systems. The wave-mechanical treatment, which is applied to hydrogen atom, is also used for hydrogen-like or closely related atoms. The Schrodinger's wave equation is expressed in the form

$$\nabla^2\Psi + \frac{8\pi^2 m}{h^2}(E - V)\Psi = 0 \qquad ...(1)$$

Dividing equation (1) by m, we get

$$\frac{1}{m}\nabla^2\Psi + \frac{8\pi^2}{h^2}(E - V)\Psi = 0 \qquad ...(2)$$

In the hydrogen atom, there are only two particles, the electron and the nucleus. For such a system, it will be convenient to express the equation (2) in the form,

$$\frac{1}{m^2}\nabla_1^{\,2}\,\Psi r + \frac{1}{m^2}\nabla_2^{\,2}\,\Psi r + \frac{8\pi^2}{h^2}(E - V)\Psi = 0 \qquad ...(3)$$

where m_1 is the mass of the electron and m_2 the mass of the nucleus

Transformation of Co-ordinates : The total energy, E, in equation (3) has to parts, (i) the translation motion of the atom as a whole, and (ii) the energy of the electron with respect to the proton. It is this latter portion of the energy in which one is interested. This leads, again to the problem of separation of variables. In order to obtain the desired equation, it becomes necessary to separate out and reject the translational portion of the total wave function.

In order to carry out this particular separation, it becomes necessary to introduce a new set of variables x, y, and z, which are Cartesian co-ordinates of the centre of mass of hydrogen atom, and the variable, r, 0 and f,, which are polar co-ordinates of the electron with respect to the nucleus. For the hydrogen atom, the Cartesian co-ordinates of the centre of mass will be given by

$$x = \frac{m_1x_1 + m_2x_2}{m_1 + m_2} \quad ...(4)$$

$$y = \frac{m_1y_1 + m_2y_2}{m_1 + m_2} \quad ...(5)$$

$$z = \frac{m_1z_1 + m_2z_2}{m_1 + m_2} \quad ...(6)$$

and the transformation to spherical co-ordinates can be seen from Fig.5.1 to be

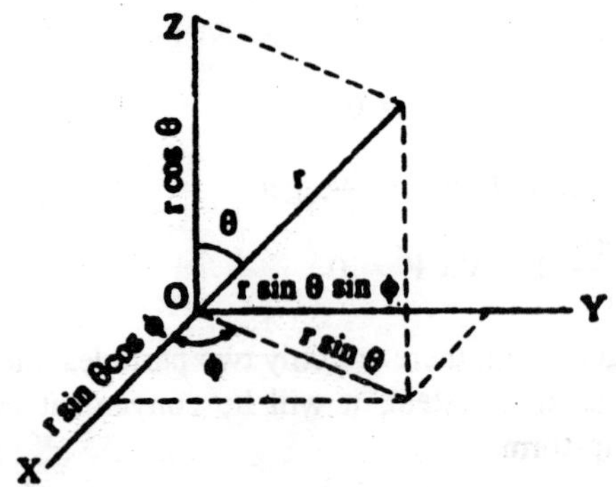

Fig. 5.1

$$r \sin\theta \cos\phi = x_2 - x_1 \quad ...(7)$$

$$r \sin\theta \sin\phi = y_2 - y_1 \quad ...(8)$$

$$r \sin\theta = z_2 - y_1 \quad ...(9)$$

Making substitutions of equations (4), (5), (6), (7), (8), and (9) in (3), we get

$$\frac{1}{m_1 + m_2}\left(\frac{\partial^2 \Psi T}{\partial x^2} + \frac{\partial^2 \Psi T}{\partial y^2} + \frac{\partial^2 \Psi r}{\partial z^2}\right) + \frac{m_1 + m_2}{m_1 m_2}\left[\frac{1}{r^2}\frac{\partial}{\partial r}\left(r^2 \frac{\partial \Psi T}{\partial r}\right)\right.$$

$$\left. + \frac{1}{r^2 \sin^2\theta}\frac{\partial^2 \Psi T}{\partial \phi^2} + \frac{1}{r^2 \sin\theta}\left(\sin\theta \frac{\partial \Psi r}{\partial \theta}\right)\right]$$

$$+ \frac{8\pi^2}{h^2}(E - V)\Psi r = 0 \quad ...(10)$$

The wave function Ψ_T is a function of the variables x, y, z, r, θ and ϕ, and energy E possesses the translational energy of the atom as well as the energy of the electron with respect to the nucleus. In the usual manner, the total wave function Ψ_T of assumed to be expressible as the product of the two wave functions such that

$$\Psi_T = F_{x\,y\,z}\, \Psi_{r\,\theta\,\phi} \quad ...(11)$$

For sake of convenience, we will express F and Ψ in place of F_{xyz} and $\Psi_{r\theta\phi}$. When equation (11) is substituted in equation (10), it is found that the following two equation are obtained.

$$\text{(i)} \quad \frac{\partial^2 F}{\partial x^2} + \frac{\partial^2 F}{a y^2} + \frac{\partial^2 F}{a z^2} + \frac{8\pi^2 (m_1 + m_2)}{h^2} E_{trans} \quad ...(12)$$

$$\text{(ii)} \quad \frac{1}{r^2}\frac{\partial}{\partial r}\left(r^2 \frac{\partial \Psi}{\partial r}\right) + \frac{1}{r^2 \sin^2\theta \partial\phi^2}\frac{\partial^2 \Psi}{\partial \phi^2} + \frac{1}{r^2 \sin\theta}\cdot\frac{\partial}{\partial \theta}\left(\sin\theta \frac{\partial \Psi}{\partial \theta}\right)$$

$$+ \frac{8\pi^2}{h^2}\mu[E - V]\Psi = 0 \quad ...(13)$$

where μ is the reduced mass and is given by

$$\mu = \frac{m_1 m_2}{m_1 + m_2}$$

Equation (12) contains only the variable x, y, and z and contains no potential energy term.

This is identical to the wave equation for a free particle and therefore represents the translational energy of the atom as a whole. Equation (3), which relates the electron to the proton is the equation of interest to us. Well now consider equation (13) only.

Separation of variables : Equation (13) is a second order partial differential equation. This contains three variables. In order to separate the variables it becomes necessary to assume that Ψ may be represented by the product of three wave functions, each having only one of the three variables, r, θ and ϕ. If we let

$$\Psi = R_{(r)}\, T_{(\theta)}\, F_{(\phi)}$$

and make this substitution into equation (13) we get

$$\frac{1}{r^2}\frac{\partial}{\partial\theta}\left(r^2\frac{\partial}{\partial r}R_{(r)}\,T_{(\theta)}\,F_{(\phi)}\right)+\frac{1}{r^2\sin^2\theta}\cdot\frac{\partial^2}{\partial\phi^2}R_{(r)}\,T_{(\theta)}\,F_{(\phi)}$$

$$+\frac{1}{r^2\sin\theta}\frac{\partial}{\partial\theta}\left[\sin\theta\frac{\partial}{\partial\theta}R_{(r)}\,T_{(\theta)}\,F_{(\phi)}\right]+\frac{8\pi^2\mu}{h^2}[E-V]R_{(r)}\,T_{(\theta)}\,F_{(\phi)}=0$$

which, on dividing by $R_{(r)}\,T_{(\theta)}\,F_{(\phi)}$ gives

$$\frac{1}{r^2R_{(r)}}\frac{\partial}{\partial r}\left(r^2\frac{\partial R_{(r)}}{\partial r}\right)+\frac{1}{F_{(\phi)}\,r^2\sin^2\theta}\frac{\partial^2F_{(\phi)}}{\partial\phi^2}$$

$$\frac{1}{T_{(\theta)}\,r^2\sin\theta}\frac{\partial}{\partial\theta}\left(\sin\theta\frac{\partial T_{(\theta)}}{\partial\theta}\right)+\frac{8\pi^2\mu}{h^2}[E-V]=0$$

If we multiply by $r^2\sin^2\theta$, we get

$$\frac{\sin^2\theta}{R_{(r)}}\frac{\partial}{\partial r}\left(r^2\frac{\partial R_{(r)}}{\partial r}\right)+\frac{1}{F_{(\phi)}}\frac{\partial^2F_{(\phi)}}{\partial\phi^2}+\frac{\sin\theta}{T_{(\theta)}}\frac{\partial}{\partial\theta}\left(\sin\theta\frac{\partial T_{(\theta)}}{\partial\theta}\right)$$

$$+\frac{8\pi^2\mu r^2\sin^2\theta}{h^2}[E-V]=0$$

$$\Rightarrow\quad\frac{\sin^2\theta}{R_{(r)}}\frac{\partial}{\partial r}\left(r^2\frac{\partial R_{(r)}}{\partial r}\right)+\frac{\sin\theta}{T_{(\theta)}}\frac{\partial}{\partial\theta}\left(\sin\theta\frac{\partial T_{(\theta)}}{\partial\theta}\right)$$

$$+\frac{8\pi^2\mu r^2\sin^2\theta}{h^2}[E-V]=-\frac{1}{F_{(\phi)}}\frac{\partial^2F_{(\phi)}}{\partial\phi^2}\quad\text{...(14)}$$

The left side of equation (14) has only the variables r and θ, whereas the right side of the equation has only the variable ϕ. If we put the right side equal to m^2, equation (14) modifies to

$$\frac{\sin^2\theta}{R_{(r)}}\frac{\partial}{\partial r}\left(r^2\frac{\partial R_{(r)}}{\partial r}\right)+\frac{\sin\theta}{T_{(\theta)}}\frac{\partial}{\partial\theta}\left(\sin\theta\frac{\partial T_{(\theta)}}{\partial\theta}\right)$$

$$+\frac{8\pi^2\mu r^2\sin^2\theta}{h^2}[E-V]\quad\text{...(15)}$$

where $$\frac{1}{F_{(\phi)}}\frac{d^2F_{(\phi)}}{d\phi^2} = -m^2 \qquad ...(16)$$

Equation (15) contains two variables r and ϕ. The problem now is to vary out the separation of the remaining two variables r and ϕ. By dividing equation (15) by $\sin^2\theta$, we get

$$\frac{1}{R_{(r)}}\frac{\partial}{\partial r}\left(r^2\frac{\partial R_{(r)}}{\partial r}\right)+\frac{1}{T_{(\theta)}\sin\theta}\frac{\partial}{\partial\theta}\left(\sin\theta\frac{\partial T_{(\theta)}}{\partial\theta}\right)$$

$$-\frac{m^2}{\sin^2\theta}+\frac{8\pi^2\mu r^2}{h^2}[E-V] = 0$$

or on rearranging

$$\frac{1}{R_{(r)}}\frac{\partial}{\partial r}\left(r^2\frac{\partial R_{(r)}}{\partial r}+\frac{8\pi^2\mu r^2}{h^2}[E-V]\right)$$

$$= \frac{m^2}{\sin^2\partial}-\frac{1}{T_{(\theta)}\sin\theta}\frac{\partial}{\partial\theta}\left(\sin\theta\frac{\partial T_{(\theta)}}{\partial\theta}\right) \qquad ...(17)$$

As each side of the equation (17) contains only one variable, they both must be equal to the same constant. If the right side of the equation is equal to the constant β, and this gives on multiplication by $T_{(\theta)}$

$$\frac{m^2T_{(\theta)}}{\sin^2\theta}-\frac{1}{\sin\theta\,\theta\,d\theta}\frac{d}{}\left(\sin\theta\frac{dT_{(\theta)}}{d\theta}\right)-\beta T_{(\theta)} = 0 \qquad ...(18)$$

This is the desired form of the $T_{(\theta)}$ equation. The remaining part of the original equation is the R equation.

$$\frac{1}{r^2}\frac{d}{dr}\left(r^2\frac{dR_{(r)}}{dr}\right)-\frac{\beta}{r^2}R_{(r)}+\frac{8\pi^2\mu}{h^2}[E-V]R_{(r)} \qquad ...(19)$$

Thus, the three variables have been successfully separated and the three independent total differential equation that result are.

(i) $$\frac{d^2F_{(\theta)}}{d\phi^2}+m^2F_{(\theta)} = 0 \qquad ...(20)$$

(ii) $$\frac{1}{\sin\theta}\frac{d}{d\theta}\left(\sin\theta\frac{dT_{(\theta)}}{d\theta}\right)-\frac{m^2T_{(\theta)}}{\sin^2\theta}+\beta R_{(\theta)} = 0 \qquad ...(21)$$

(iii) $$\frac{1}{r^2}\frac{d}{dr}\left(r^2\frac{dR_{(r)}}{dr}\right)-\frac{\beta}{r^2}R_{(r)}+\frac{8\pi^2\mu}{h^2}[E-V]R_{(r)}=0 \quad ...(22)$$

Solution of F(f) equation.

Equation (20) is $\dfrac{d^2F_{(\phi)}}{d\phi^2}+m^2F_{(\phi)}=0$

Equation (20) is of the same form as the wave equation for the particle in a box. In terms of sine and cosine, its solution is

$$F_m(\phi) = A \sin m\phi + B \cos m\phi$$

In order for a wave function to be acceptable, it should be of the well behaved class. One of the requirements of such function is that it must be single valued. In order to meet this restriction, the function $F_m(\phi)$ should have the same value for $\phi = 0$, as it has for $\phi = 2\pi$. For the case of $\phi = 0$, it follows that

$$F_m(\phi) = A \sin m\phi + B \cos 0 = B$$

and when $\phi = 2\phi$, we get

$$F_m(2\pi) = A \sin m\, 2\pi + B \cos m\, 2\pi$$

As the value of $F_m(\phi)$ must be the same under both of these conditions, it becomes necessary that

$$B = A \sin m\, 2\pi + B \cos m\, 2\pi \quad ...(23)$$

This identity is true only if m is zero or has a positive or negative integral value. Such characteristics are those that would be expected of a quantum number, and the particular restrictions on m indicate that it is the analog of the magnetic quantum number of the Bohr-Sommerfeld model.

Very often, in the treatment of the hydrogen atom, the exponential solution to the $F_{(\phi)}$ equation

$$F_{(m)}(\phi) = C_e^{\pm im\phi} \quad ...(24)$$

is used. In order to evaluate the constant, C it would be most convenient to select C in such a manner that the wave function $F(\phi)$, will be a normalised wave function. This requires that

$$\int_0^{2\pi} F_{(\phi)}\, F_{(\phi)}{}^*\, d\phi = 1$$

which leads to

$$\int_0^{2\pi} C^2\, e^{\pm im\phi}\, e^{\pm im\phi}\, d\phi = 1$$

$$\Rightarrow \quad C^2 \int_0^{2\pi} d\phi = 2\pi\, C^2 = 1$$

The value of the constant, C, that gives a normalised wave-function is seen to be C = 1/(2π), and the final normalised wave-function will be

$$F_m(\phi) = \frac{1}{\sqrt{(2\pi)}} e^{\pm im\phi} \text{ where } m = 0\ 1, \pm 2 \qquad ...(25)$$

F_m (ϕ) must be single valued, and therefore F_m(ϕ) must have the same value after any number of whole revolutions, and hence

$$e^{im\phi} = e^{im\ (\phi + 2\pi n)}$$
$$= e^{i\ (m\phi + 2\pi nm)}$$

where n is whole number and therefore the product *nm* must be whole number, and m must be whole number. Possible values of m are 0, ± 1, ± 2, ± 3,..., m is known as the *magnetic quantum number.* It is concerned with the behaviour of the electrons in the atom when it is placed in a magnetic field.

Solution of Yq) Dependent Equation : Equation (21) is

$$\frac{1}{\sin\theta}\frac{d}{d\theta}\left(\sin\theta\frac{dT_{(\theta)}}{d\theta}\right) - \frac{m^2 T_{(\theta)}}{\sin^2\theta} + \beta T_{(\theta)} = 0 \qquad ...(26)$$

Suppose z =n cos q, which varies between the limits + 1 and – 1. Replace $T_{(\theta)}$ by P where P is a function of z. The following three relations may be written down.

$$z^2 = 1 - \sin^2\theta$$

$$\frac{dT_{(\theta)}}{d\theta} = \frac{dP}{dz}\frac{dz}{d\theta} = -\frac{dP}{dz}\sin\theta \qquad ...(27)$$

$$\frac{d^2T_{(\theta)}}{d\theta^2} = \frac{dP}{dz}\frac{d^2z}{d\theta^2} + \left(\frac{dz}{d\theta}\right)^2\frac{d^2P}{dz^2} \qquad ...(28)$$

$$\text{Now } \frac{1}{\sin\theta}\frac{d}{d\theta}\left(\sin\theta\frac{dT_{(\theta)}}{d\theta}\right) = \frac{1}{\sin\theta}\sin\theta\frac{d^2T_{(\theta)}}{d\theta^2} + \frac{\cos\theta}{\sin\theta}\frac{dT_{(\theta)}}{d\theta}$$

$$\frac{dP}{dz}\frac{d^2z}{d\theta} + \left(\frac{dz}{d\theta}\right)^2\frac{d^2P}{dz^2} + \frac{\cos\theta}{\sin\theta}\frac{dP}{dz}\frac{dz}{d\theta}$$

$$\text{But } \frac{dz}{d\theta} = -\sin\theta \qquad [\because z = \cos\theta]$$

and $$\frac{d^2z}{d\theta^2} = -\cos\theta = z$$

Hence $$\frac{1}{\sin\theta}\frac{d}{d\theta}\left(\sin\theta\frac{dT_{(\theta)}}{d\theta}\right) = -z\frac{dP}{dz} + (1-z^2)\frac{d^2P}{dz^2} - z\frac{dP}{dz}$$

$$= (1 - z2)\frac{d^2P}{dz^2} - 2z\frac{dP}{dz}$$

$$= \frac{d}{dz}\left[(1-z^2)\frac{dP}{dz}\right]$$

Therefore,

$$\frac{1}{\sin\theta}\frac{d}{d\theta}\left(\sin\theta\frac{dT_{(\theta)}}{d\theta}\right) - \frac{m^2T_{(\theta)}}{\sin\theta} + \beta T_{(\theta)} = \frac{d}{dz}$$

$$\left[(1-z^2)\frac{dP}{dz} + \left(\beta - \frac{m^2}{1-z^2}\right)P\right]$$

and $$\frac{d}{dz}\left[(1-z^2)\frac{dP}{dz}\right] + \left(\beta - \frac{m^2}{1-z^2}\right)P = 0 \qquad ...(29)$$

$$\Rightarrow \quad (1-z^2)\frac{d^2P}{dz^2} - 2z\frac{dP}{dz} + \left(\beta - \frac{m^2}{1-z^2}\right)P = 0 \qquad ...(30)$$

This last relation is of the form of differential equation which is so-called associated Lagendre functions.

Legendre Polynomials and Associated Legendre Functions The Legendre polynomials satisfy the differential equation

$$\frac{d}{dz}\left[(1-z^2)\frac{dP_I(z)}{dz}\right] + I(I+1)P_I(z) = 0$$

where $P_l(z)$ is the Legendre polynomial of degree l. The properties of polynomial restrict I in such a way that it must be zero or an integer.

$$P_t(z) = \frac{1}{z^I I!}\frac{d^I(z^2-1)}{dz^I}$$

When we know two adjacent polynomials, one can calculate others from the recursion formula

$$(I + 1)P_{l+1}(z) - (2l + 1)z P_l(z) + I P_l - 1(z) = 0$$

The associated functions of degree l and | m |, where l = 0, 1, 2, ... and | m | = 0, 1, 2, ..., l, are defined in terms of the Legendre polynomials by

$$P^{l|m|}(z) = (1 - z^2)^{|m|/2} \frac{d^{|m|}}{dz^{|m|}} P_l(z)$$

This satisfies the differential equation

$$(1 - z^2)$$

$$\frac{d^2 P_l^{|m|}(z)}{dz^2} - 2z\frac{d P_l^{|m|}}{dz} + \left[1(1+1 - \frac{m^2}{1-z^2}\right] P_l^{|m|^2}(z) = 0 \quad ...(31)$$

Solution of $T_{(\theta)}$ *dependent equation. Comparing* equation (31), the differential equation which is satisfied by the associated Legendre function, with equation (30), from transformed $T_{(\theta)}$ equation), we see that they are of the same form. We need only to identify P (z) with $P_l^{|m|}(z)$ and β with l (l + 1). We need now have

$$T_{(\theta)} = N_\theta P^{l|m|}(z) \quad ...(32)$$

and

$$\beta = l(l + 1) \quad ...(33)$$

where l is known as the azimuthal quantum number. Here l ≥ m, (m + 1), (m + 2), (m + 3), ...

One should now normalise $T_{(\theta)}$. One knows that since the variable is z = cos q, the electron is confined between z = – 1 and z = + 1. The value of the integral is

$$\int_{-1}^{+1} P_l^{|m|}(z) P\gamma^{|m|}(z)\, dz = \frac{2}{(2l+1)} \frac{(l+|m|!}{(l-m)!} \delta u' \quad ...(34)$$

has been determined. The symbol. The symbol $\delta u'_l$ the Kronecker delta, has the value zero when l ≠ l′ and unity when l = l′.

Variation from z = – 1 to z = + 1, corresponds to variation of θ from zero to π. The normalisation integral,

$$\int_\theta^\pi T^*_{(\theta)ml} T_{(\theta)ml} \sin\theta\, d\theta$$

becomes

$$N_\theta^2 \int_{-1}^{+1} P_l^{|m|}(z) P_l'^{|m|}(z)\, dz = 1 \quad ...(35)$$

in terms of the associated Legendre functions. Substituting for the integral its value in terms of l and m from equation (34) in (35) we find

$$N_\theta^2 = \frac{(2l+1)(l-|m|)!}{2(l+|m|)!}$$

Substituting above equation in (32), we get

$$T_{(\theta)} = \frac{(2l+1)(l-|m|!}{2(l+|m|)!} P_l^{|m|} \cos\theta \qquad ...(36)$$

The solution of r Dependent Equation : It now remains to solve the equation describing that ρ dependence of the hydrogen atom wave function

$$\frac{1}{r^2}\frac{d}{dr}\left(r^2 \frac{dR_{(r)}}{dr}\right) - \frac{\beta}{r^2} R_{(r)} + \frac{8\pi^2\mu}{h^2}[E-V]R_{(r)} = 0 \qquad ...(37)$$

Here we know that $\beta = l(l+1)$ and E is negative and

$$V = \frac{Z_{e^2}}{r}.$$

Suppose we make the following substitutions

$$\alpha^2 = \frac{8\pi^2\mu E}{h^2}$$

and $$l = r\frac{4\pi^2\mu Ze^2}{h^2\alpha}$$

and $\rho = 2\alpha r$, where r is a new independent variable lying between 0 and ∞.

Suppose $S(\rho) = R(\rho)$. Then, $$\frac{dS}{d\rho} = \frac{dS}{dr}.\frac{dr}{d\rho} = 2\alpha$$

Putting $\beta = l(l+1)$ and making the appropriate substations, the r dependent equation (37) modifies to

$$\frac{1}{\rho^2}\frac{d}{d\rho}\left(\rho^2 \frac{dS}{d\rho}\right) + \left[\frac{-l(l+1)}{\rho^2} \frac{1}{4} + \frac{\lambda}{\rho}\right] S = 0 \qquad ...(39)$$

For large values of r, equation (39) becomes approximately as

$$\frac{d^2S}{d\rho^2} = \frac{S}{4} \qquad ...(40)$$

Solutions of this equation are $S = e^{\rho/2}$ and $S = e^{-\rho/2}$, only the second is suitable for wave functions. Assume that the solution of the complete equation is of the following form.

$$S(\rho) = e^{-\rho/2} F, \leq \rho £ \infty.$$

Then $\dfrac{dS}{d\rho} = e^{-r/2}\dfrac{dP}{d\rho} - \dfrac{1}{2}e^{-r/2}\ F$ ...(41)

and $\dfrac{dS}{d\rho} = e^{-r/2}\dfrac{d^2F}{d\rho^2} - \dfrac{e^{-\rho/2}}{2}\dfrac{dF}{d\rho} - \dfrac{e^{-\rho/2}}{2}\dfrac{dF}{d\rho} + \dfrac{e^{-\rho/2}}{4}F$

...(42)

On making the appropriate substitutions in equation (39).

$$\frac{d^2F}{d\rho^2} + \frac{dF}{d\rho}\left(\frac{2}{\rho} - 1\right) + \left(\frac{\lambda}{\rho} - \frac{l(l+1)}{\rho^2} - \frac{1}{\rho}\right)F = 0 \qquad ...(43)$$

Coefficients of dF/dr and F posses singularities at the origin, which is a regular point. Therefore a power series can be substituted for F beginning with a non-vanishing constant term.

Let F $= r'L$ where $L = a_0 + a_1\rho + a_2\rho^2 + a_3\rho^3 + ...$ ...(44)

Then $\dfrac{dF}{d\rho} = \rho^2\dfrac{dL}{d\rho^2} + s\rho^{2-1}L$ ...(45)

and $\dfrac{d^2F}{d\rho^2} = \rho^2\dfrac{d^2L}{d\rho^2} + 2s\rho^{2-1}\dfrac{dL}{d\rho} + (s-1)\rho^{s-1}L$ ...(46)

Multiplying equation (43) by ρ^2,

$$\rho^2\frac{d^2F}{d\rho^2} + (2\rho - \rho^2)\frac{dF}{d\rho} + (\lambda - 1)\rho E - l(l+1)F = 0 \qquad ...(47)$$

Substituting for $\dfrac{d^2F}{d\rho^2}$ and $\dfrac{dF}{d\rho}$,

$$\rho^{s+2}\frac{d^2L}{d\rho^2} + 2s\rho^{s+2}\frac{dL}{d\rho} + s(s-1)\rho^2L\ \ 2\rho^{s+1}\frac{dL}{d\rho} + 2s\rho^sL$$

$$-\ \rho^{s+2}\frac{dL}{d\rho} - s\rho^{s+1}L + (\lambda - 1)\rho^{s+1}L - l(l+1)\rho'L = 0 \qquad ...(48)$$

The equation is an identity in r, and therefore coefficients of individual powers of s get vanished. Taking the co-efficients of ρ^2,

$$\{s\ (s-1) + 2s - l)\ l + 1)\}\ a_0 = 0$$

Since $a_0 \neq 0$, therefore $\{s\ \{s - 1) + 2s - l\ (l + 1)\} = 0$,

...(49)

Therefore either s = l or s = – (l + 1) but the second value does not lead to an acceptable wave function.

Taking s = l,

$$F(\rho) = \rho^{l} L \qquad \text{[Use equation (44)]}$$

Substituting l for s and dividing by ρ^{l+1} in equation (48).

$$\rho \frac{d^2L}{d\rho^2} + \{2(l+1) - \rho\} \frac{dL}{d\rho} + (\lambda - 1 - l) L = 0 \qquad ...(49)$$

Suppose

$$L = a_0 + a_1\rho + a_2\rho + a_3\rho + ... \qquad ...(51)$$

$$\frac{dL}{d\rho} = a_1 + 2a_2\rho + 3a_2\rho^3 + ... \qquad ...(52)$$

and

$$\frac{d^2L}{d\rho^2} = 2a_2 + 6a_3\rho^3 + ... \qquad ...(53)$$

Substituting equations (51), (52) and (53) in (5(i), we get

$$2a_1\rho + 6a_2\rho^2 + 12a_3\rho^3 + ... + 2(l+1)(a_1 + 2a_1\rho + 3a_2\rho^2 + ... = 0$$

$$- a_1\rho - 2a_2\rho^2 - 3a_3\rho^3 - ...) + (l - 1 - 1)(a_0 + a_1\rho + a_2\rho^2 + ...) = 0.$$

The equation is an identity in ρ, and therefore, coefficients of powers of ρ must vanish individually. Thus.

$$(l - 1 - 1) a_0 + 1 \times 2(l+1) a_1 = 0$$

and $$(l - 1 - 1) a_1 + (2 \times 2(l+1) + 1 \times 2\} a_2 = 0$$

and $$(l - 1 - 2) a_2 + \{3 \times 2(l+l) + 2 \times 3\}\} a_3 = 0$$

$$(l - 1 - 1 - v) a_v + \{2(v+1)(l+1) + v(v+1)\} a_v + 1 = 0.$$

If the solution of the main equation is acceptable as a wave function, the series should break of after a finite number of terms (so that Ψ approaches zero at infinity). If it breaks off after the term v c_v = rn′, then we get

$$(l - 1 - n') = 0 \qquad ...(54)$$

where n′ has an integral value, 0 1, 2,...

Suppose l = n, then ...(55)

As l is an integer, n should be an integer of series 1, 2, 3,...n is known as the principal quantum number. Thus, the three quantum numbers defining the state of hydrogen atom are as follows.

Magnetic quantum number : *m* = 0, ± 1, ± 2, ± 3,...

Azimuthal quantum number :*l* ³ m, (m + 1), (m + 2), (m + 3)...

Principal quantum number : *n* ³ l + 1.

It becomes more convenient for the interpretation of spectra to rewrite these numbers as.

n = 1, 2, 3, ...

l = 0, 1, 2, 3, ...(n – 1)

m = – l, – l + 1, ... – 1, 0, + 1, ... 1, ... + l – 1, + l

Evaluation of the total energy E of an orbital of the hydrogen atom:

Since $\lambda = \frac{4\pi^2\mu Ze^2}{h^2\alpha}$ [Equation (38)] ...(56)

and $\alpha^2 = \frac{8\pi^2\mu E}{h^2}$ [Equation 38)] ...(57)

Squaring equation (56), we get

$$\lambda^2 = \frac{16\pi^4\mu^2 Z^2 e^4}{h^4\alpha^2}$$

$$\Rightarrow \quad \alpha^2 = \frac{16\pi^4\mu^2 Z^2 e^4}{h^4\lambda^2} \quad \text{...(58)}$$

From equation (57) and (58), we get

$$-\frac{8\pi^2\mu E}{h^2} = \frac{16\pi^4\mu^2 Z^2 e^4}{h^4\lambda^2}$$

$$\Rightarrow \quad E = -\frac{2\pi^2\mu Z^2 e^4}{h^2\lambda^2}$$

From equation (55), we have

n = l

$$\therefore \quad E = \frac{2\pi^2\mu Z^2 e^4}{n^2 h^2} \quad \text{...(59)}$$

For hydrogen atom Z = 1, it means that

$$E = -\frac{2\pi^2\mu e^4}{n^2 h^2} \quad \text{...(60)}$$

From equation (60), it follows that the total energy obtained by the Schrodinger treatment is seen to be the same as that obtained by the Bohr theory.

The Space Wave Function Y for the Electron in the Hydrogen Atom : The possible values of Ψ for the electron in the hydrogen atom are given by the product of three solutions derived above.

$$\Psi = e^{-r/2} \rho^{l} L(\rho) P_{l}^{m} (\cos \theta) e^{\pm \sin \phi} \qquad ...(61)$$

Equation (61) does not include the spin of the electron which is governed by a fourth quantum number, called spin quantum number. *The wave function which does not include the spin of the electron will be termed as the space wave function.*

Computations of a large number of wave functions corresponding to different values of the quantum numbers n, l, m have been made for the hydrogen atom.

The results have been tabulated in details by Pauling and others. For convenience of reference the separated functions $F_{(\phi)}$, $T_{(\theta)}$ and $R_{(r)}$ of the hydrogen atom are set out in Table i, ii and iii. The Table, IV gives the space wave function for the hydrogen atom in any given state which is obtained by multiplying together appropriate separated functions from Tables i, ii, and iii obtained by multiplying together appropriate separated functions from Tables i, ii, and iii.

Table 1 : Functions $F_{(\phi)}$ for the hydrogen atom.

m	$F_{(\phi)}$ *symbol*	$F_{\phi} = \frac{1}{\sqrt{2}} e^{\pm im\phi}$
0	$F_0(\phi)$	$\frac{1}{\sqrt{2\pi}}$
1	$F_1(\phi)$	$\frac{1}{\sqrt{2\pi}} e^{i\phi}$ or $\frac{2}{\sqrt{\pi}} \cos \phi$
– 1	$F_{-1}(\phi)$	$\frac{1}{\sqrt{2\pi}} e^{-i\phi}$ or $\frac{1}{\sqrt{\pi}} \sin \phi$
2	$F_2(\phi)$	$\frac{1}{\sqrt{2\pi}} e^{2i\phi}$ or $\frac{1}{\sqrt{\pi}} \cos 2\phi$
– 2	$F_{-2}(\phi)$	$\frac{1}{\sqrt{2\pi}} e^{2i\phi}$ or $\frac{1}{\sqrt{\pi}} \sin 2\phi$

Table 2 : Function $T_{(\theta)}$ for the hydrogen atom.

l	*m*	$T_{(\theta)}$ *symbol*	$T_{(\theta)}$ *in terms of θ*
0	0	T_{00}	$\frac{\sqrt{2}}{2}$
1	0	T_{10}	$\frac{\sqrt{6}}{2}\cos\theta$
1	± 1	$T_{1\pm1}$	$\frac{\sqrt{3}}{2}\sin\theta$
2	0	T_{20}	$\frac{\sqrt{10}}{4}(3\cos^2\theta - 1)$
2	± 1	$T_{2\pm1}$	$\frac{\sqrt{11}}{2}\sin\theta\cos\theta$
2	± 2	$T_{2\pm2}$	$\frac{\sqrt{15}}{4}\sin^2\theta$

Table 3 : From $R_{(r)}$ for the hydrogen atom

n	*l*	*Orbital*	*R symbol*	*R in terms of r and a*
1	0	1_s	R_{10}	$\frac{1}{a^{2/2}} 2e^{-r/a}$
2	0	2_s	R_{20}	$\frac{1}{2\sqrt{2a^2}}\left(2-\frac{r}{a}\right)e^{-r/2a}$
2	2	2_p	R_{21}	$\frac{1}{2\sqrt{6\alpha^3}}\frac{r}{a}e^{-r/2a}$
3	0	3_s	R_{30}	$\frac{1}{9\sqrt{3a^3}}\left(6-4\frac{r}{a}+\frac{4r^2}{9a^2}\right)e^{-r/2a}$

Probability Distribution Curves for s and p Orbitals : For sake of convenience, the wave function is divided into two independent parts.

(a) ***Radiant Function :*** This is r dependent part. This depends on the two quantum numbers n and l. The r dependent part of the wave function may be represented in one three ways.

(i) For the given values of n and l, the different values of the variable r are plotted against the r dependent part of the wave function. This is shown in (Fig 5.2).

Table 4 : Space wave functions of the hydrogen atom

n	*l*	*m*	*Orbital*	*Wave function*
1	0	0	1_s	$\Psi_{100} = \frac{1}{\sqrt{\pi a^3}} e^{-r/a}$
2	0	0	2_s	$\Psi_{200} = \frac{1}{4\sqrt{(2\pi a^3)}}\left(2 - \frac{r}{a}\right) e^{-r/2a}$
2	1	0	2_{pz}	$\Psi_{210} = \frac{1}{4\sqrt{(2\pi a^3)}} \frac{r}{a} e^{-r/2a} \cos\theta$
2	1	± 1	2_{px}	$\Psi_{211} = \frac{1}{4\sqrt{(2\pi a^3)}} \frac{r}{a} e^{-r/2a} \sin\theta \cos\phi$
			2_{py}	$\Psi_{211} = \frac{1}{4\sqrt{(2\pi a^3)}} \frac{r}{a} e^{-r/2a} \sin\theta \cos\phi$

(ii) For the given values n and l, the different values of the variable r are plotted against the square of the r dependent part of the wave function. This is shown in (Fig. 5.3.).

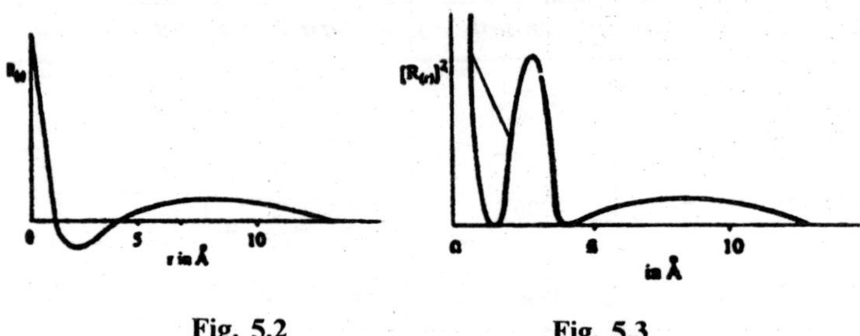

Fig. 5.2 Fig. 5.3

(iii) The different values of the variable, r are plotted against the radial distribution function, $4\pi r^2 [R_{(r)}]^2$. This is shown in (Fig. 5.4).

The variation of electron density with distance is best illustrated by plotting *radial distribution function* $4\pi r^2 [R_{(r)}]^2$ versus distance r from the nucleus, $4\pi r^2$ is the surface area of a sphere of radius r. Such a graph shows the probability of finding the electron on the surface of sphere at a distance r from the nucleus. The net result for a 1s electron in the probability curve is shown in (Fig. 5.4).

From the above curve we find that the probability of finding the electron is greatest at a distance r_0 from the nucleus. This distance

(0.529Å) is in close agreement with the calculated value for the first circular orbit for hydrogen.

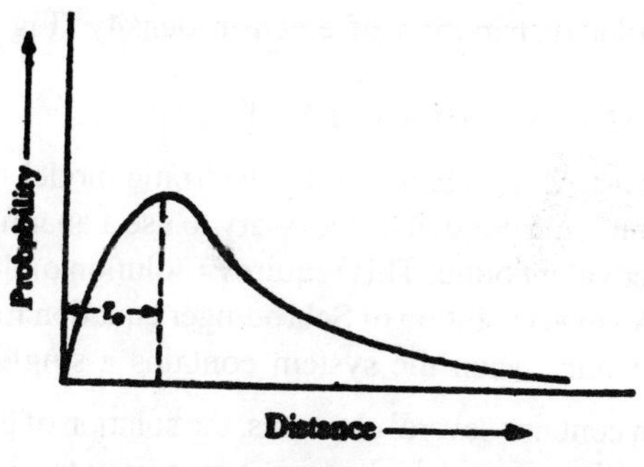

Fig. 5.4

Even after the maximum has passed, there is finite though very small, electron probability even at infinite distance from the nucleus.

(b) *Angular dependent function :* The polar function may be represent by a two-dimensional graph in which θ dependent part of the function $[T_{(\theta)}]$ only is concerned. There are two methods.

1. A polar graph is drawn by plotting the $T_{(\theta)}$ against different values of θ for given values of l and m (Fig. 5.5).

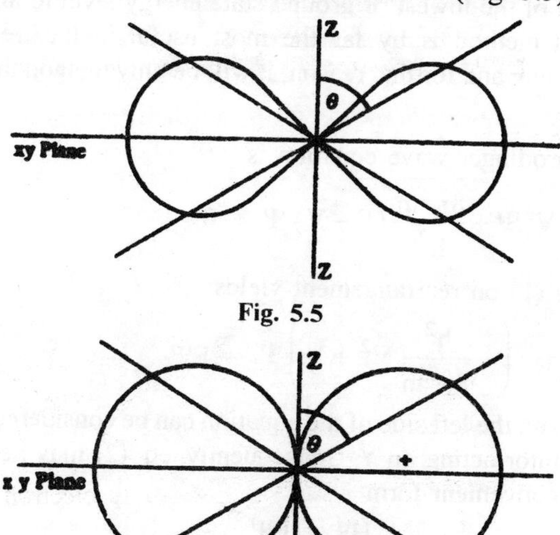

Fig. 5.5

Fig. 5.6

2. A polar graph is also drawn by plotting $(T_\theta)^2$ against different values of θ for given values of l and m. This graph is a map of the angular distribution of electron density (Fig. 5.6).

WAVE MECHANICS AND VALENCE

Introduction : Although, a simple electronic model is adequate for the treatment of an ionic bond, it is necessary to use a quantum mechanical approach to the covalent bond. This requires a solution of the Schrodinger wave equation. An exact solution of Schrodinger equation for the hydrogen atom is possible only when the system contains a single electron.

If the system contains several electrons, the solution of the approximate Schrodinger equation for a polyelectron system is extremely difficult. In such a case we have to take recourse to some approximate producers. There are two rather common approximation methods used in wave mechanics. These are the variation method and perturbation method.

The Variation Method

According to the variation method, *the wave function which gives the lowest energy of the system is closest to the accurate have function of the system.* This method has been mainly used for the approximate determination of the lowest or ground state energy level to any system. The variation method is by far the most useful in the treatment of chemical bonding and for that reason, it will be only method that we will discuss here.

This Schrodinger wave equation is

$$\nabla^2\Psi + \frac{8\pi^2 m}{h^2}(E - V)\Psi = 0 \qquad ...(1)$$

Equation (1) on rearrangement yields

$$\left(-\frac{h^2}{8\pi^2 m}\nabla^2 + V\right)\Psi = E\Psi \qquad ...(2)$$

The term on the left side of the equation can be considered, together, to be an operator acting on Y. Consequently, eq. (2) may be expressed in the more convenient form

$$H\Psi = E\Psi \qquad ...(3)$$

where H represents this operator and is known as Hamiltonian operator. If we multiply both sides of equation (3) by Ψ^* and integrate over the configuration space, we will obtain the expression

$$\int \Psi^* H\Psi \, d\tau = E \int \Psi\Psi^* \tau \qquad ...(4)$$

$$\Rightarrow \; E = \frac{\int \Psi^* H\Psi \, d\tau}{\int \Psi\Psi^* \, d\tau} \qquad ...(5)$$

Application of Variation Method

We shall now apply this principle in calculating the energy of a hydrogen atom.

Hydrogen Atom

We have already stated that in selecting the proper wave function use must be made of some of the physical conditions. In the case of the hydrogen atom we are to assume that

1. The function is radially symmetrical.
2. The electron is inside the atom and so the function rapidly dies away as its distance from the nucleus increases, and
3. Positive charge of the nucleus attracts the electrons and hence the function expressing the probability of the electron has the maximum value at the nucleus.

From assumptions (2) and (3), it is evident that the function is of the type Ψ (r) = e^{-ct}, where c is a constant.

We have $\quad H\Psi = E\Psi$

and $\quad \Psi H\Psi = E\Psi^2$

or
$$E = \frac{\int \Psi H \Psi \, d\tau}{\int \Psi^2 \, d\tau} \qquad ...(6)$$

Here the function is not normilised. For the H-atom,

$$H = -\frac{h^2}{8\pi^2 m}\nabla^2 - \frac{e^2}{r} \qquad ...(7)$$

Again since the postulated function is radially symmetrical, it is independent of θ and ϕ and hence.

$$\nabla^2 = \frac{1}{r^2}\frac{d}{dr}\left(r^2 \frac{d}{dr}\right) \qquad ...(8)$$

or
$$H\Psi = \left(-\frac{h^2}{8\pi^2 m}\nabla^2 - \frac{e^2}{r}\right)\Psi \qquad ...(9)$$

Let $\Psi = e^{-er}$ be the trial wave function

$$\therefore \qquad H\Psi = \left(-\frac{h}{8\pi^2 m}\nabla^2 - \frac{e^2}{r}\right)e^{-cr} \qquad ...(10)$$

But $$\nabla^2 e{-}cr = \frac{1}{r^2}\frac{d}{dr}\left(r^2 \frac{d}{dr}e^{-cr}\right)$$

[From equation (8)] ...(11)

$$\nabla^2 e^{-cr} = \frac{1}{r^2}\frac{d}{dr}\left[r^2(-c)e^{-cr}\right]$$

$$= +\frac{c^2}{r^2}e^{-cr}r^2 - \frac{2cr}{r^2}e^{-cr} \qquad ...(12)$$

$$= c^2 e^{-cr} - \frac{2cr}{r^2}e^{-cr} \qquad ...(13)$$

Now $$E = \frac{\int \Psi H \Psi d\tau}{\int \Psi^2 d\tau} = \frac{\int e^{-cr}He^{-cr}d\tau}{\int e^{-2cr}d\tau} \qquad ...(14)$$

$$E = \frac{\int e^{-cr}\left\{-\frac{h^2c^2}{8\pi^2 m}e^{-cr} + \frac{2h^2c\,e^{-cr}}{8\pi^2 mr} - \frac{e^2}{r}e^{-cr}\right\}d\tau}{\int e^{-2cr}d\tau} \qquad ...(15)$$

Again dt = r^2 since ϕ and θ do not come in for a radially symmetrical function.

$$E = \frac{-\frac{h^2c^2}{8\pi^2 m}\int e^{-2cr}r^2\,dr + \frac{2h^2c}{8\pi^2 m}\int c^{-2cr}\,rdr - e^2\int e^{-2cr}\,r\,dr}{\int e^{-2cr}r^2 dr} \qquad ...(16)$$

$$\Rightarrow \qquad E = \frac{-\frac{h^2c^2}{8\pi^2 m}\frac{2!}{(2c)^3} + \frac{2h^2}{8\pi^2 m}\frac{1!}{(2c)^2} - e^2\frac{1!}{(2c)^2}}{2!(2c)^3}$$

$$= \left\{-\frac{h^2c^2}{8\pi^2 m}\frac{2}{(2c)^3} + \frac{2h^2c}{8\pi^2 m(2c)^2} - \frac{e^2}{(2c)^2}\right\}\frac{(2c)^3}{2}$$

$$= -\frac{h^2c^2}{8\pi^2 m} + \frac{2h^2c^2}{8\pi^2 m} - e^2c$$

$$= \left(\frac{h^2c^2}{8\pi^2 m} - e^2c\right) \qquad ...(17)$$

At this stage will apply the variation principle. If the right wave function has been picked up, then dE/dc will be = 0. E is minimum.

$$\Rightarrow \quad \frac{dE}{dc} = \frac{d}{dc}\left(\frac{h^2c^2}{8\pi^2 m} e^2 c\right) \qquad ...(18)$$

$$\Rightarrow \quad = \frac{2ch^2}{8\pi^2 m} - e^2 = 0$$

$$\Rightarrow \quad c = \frac{8\pi^2 me^2}{2h^2} = \frac{4\pi^2 me^3}{h^2} = \frac{1}{a_0} \qquad ...(19)$$

and hence E = $\Psi^{-r/a0} = 0$

$$\text{and} \qquad E = \frac{h^2}{8\pi^2 m}\left(\frac{4\pi^2 me^2}{h^2}\right)h^2 - \left(\frac{4\pi^2 me^2}{h^2}\right)$$

$$\Rightarrow \quad E = \frac{2\pi^2 me^4}{h^2} - \frac{4\pi^2 me^4}{h^2} = -\frac{2\pi^2 me^4}{h^2} \qquad ...(20)$$

which is the ground energy state when n = 1.

Again we may assume $\Psi = e^{cr^2}$, which is also spherically symmetrical and varying exponentially with r. But it yields an energy value slightly greater than $\Psi = e^{-cr}$, the exact one.

Greater flexibility can, however, be introduced by representing $\Psi = e^{-cr^2}$ not knowing what power of r is actually involved. In such a case there are two adjustable parameters and the variation principle has to be applied to each case.

$$\frac{dE}{dc} = 0 \text{ and } \frac{dE}{dn} = 0 \qquad ...(21)$$

THE PERTURBATION METHOD

This perturbation principle is another way for approximate solution of the problem. This method is based on the principle that the actual problem can be treated as a slight modification or perturbation of another problem capable of exact solution. If HY= Ef represents the actual Schrodinger equation to be solved then it is assumed that an exact solution is available for $H_0Y_0 = E_0Y_0$, where H_0 differs only slightly from the Hamiltonian H for the actual problem.

This method may be illustrated by taking the state of any system (say that of H-atom) characterised by the wave function Ψ. When acted on by external electrical field its electronic state will be disturbed. Let the wave function of the perturbed state by $\Psi + \phi$.

The Hamiltonian will also change slightly due to the change in potential energy. Let H be the Hamiltonian in the unperturbed state and (H′ + H) that of the perturbed state. H′ is called the Schrodinger equation they must form an orthogonal set, so f must be represented as

$$\phi = c_1\Psi_1 + c_2\Psi_2 + ... \quad ..(1)$$

where c′ s and Ψ_1, etc., are very small quantities,

Let us consider a perturbed state of the system Ψ. The wave function in the perturbed state is given by Ψ_2, + ϕ, *i.e.*,

$$\Psi = \Psi_2 + c_2\Psi_1 \quad c_3\Psi_3 \quad ...(2)$$

Here $c_2\Psi_2$ does not appear in the perturbed stated of Ψ_2 because $c_2\Psi_2$ makes no contribution. By Schrodinger equation $H\Psi = E\Psi$.

In the present case $(H + H')\phi = E\Psi$, where E is energy in he perturbed state.

$$\Rightarrow \quad (H + H')\,\Psi_2 + c_1\Psi_1 + c_3\Psi_3) = E\,(\Psi_2 + c_1\Psi_1 + c_3\Psi_3)$$

$$\Rightarrow \quad H\Psi_2 + c_1H\Psi_1 + c_3H\Psi_3 + H'\Psi_2 + H'c_1\Psi_1 + H'c_3\Psi_3$$

$$\Rightarrow \quad E\Psi_2 + Ec_1\Psi_1 + Ec_3\Psi_3 \quad ...(3)$$

Since c_1, c_3, and H′ are small quantities, the terms

$H'c_1\Psi_1$ and $H'c_3\Psi_3$ are neglected and we get

$$E\Psi_2 - H\Psi_2 + Ec_1\Psi_1 - c_1H\Psi_1 + Ec_3\Psi_3 - c_3H\Psi_3 = H'\Psi_2 \quad ...(4)$$

Multiplying by Ψ_2 and integrating on over the entire configurations space, we get

$$\int\Psi_2\,(E - H)\,\Psi_2 d\tau + c_1 \int\Psi_2\,(E - H)\,\Psi_2 d\tau$$

$$\Rightarrow \quad + c_3 \int\Psi_2\,(E - H)\,\Psi_3 d\tau = \int\Psi_2\,H'\Psi_2 d\tau \quad ...(5)$$

$$\Rightarrow \quad \int\Psi_2\,E\Psi_2 d\tau - \int\Psi_2 H\Psi_2\,d\tau = \int\Psi_2\,H'\Psi_2 d\tau \quad ...(6)$$

$$\Rightarrow \quad \int\Psi_2\,E\Psi_2 d\tau = \int\Psi_2\,H\Psi_2\,d\tau + \int\Psi_2 H'\Psi_2\,d\tau$$

$$\Rightarrow \quad E = E_0 + \int\Psi_2\,H'\Psi_2\,d\tau \quad ...(7)$$

where E_0 is the energy of the unperturbed state.

$H' = eEX$ where e is the electronic charge, E is strength of the electric field and X the coordinate of the system in the direction of the field.

$$\therefore \quad E = E_0 + E_e \int \Psi_2 X' d\tau \qquad ...(8)$$

All the terms of the right hand side can be known and hence E can also be found out.

The wave function of the perturbed state is

$$(E - H - \Psi_2 + c_1 (E - H) \Psi_1 + c_3 (E - H) \Psi_3 = H'\Psi_2$$

$$\Rightarrow \int \Psi_1 (E - H) \Psi_2 d\tau + c_1 \int \Psi_1 (E - H) \Psi_1 d\tau + c_3 \int \Psi_1 (E - H) \Psi_3 d\tau$$

$$= \int \Psi_1 H' \Psi_2 \, d\tau$$

$$\Rightarrow \quad c_1 \int \Psi_1 E \Psi_1 \, d\tau - c_3 \int \Psi_1 H \Psi_1 \, d\tau = \int \Psi_1 H' \Psi_2 \, d\tau$$

$$\Rightarrow \quad c_1 E + c_1 \int \Psi_1 H \Psi_1 \, d\tau = \int \Psi_1 H' \Psi_2 \, d\tau .$$

All the terms being known, c1 can be found out. Similarly, multiplying eq. (3) by Y_3, c_3 can be known

$\therefore \quad \Psi = \Psi_2 + c_1\Psi_1 + c_3\Psi_3$ can be known since Ψ_1, Ψ_2, Ψ_3 can be easily known.

The application of this method has been used in connection with the energy change involved when an atom in a molecule is replaced by another atom, as in butadene C = C – C = C. If this be changed into C = N – C = C, then the energy state at C_2 atom will be disturbed or perturbed. From the amount of perturbation, energy can be calculated.

THE SECULAR EQUATIONS

In many examples, it is generally desirable to express the variation function in term of a set of functions, ϕ. These functions are of class Q and are generally normilised. Thus,

$$\Psi = a_1\Psi_1 + a_2\Psi_2 + ... + a_n\phi_n$$

where a_2, a_2, ... an are arbitrary parameters, their values can be varied to give a minimum in the energy. For sake of simplicity, we will consider the variation functions in terms of two functions only,

i.e., $\quad \Psi = a_1\Psi_1 + a_3\phi_3$

From the variation method, we know

$$E = \frac{\int \Psi^* H \Psi \, d\tau}{\int \Psi \Psi^* \, d\tau}$$

Substituting equation (1) into (2), we get ...(1)

$$E = \frac{\int (a_1\phi_1{}^* + a_2\phi_2{}^*) H (a_1\phi_1 + a_2\phi_3) \, d\tau}{\int (a_1\phi_1{}^* a_2\phi_2{}^*)(a_1\phi_1 + a_2\phi_2) \, d\tau} \quad ...(2)$$

$$\Rightarrow \quad E \int (a_1\phi_1{}^* + a_2\phi_2{}^*)(a_1\phi_1 + a_2\phi_2) \, d\tau$$

$$= \int (a_1\phi_1{}^* + a_2\phi_2{}^*)(a_1\phi_1 + a_2\phi_2) \, d\tau$$

$$E \left(a_1{}^2 \int \phi_1{}^*\phi_1 \, d\tau + 2a_1a_2 \int \phi_1{}^*\phi_2 \, d\tau + 2a_2{}^2 \int \phi_2{}^*\phi_2 \, d\tau\right)$$

$$= a_1{}^2 \int \phi_1{}^* H\phi_1 \, d\tau + 2a_1a_2 \int \phi_1{}^* H\phi_2 \, d\tau + a_2{}^2 \int \phi_2{}^* H\phi_2 \, d\tau$$

Fig. 3.10...(3)

In order to get the minimum value of E, it is required to minimise E with respect to both a1 and a_3. For this, differentiation is used with respect to both a_1 and a_3. Differentiating eq. (3) with respect to a_1, we get

$$E = \left[2a_1 \int \phi_1{}^*\phi_1 \, d\tau + 2a_2 \int \phi_1{}^*\phi_2 \, d\tau\right]$$

$$+ \frac{\partial E}{\partial a_1}\left[a_1{}^2 \int \phi_1{}^*\phi_1 \, d\tau + 2a_1a_2 \int \phi_1{}^*\phi_2 \, d\tau + a_2{}^2 \int \phi_2{}^*\phi_2 \, d\tau\right]$$

$$+ a_2{}^2 \int \phi_2{}^*\phi_2 \, d\tau\Big] = 2a_1 \int \phi_2{}^* H\phi_1 \, d\tau + 2a_2 \int \phi_1{}^* H\phi_2 \, d\tau$$

...(4)

If we differentiate equation (3) with respect to a_2, we get an equivalent equation to (4). According to the rules of differential calculus, the minimum value of E with respect to a_1 and a_2 is obtained by putting

$$\left(\frac{\partial E}{\partial a_1}\right)_{a_2} = \left(\frac{\partial E}{\partial a_2}\right)_{a_1} = 0 \quad ...(5)$$

For the sake of simplicity, it is desirable to introduce the symbolism

$$H_{IJ} = \int \phi_I{}^* H\phi_J \, d\tau \ \text{a}$$

$$\text{nd SIJ} = \int \phi_I{}^* \phi_J \, d\tau \quad ...(6)$$

By applying the condition of equation (5), *i.e.*, $\partial E/\partial a_1 = 0$ to equation (4), we obtain an equation which on substitution according to equation (6) yields a new equation.

$$(H_{11} - ES_{11})\, a_1 + (H_{12} - ES_{12})\, a_2 = 0 \quad ...(7)$$

The equivalent equation will be obtained by applying the condition

$\partial E/\partial a_2 = 0$, *i.e.*,

$$(H_{21} - ES_{21})\, a_1 + (H_{22} - ES_{22})\, a_2 = 0 \qquad ...(8)$$

Equations (7) and (8) are known as *secular equations.* These equations are of the form

$$ax + by = 0$$

$$cx + dy = 0$$

The above two equations can be expressed in the determinatal form as

$$\begin{vmatrix} a & b \\ c & d \end{vmatrix} = 0$$

Obviously, the above condition can be applied to equations (7) and (8), we get

$$\begin{vmatrix} H_{11} - ES_{11} & H_{12} - ES_{12} \\ H_{21} - ES_{11} & H_{22} - ES_{22} \end{vmatrix} = 0 \qquad ...(9)$$

In order to express Ψ in terms of n independent terms, the secular determinant of equation (9) becomes as

$$\begin{vmatrix} H_{11} - ES_{11} & H_{12} - ES_{12} ... H_{1^n} - ES_{1^n} \\ H_{21} - ES_{21} & H_{22} - ES_{22} ... H_{2^n} - ES_{2^n} \end{vmatrix} = 0$$

THE VALENCE BOND THEORY

The valence bond theory was put forward by Heitler and London in 1927 to explain the stability of a covalently bonded molecule. This theory is based on the following two principles.

Principle I: If two unconnected systems can be separately described by wave function Ψ_n and Ψ_b respectively, then the wave function for the combined system will be

$$\Psi = \Psi_a \Psi_b \qquad ...(1)$$

Principle II: If Ψ_1, Ψ_2, ... Ψ_n are several wave functions for the same system, then the true wave function Ψ, is obtained by combining the several wave functions linearly, *i.e.*,

$$\Psi = c_1\Psi_1 + c_2\Psi_2 + c_3\Psi_3 + ...\ c_n\Psi_n \qquad ...(2)$$

where c_1, c_2, c_3, ... cn are various coefficients.

Equation (2) is based on the law of linear combination.

From the above two principles, we shall deduce the formation of hydrogen molecule. Formation of Hydrogen Molecule

The hydrogen molecule consists of two positively charged nuclei, a and b and two electrons, 1 and 2 as shown in Fig. 5.7.

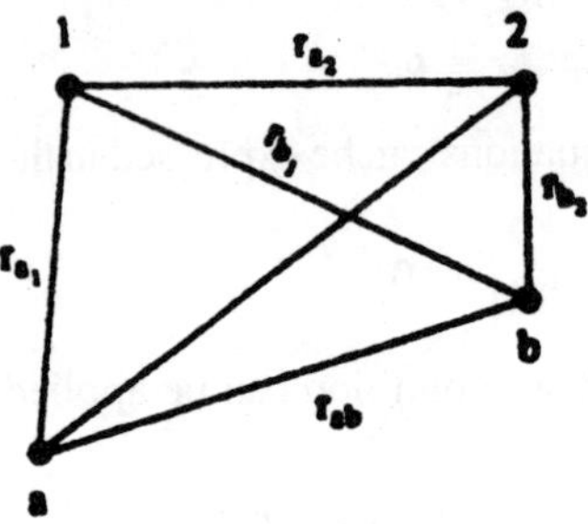

Fig. 5.7

When the two atoms are far apart, their mutual potential energy is nil, *i.e.*, the potential V, is zero. Thus, the system under consideration is equivalent to two separate hydrogen atoms and if these atoms be in their normal state, the appropriate wave functions (excluding spin) will be the hydrogen 1 s wave functions. These orbitals may be represented by $\Psi_{a(1)}$ and $\Psi_{b(2)}$, which implies that electron (1) is associated with nucleus (a) and electron (2) with nucleus (b). Since $\Psi^2_{a(1)}$ and $\Psi^2_{b(2)}$ yield the simultaneous probabilities of electron (1) being in the field of nucleus (a)] and the electron (2) on nucleus (b), then

$$\Rightarrow \quad \Psi_1^2 = \Psi^2_{a(1)}, \Psi^2_{b(2)}$$

$$\Rightarrow \quad \Psi_1 = \Psi_{a(1)} \Psi_{b(2)} \qquad \text{[Applying principle I]} \quad ...(3)$$

Similarly, the simultaneous probabilities of electron (2) one nucleus (a) and electron (1) on (b) may be written as

$$\Rightarrow \quad \Psi^2_{11} = \Psi^2_{a(2)} \Psi^2_{b(1)}$$

$$\Rightarrow \quad \Psi_{11} = \Psi_{a(2)} \Psi_{b(1)} \qquad ...(4)$$

When two atoms of hydrogen are brought together to form a molecule, a reasonably simple wave function would be the product of two 1 s atomic wave functions as presented by equations (3) and (4) respectively. Since electrons (1) and (2) cannot be distinguished, the wave function $\Psi_{a(2)} \Psi_{b(1)}$

is as much acceptable as $\Psi_{a(1)}\ \Psi_{b(2)}$. Presumably the true wave function should have the characteristics of both. One must assume, therefore, that the complete wave function is of the type.

$$\Psi = c_1\Psi_{(I)} + c_2\Psi_{(II)}$$

Substituting equations (3) and (4) in the above equation, we get

$$\Psi = c_1\ \Psi_{a(1)}\ \Psi_{b(2)} + c_2\ \Psi_{a(2)}\ \Psi_{b(1)} \qquad ...(5A)$$

As there are two electrons in the hydrogen molecule, it means that there should be two Laplacian's operators ∇_1^2 and ∇_2^2 and the Schrodinger's wave equation will be given by

$$(\nabla_1^2 + \nabla_2^2)\Psi + \frac{8\pi^2 m}{h^2}(E - V)\Psi$$

Using the symbolism of Fig. 1, the Hamiltonian operator of the hydrogen molecule will be

$$H = \left[-\frac{h^2}{8\pi^2 m}(\nabla_1^2 + \nabla_2^2) - \frac{e^2}{r_{a(1)}} - \frac{e^2}{r_{b(1)}} - \frac{e^2}{r_{a(2)}} - \frac{e^2}{r_{b(2)}} + \frac{e^2}{r_{12}} + \frac{e^2}{r_{ab}}\right]$$

$$\text{where } V = -\frac{e^2}{r_{a(1)}} - \frac{e^2}{r_{b(1)}} - \frac{e^2}{r_{a(2)}} - \frac{e^2}{r_{b(2)}} + \frac{e^2}{r_{12}} + \frac{e^2}{r_{ab}}$$

The integral H11 is now defined

$$H_{11} = \iint \Psi_1 H \Psi_1\ d\tau\ d\tau_2$$

where the double integral results in from a consideration of the co-ordinates of both electrons (1) and (2). On introducing the Hamiltonian operator, the integral becomes

$$H_{11} = \iint \Psi_{a(1)}\Psi_{b(2)}$$

$$\left\{\frac{-h^2}{8\pi^2 m}\left(\nabla_1^2 + \nabla_2^2\right) - \frac{e^2}{r_{a(1)}} - \frac{e^2}{r_{b(2)}} - \frac{e^2}{r_{a(2)}} - \frac{e^2}{r_{b(2)}} + \frac{e^2}{r_{12}} + \frac{e^2}{r_{ab}}\right\}$$

$$\Psi_{a(1)}\ \Psi_{b(2)}\ dt_1\ dt_2$$

Now $$\left(-\frac{h^2}{8\pi^2 m}\nabla_1^2 - \frac{e^2}{r_{b(2)}}\right)\Psi_{a(1)} = E_0\ \Psi_{a(1)}$$

and $$\left(-\frac{h^2}{8\pi^2 m}\nabla_1^2 - \frac{e^2}{r_{a(1)}}\right)\Psi_{b(2)} = E_0\ \Psi_{b(2)}$$

where E_0 is the group state energy of the hydrogen atom. Therefore, the integral H_{12} can be expressed more simply as,

$$H_{11} = \iint \Psi_{a(1)} \Psi_{b(2)} \left(2E_0 + \frac{e^2}{r_{ab}} + \frac{e^2}{r_{12}} + \frac{e^2}{r_{a(2)}} - \frac{e^2}{r_{b(1)}}\right) \Psi_{a(1)} \Psi_{b(2)} d\tau_1 d\tau_2$$

Or since the 1s wave functions are assumed to be normilised

$$H_{11} = 2E_0 + \frac{e^2}{r_{ab}} + J_1 - 2J_2$$

Here, $$J_1 = e^2 \iint \frac{1}{r_{12}} \left[\Psi_{a(1)} \Psi_{b(2)}\right]^2 d\tau_1 \, d\tau_2$$

and because of the equivalence of the two electrons

$$J_2 = e^2 \iint \frac{1}{r_{b(1)}} \left[\Psi_{a(1)} \Psi_{b(2)}\right]^2 d\tau_1 \, d\tau_2 .$$

The solution of the integrals is even more complicated than that for the corresponding integrals of the hydrogen molecules ion, and for this reason they will not be treated further here. However, if will again be useful to consider the basic form of the energy expression.

By the same general argument as was used for H_{11}, if can be shown that

$$H_{12} = 2E_0 S_{12} + \frac{e^2}{r_{ab}} + K_1 - 2K_2$$

where $$K_1 = e^2 \iint \frac{1}{r_{12}} \left[\Psi_{a(1)} \Psi_{b(2)} \Psi_{a(2)} \Psi_{b(1)}\right] d\tau_1 \, d\tau_2$$

and again because of the equivalence of the two electrons.

$$K_2 = e^2 \iint \frac{1}{r_{a(1)}} \left[\Psi_{a(1)} \Psi_{b(2)} \Psi_{a(2)} \Psi_{b(1)}\right] d\tau_1 \, d\tau_2$$

$$= e^2 \iint \frac{1}{r_{b(2)}} \left[\Psi_{a(1)} \Psi_{b(2)} \Psi_{a(2)} \Psi_{b(1)}\right] d\tau_1 \, d\tau_2 .$$

If these results are now substituted into the expression for the energy states of the hydrogen molecule and then rearranged, we can obtain

$$E_S - 2E_0 = \frac{e^2}{r_{ab}} + \frac{J_1 - 2J_2 + K_1 - 2K_2}{1 + S_{12}}$$

and $$E_A - 2E_0 = \frac{e^2}{r_{ab}} + \frac{J_1 - 2J_2 + JK_1 - 2K_2}{1 - S_{12}}$$

which correspond to the symmetric and antisymmetric wave functions respectively.

The potential energy curves resulting from both the symmetric and the antisymmetric states are shown in Fig. 5.8. By setting the ground state energies of the isolated hydrogen atoms equal to zero, *i.e.*, $E_0 = 0$ thus, the resultant potential energy curve represents the interaction energy between the hydrogen atoms as they form the hydrogen molecule.

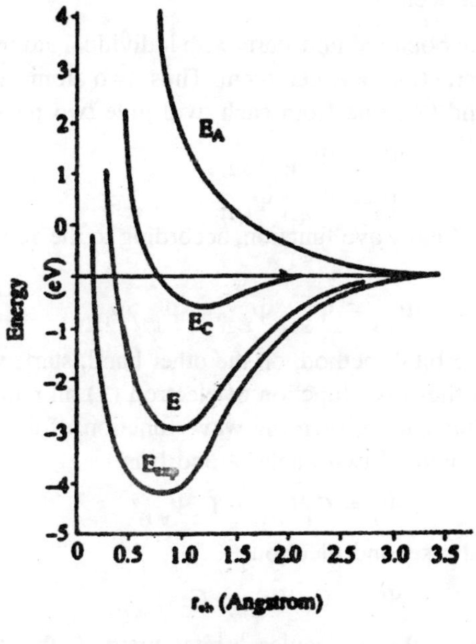

Fig. 5.8

It is significant that the potential minimum using this trial function is considerably closer to the experimental potential minimum than that obtained by ignoring the indistinguishability of the electrons.

The symmetric state exhibits a deep minimum in the curve, thereby indicating a stable molecular species. On the other hand the antisymmetric state fails to exhibit a minimum, and thus it represents an unstable state.

It is important to note that potential minimum using this trial function is very much closer to the experimental potential minimum than that obtained by neglecting the concept by the indistinguishability of the electrons.

MOLECULAR ORBITAL METHOD

The valence bond and molecular orbital methods are the two basic approaches to the quantum chemistry of molecules. Whereas the valence bond theory retains the individuality of the atoms composing a molecule, the molecular orbital theory attempts to construct molecular orbitals in which the individually of the atoms has disappeared completely, *i.e.*, molecular orbitals are polycentric, whereas orbitals in valence bond method are monocentric.

The valence bond method starts with individual atoms and takes into account the interaction between them. Thus, two atoms a and b and two electrons (1) and (2), one from each, will give two possible functions.

$$\Psi_1 = \Psi_{a(1)} \Psi_{b(2)}$$

and

$$\Psi_2 = \Psi_{a(2)} \Psi_{b(1)}.$$

Then, the final wave function, according to the valence bond method, is

$$\Psi_{VB} = \Psi_{a(1)} \Psi_{b(2)} + \Psi_{a(2)})\Psi_{b(1)}$$

The molecular orbital method, on the other hand, starts with two nuclei only. If $\Psi_{a(1)}$ is the wave function of electron (1) on nucleus a and $\Psi_{b(t)}$ that of (1) on nucleus b, then the wave function of the single electron moving in the field of two nuclei a and b is

$$\Psi_1 = C_1\Psi_{a(1)} + C_2\Psi_{b(1)}.$$

Similarly, for the second electron

$$\Psi_2 = C_1 \Psi_{a(2)} + C_2 \Psi_{b(1)}.$$

According to the molecular orbital method, the combined wave function for the two is the product of two wave functions, Ψ_1 and Ψ_1, *i.e.*,

$$\Psi_{MO} = \Psi_1\Psi_2 = [C_1\Psi_{a(1)} + C_2\Psi_{b(1)}] [C_1\Psi_{a(2)} + C_2\Psi_{b(2)}]$$

We will now discuss the application of molecular orbital method to the hydrogen molecule ion.

Hydrogen Molecule Ion

In order to illustrate molecular orbital theory, we will consider the simplest possible case of the hydrogen molecule ion. If one is interested in determining the group state energy of this species, one can construct trial wave function from a linear combination of the 1s orbitals of the hydrogen atoms. Initially it is imagined that the nuclei are separated by an infinite distance leading to the following arrangement.

$$H^- \qquad\qquad H^+$$
$$(a) \qquad\qquad (b)$$

In this case, these is an electron on atom a, and in the ground state the molecular orbital will be represented by the atomic orbital.

$$\Psi_a = \Psi_{1s(a)}.$$

If one assumes that the electron is associated with electron b, then the new situation arises as depicted below.

$$H+ \qquad\qquad H.$$
$$(a) \qquad\qquad (b).$$

The ground state molecular orbital for the above mentioned case will be the atomic orbital

$$\Psi_b = \Psi_{1s(b)}.$$

If the two nuclei are permitted to come together, it would be justified to consider that the resultant one electron molecular orbital will be characteristic of the two atomic orbitals.

This leads to the approximation of linear combination of atomic orbitals (LCAO). Thus, one can say that the one-electron molecular orbital for the first electron is

$$\Psi_1 = a_1\Psi_a + a_2\Psi_b \qquad ...(1)$$

As there is one electron in hydrogen molecule ion, the total wave function

$$\Psi_{MO} = \Psi_1, \textit{ i.e.,}$$
$$\Psi_{MO} = \Psi_1 = a_1\Psi_a + a_2\Psi_b \qquad ...(2)$$

As the molecular orbital of one electron in the hydrogen molecule ion is here represented as a linear combination of two independent terms, the scalar determinant will take the form as given below.

$$\begin{vmatrix} H_{aa} - ES_{aa} & H_{ab} - ES_{ab} \\ H_{ba} - ES_{ba} & H_{bb} - ES_{bb} \end{vmatrix} = 0.$$

As the hydrogen atoms, and therefore, the ground state atomic orbitals are identical, it should be apparent that

$$H_{aa} = H_{bb}$$
$$H_{ba} = H_{ab}$$
$$S_{ba} = S_{ab}.$$

Further, if one uses normilised wave functions, $S_{aa} = S_{bb} = 1$. Consequently, the secular determinant will now be reduced to

$$\begin{vmatrix} H_{aa} - E & H_{ba} - ES \\ H_{ba} - ES & H_{aa} - E \end{vmatrix} = 0 \text{ and this leads to the expression,}$$

$$(H_{aa} - E)^2 - (H_{ba} - ES)^2 = 0.$$

If one solves the above expression by means of the quadratic equation, two roots are obtained.

$$E_{Sm} = \frac{H_{aa} + H_{ba}}{I + S} \qquad \text{...(2A)}$$

and
$$E_A = \frac{H_{aa} + H_{ba}}{I - S}$$

where E_S and E_A denote symmetric and antisymmetric energy states, respectively.

Bonding and Antibonding Orbitals

One is aware of the fact that the original 1s energy states of the two hydrogen atoms are degenerate, but on combination, they split into two new energy states, one of lower energy and the other of higher energy than the original 1s states. This is depicted in Fig. 5.9.

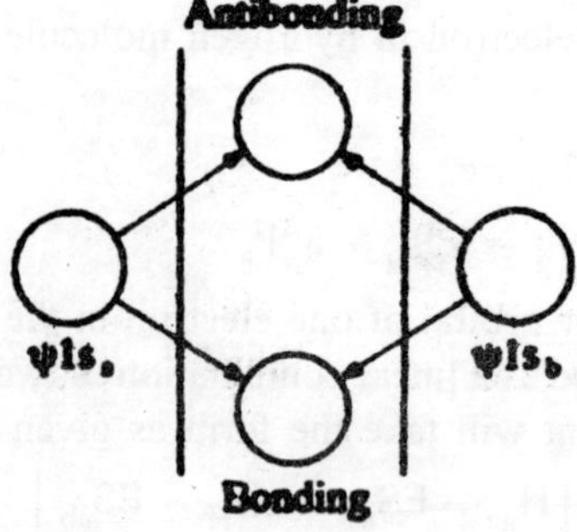

Fig. 5.9

In terms of molecular orbital theory, lower-energy orbital is termed as a *bonding orbital* and the higher energy orbital is called an *antibonding orbital. However*, both orbitals can accommodate, a pair of electrons but the bonding orbital, since it is of lower energy, will be filled first. Thus for the hydrogen molecule ion, the electron will first enter the bonding orbital.

Electron distribution in the hydrogen molecule ion : It is obvious that one gets interested to know the electron distribution in the hydrogen molecule ion for both the symmetric and the antisymmetric states. From equation (2), the wave function for the hydrogen molecule ion is

$$\Psi_{MO} - a_1\Psi_a + a_2\Psi_b$$

As described earlier, the two energy states are obtained, thus one can conclude that two wave functions must exist, *i.e.*, one for the symmetric state and the other for antisymmetric state. These can be determined from the general expression relating a_1 to a_2.

$$(H_{11} - ES_{11})\, a_2 + (H_{21} - ES_{21})\, a_2 - 0.$$

In terms of our present symbolism, this can be put as

$$(H_{aa} - ES_{aa})\, a_1 + (H_{ba} - ES_{ba})\, a_2 = 0.$$

In order to achieve the symmetric solution, it becomes necessary to substitute E_S for E in this equation and for this antisymmetric solution, it is necessary to substitute E_A for E. If this is done, and the resultant expression is solved, it is found, respectively, that $a_1 = a_1$ and $a_1 = - a_2$.

This , then, leads to the molecular orbital wave functions

$$\Psi_S = N_S\,(\Psi_a + \Psi_b)$$

$$\Psi_A = N_A\,(\Psi_a - \Psi_b)$$

where N is a normalising constant.

In order to get the final wave functions, it becomes necessary to carry out the normalisation of Ψ_S are Ψ_A. This can be achieved simply for both the symmetric and antisymmetric functions. In order to illustrate the procedure, one can do it for a symmetric function. On knows that for a normilised function.

$$\Rightarrow \quad \int \Psi_S{}^* \Psi_S{}^* d\tau = 1$$

$$\Rightarrow \quad \int N_S{}^2\, (\Psi_a + \Psi_b)^* d\tau = 1$$

On expansion this expression becomes as

$$N_S{}^2 \left[\int \Psi_a{}^2\, d\tau + 2 \int \Psi_a \Psi_b\, d\tau + \int \Psi_b{}^2\, d\tau \right] = 1 \qquad ...(3)$$

The complex conjugate forms have been neglected here because both Ψ_a and Ψ_b are real. Now if one has originally chosen Ψ_a to be normilised wave unction, it follows that

$$\int \Psi_a^2\, d\tau = \int \Psi_b^2\, d\tau = 1$$

and by definition,

$$\int \Psi_a \Psi_b\, d\tau = S_{ab}.$$

Thus, equation (3) becomes

$$N_S^2 (1 + 2S_{ab} + 1) - 1$$

$$\Rightarrow \qquad N_S = \frac{1}{\sqrt{(2+2S_{ab})}}.$$

In the same manner, it can be proved that

$$N_A = \frac{1}{\sqrt{(2+2S_{ab})}}.$$

This now gives us the normilised wave functions

$$\Psi_S = \frac{1}{\sqrt{(2+2S_{ab})}}(\Psi_a + \Psi_b)$$

and

$$\Psi_A = \frac{1}{\sqrt{(2-2S_{ab})}}(\Psi_a - \Psi_b).$$

For the wave functions one can determine the distribution of electron charge in the molecule, and from the expression for the energy states, one can calculate the molecular energy levels. Considering the charge distribution first, one can see that if S_{ab} is sufficiently near to zero then

$$\Psi_S^2 = \frac{1}{2}[\Psi_a^2 + \Psi_b^2 + 2\Psi_a \Psi_b]$$

and

$$\Psi_A^2 = \frac{1}{2}[\Psi_a^2 + \Psi_b^2 + 2\Psi_a \Psi_b].$$

From the above expression it follows that the symmetric function results in an increase in electron charge density in the region of overlap between two atoms over that of the individual atoms as described by the function Ψ_a^2 and Ψ_b^2 in Fig. 5.10(a). On the other hand, the antisymmetric function is represented graphically in Fig 5.10(b). The dotted lines represent the charge densities of the individual atoms when separated to infinity, and the heavy lines represent the electron charge distribution in the hydrogen molecule ion along a line passing through the nuclei.

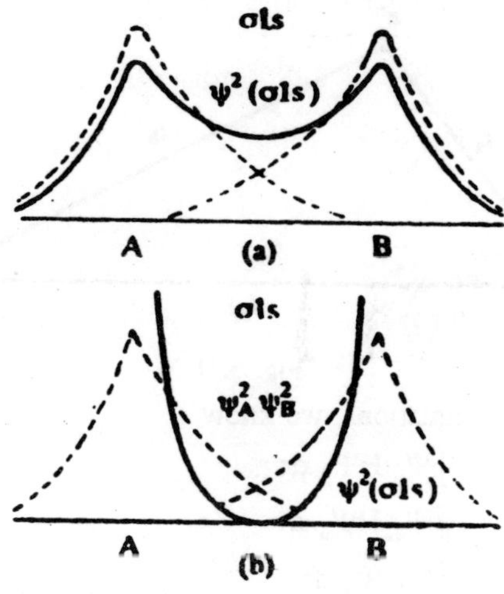

Fig. 5.10

It is evident that the bonding orbital favours a charge distribution which is concentrated between the nuclei, whereas the antibonding orbital tends to decrease the charge density in this region and to concentrate it around the individual atoms. As one considers the formation of a covalent bond to be associated with an electron charge build up in the region of the bond, it means that only the symmetric function would result in the formation of a stable molecule.

Stability of the Hydrogen Molecule Ion

If one assumes that the hydrogen molecules ion is a stable species, one will expect a potential energy diagram as a function of distance of the separation of the two nuclei to show a minimum at some equilibrium separation of the two atoms, a and b. If one could evaluate the expression for the energy of the molecule as a function of the internuclear distance, it becomes possible to draw the potential energy diagram.

Actually, two potential energy diagrams will be obtained. One for bonding orbital and the other for antibonding orbital. In both E_S and E_A, the same integrals will appear but the energies will be different due to the signs of various terms. From Fig. 5.11, it follows that the Hamiltonian operator for the hydrogen molecule ion is

$$H = -\frac{h^2}{8\pi^2 m}\nabla^2 - \frac{e^2}{r_a} - \frac{e^2}{r_b} + \frac{e^2}{r_{ab}}. \quad ...(4)$$

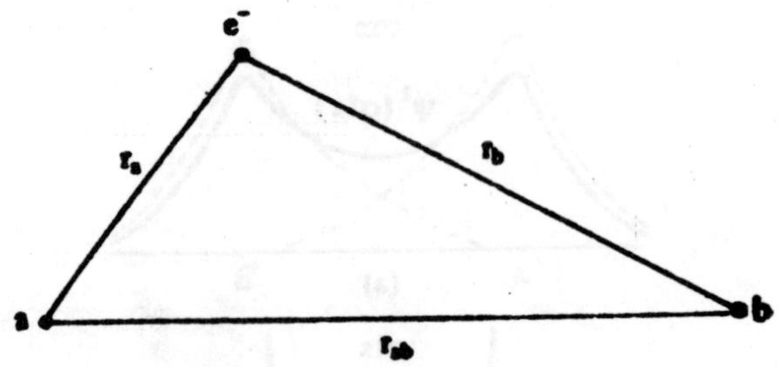

Fig. 5.11

From Secular equations, we know

$$H_{aa} = \int \Psi_a H \Psi_a \, d\tau \qquad ...(5)$$

$$H_{ba} = \int \Psi_b H \Psi_a \, d\tau \qquad ...(6)$$

and

$$S_{ba} = \int \Psi_b \Psi_a \, d\tau \qquad ...(7)$$

Substituting equation (4) into equation (5), we have

$$H_{aa} = \int \Psi_{1Sa} \left[\frac{h^2}{8\pi^2 m} \nabla^2 - \frac{e^2}{r_b} - \frac{e^2}{r_b} + \frac{e^2}{r_{ab}} \right] \Psi_{iSa} \, d\tau \qquad ...(8)$$

$$= \int \Psi_{1Sa} \left[E_0 - \frac{e^2}{r_b} + \frac{e^2}{r_{ab}} \right] \Psi_{iSa} \, d\tau \qquad ...(9)$$

where $E_0 = \frac{-h^2}{8\pi^2 m} \nabla^2 - \frac{e^2}{r_a}$, it is the ground state energy of the hydrogen atom. As E_0 and r_{ab} are both constants, it is possible to remove them from equation (9) under the integral sign, giving

$$H_{aa} = E_0$$

$$\int \Psi_{iSa} \Psi_{iSa} \, d\tau + \frac{e^2}{r_{ab}} \int \Psi_{iSa} \Psi_{iSa} \, d\tau - e^2 \int \frac{1}{r_b} \Psi_{iSa} \Psi_{iSa} \, d\tau \qquad ...(10)$$

As the 1s wave functions are normilised, it shows that

$$\int \Psi_{iSa} \Psi_{iSb} \, d\tau = 1 \qquad ...(11)$$

Substituting equation (11) in (10), we get

$$H_{aa} = E_0 + \frac{e^2}{r_{ab}} - J$$

where J denotes the integral

$$J = e^2 \int \frac{1}{r_b} \Psi_{iSa} \Psi_{iSb} \, d\tau \quad ...(12)$$

Similarly, one can introduce the Hamiltonian operator in the same manner as discussed above, yielding the equation similar, *i.e.*,

$$H_{ba} = E_0$$

$$\int \Psi_{iSa} \Psi_{iSb} \, d\tau + \frac{e^2}{r_{ab}} \int \Psi_{iSa} \Psi_{iSb} \, d\tau - e^2 \int \frac{1}{r_b} \Psi_{iSa} \Psi_{iSb} \, d\tau \quad ...(13)$$

We have

$$S_{ba} = \int \Psi_b \Psi_a \, d\tau \quad ...(14)$$

Substituting equation (14) in (13), we get

$$H_{ba} = E_0 \, S_{ab} + \frac{e^2}{r_{ab}} S_{ab} - K \quad ...(15)$$

where K denoted the integral

$$K = e^2 \int \frac{1}{r_b} \Psi_{iSa} \Psi_{iSb} \, d\tau \quad ...(16)$$

Both integrals J and K are difficult to evaluate, but these can be helpful to see how these will affect the energy of the molecule.

Substituting equation (11) and (15), we obtain for the symmetric state.

$$E_S = \frac{H_{aa} + H_{ba}}{1 + S_{ba}}$$

$$= \frac{E_0 + \frac{e^2}{r_{ab}} - J + E_0 \, S_{ba} + \frac{e^2}{r_{ab}} S_{ba} - K}{1 + S_{ba}}$$

or

$$E_S - E_0 = \frac{e^2}{r_{ab}} - \frac{J - K}{1 + S_{ba}} \quad ...(17)$$

For the antisymmetric states it is found that the corresponding equation is

$$E_A - E_0 = \frac{e^2}{r_{ab}} - \frac{J - K}{1 - S_{ba}} \quad ...(18)$$

Both the equation (17) and (18) can be represented by means of a potential energy diagram such that as shown in (Fig. 5.12). Here the antisymmetric state is seen to correspond to an unstable energy state and if the electron were in the antisymmetrical orbital, one would draw the conclusion that the hydrogen molecule ion would be an unstable species. On the other hand, the symmetric energy state leads to a potential minimum and, therefore, a stable molecular species.

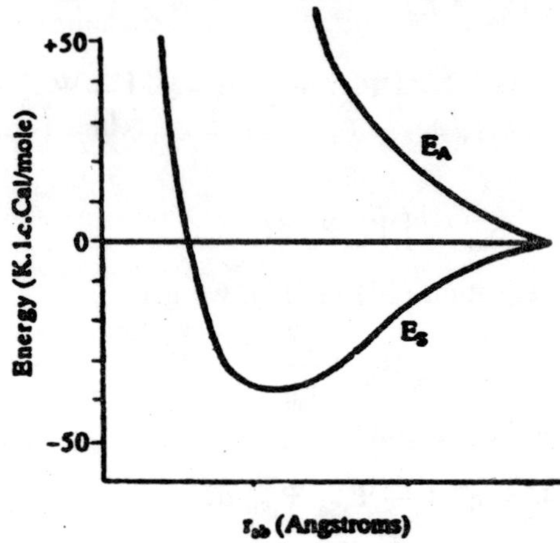

Fig. 5.12 : Potential energy diagram for the hydrogen molecule ion.

By using the particular wave function which we have chosen, the potential energy has been found to be 1.76 eV and equilibrium separation of 1.32 A°. Experimentally, the value of dissociation energy is 22.791 eV and equilibrium separation is 1.06 A°. Thus, our calculations indicate that the hydrogen molecule ion is a stable species.

HARMONIC OSCILLATOR

Many systems of interest can be approximated by harmonic oscillator for example, the vibrations of a diatomic molecule and the motions of atoms in a crystal lattice can be treated to a first approximation as motion can be treated to a first approximation as motion of a particle in a harmonic field. The linear harmonic oscillator are of great importance in connections with the study of molecular spectra. A harmonic oscillator is a particle of mass m moving in a straight line say, along the x-axis subject to a potential, $V = \frac{1}{2}kx^2$, so that the force on the particle is – kx. According to second law of motion.]

$$m = \frac{d^2x}{dt^2} = -kx$$

$$\therefore \qquad \frac{d^2x}{dt^2} = -\left(\frac{k}{m}\right)x = -\omega^2 x \qquad ...(1)$$

where $\qquad w = \sqrt{\left(\frac{k}{m}\right)}$.

The general solution for this differential equation is

$$x = A \sin(\omega t - \theta)$$

where A and θ are constants, which can be determined from the initial conditions. Equations (1) shows that the particle performs simple harmonic oscillations with frequency.

$$V = \frac{\omega}{2\pi} = \frac{1}{2\pi}\sqrt{\left(\frac{k}{m}\right)} \qquad ...(2)$$

where w refers to angular frequency of oscillations.

$$V = \int_0 dV = \int_0^x \frac{dV}{dx}\,dx$$

But $\qquad \frac{dV}{dx} = -F = -(-kx) = kx$

$$\therefore \qquad V = \int_0^x kx\,dx = \frac{1}{2}kx^2 = \frac{1}{2}m\omega^2x^2 \qquad ...(3)$$

The kinetic energy K at the displacement x is given by,

$$K = \frac{1}{2}m\left(\frac{dx}{dt}\right)^2 = \frac{1}{2}mA^2\omega^2\cos^2(\omega t + \theta)$$

$$= \frac{1}{2}m\omega^2A^2\,[1 - \sin^2(\omega t + \theta)]$$

$$= \frac{1}{2}m\omega^2(A^2 - x^2) + \frac{1}{2}m\omega^2x^2 \qquad ...(4)$$

According to classical mechanics the total energy E of the oscillator is given by

$$E = K + V$$

$$= \frac{1}{2}m\omega^2(A^2 - x^2) + \frac{1}{2}m\omega^2x^2$$

$$= \frac{1}{2}m\omega^2A^2 \qquad ...(5)$$

Equation (3) shows that the potential energy has a parabolic form. It means that the particle moves is a potential well of parabolic form.

Equation (5) shows that the total energy of a particle oscillating with a constant frequency $V = \dfrac{\omega}{2\pi}$ is proportional to the square of the amplitude. According to classical mechanics, the particle can oscillate with any amplitude, and there fore, the total energy increases, continuously with increase in the amplitude.

ONE DIMENSIONAL SIMPLE HARMONIC OSCILLATOR IN QUANTUM MECHANICS

Wave Equation for the Oscillator

The time independent Schrodinger wave equation for linear motion of a particle along the x-axis is given as,

$$-\frac{h^2}{2m}\frac{d^2\Psi}{dx^2} + V\Psi = E\Psi$$

or $$\frac{d^2\Psi}{dx^2} + \frac{2m}{h^2}(E - V)\Psi = 0 \qquad ...(1)$$

where E is the total energy of the particle, V is the potential energy, Y is the wave function for the particle, which is the function of x alone and $h = h/2\pi$.

For a linear oscillator with an angular frequency w under a restoring force proportional to the displacement x (along x axis), the potential energy is given by

$$V = \frac{1}{2}m\omega^2x^2 \qquad ...(2)$$

Substituting the value of V in equation (1) we get

$$\frac{d^2\Psi}{dx^2} + \frac{2m}{h^2}\left(E - \frac{1}{2}m\omega^2x^2\right)\Psi = 0 \qquad ...(3)$$

or $$\frac{d^2\Psi}{dx^2} + \left(\frac{2mE}{h^2} - \frac{m^2\omega^2}{h^2}x^2\right)\Psi = 0 \qquad ...(4)$$

Thus is known as Schrodinger wave equation for the given oscillator.

In order to simplify equation (4), consider a dimensionless independent variable y which is related to x by the equation,

$$y = \left(\sqrt{\frac{m\omega}{h}}\right)x \qquad ...(5)$$

so that $x = \left(\sqrt{\frac{h}{m\omega}}\right) y$

we know that,

$$\frac{d\Psi}{dx} = \frac{d\Psi}{dy}\frac{d\Psi}{dx} = \frac{d\Psi}{dy}\sqrt{\frac{m\omega}{h}}$$

and $$\frac{d^2\Psi}{dx^2} = \frac{d^2\Psi}{dy^2}\cdot\frac{dy}{dx}\sqrt{\frac{m\omega}{h}} = \frac{d^2\Psi}{dy^2}\frac{m\omega}{h} = \frac{m\omega}{h}\frac{d^2\Psi}{dy^2} \quad ...(6)$$

Substituting the values of $\frac{d^2\Psi}{dx^2}$ and x^2 is equation (4) we get

$$\frac{m\omega}{h}\frac{d^2\Psi}{dy^2} + \left(\frac{2mE}{h^2} - \frac{m^2\omega^2}{h^2}\cdot\frac{h}{m\omega}y^2\right)\Psi = 0$$

or $$\frac{m\omega}{h}\frac{d^2\Psi}{dy^2} + \left(\frac{2mE}{h^2} - \frac{m\omega}{h}y^2\right)\Psi = 0$$

or $$\frac{m\omega}{h}\frac{d^2\Psi}{h} + \frac{m\omega}{h}\left(\frac{h}{m\omega}\cdot\frac{2mE}{h^2} - y^2\right)\Psi = 0$$

or $$\frac{d^2\Psi}{dy^2} + \left(\frac{2E}{h\omega} - y^2\right)\Psi = 0 \quad ...(7)$$

or $$\frac{d^2\Psi}{dy^2} + \left(\lambda - y^2\right)\Psi = 0 \quad ...(8)$$

where $$l = \frac{2E}{h\omega}$$

Through equation (1) is in a simplified from, it is not easy to solve it. For large values of y, such that $y^2 >> l$, we may neglect l. There fore, equation (7) can be changed to the following form.

$$\frac{d^2\Psi}{dy^2} - y^2\,\Psi = 0 \quad ...(9)$$

For large values of y, the approximate solution of equation (9) is

$$\Psi = e^{-y^2/2} \quad ...(10)$$

[Since $\Psi \to 0$, when $y \to \infty$, we reject the from $e^{+y^2/2}$] If we substitute this equation in equation (9).

we get $$\frac{d^2\Psi}{dv^2} - (y^2 - 1)\,\Psi = 0 \quad ...(11)$$

For large values of y this equation is radical to equation (9). This suggest that an accurate solution of equation (7) must be of the form

$$\Psi = e^{-y^2/2} H(y) \qquad ...(12)$$

where H (y) is a finite polynomial in y.

Change of Dependent Variable from Y to (H (y) in Equation (7).

To change the differential equation (7) in to a differential equation for the dependent variable (H (y) we differentiate equation (12) with respect toy. Therefore.

$$\frac{d\Psi}{dy} = e^{-y^2/2}\frac{dH}{dy} + He^{-y^2/2}(-y)$$

$$\frac{d\Psi}{dy} = \left(\frac{dH}{dy} - Hy\right)e^{-y^2/2}$$

Differentiating the above equation with respect to y, we get

$$\frac{d^2\Psi}{dy^2} = \left(\frac{d^2H}{dy^2} - y\frac{dH}{dy} - H\right)e^{-y^2/2} + \left(\frac{dH}{dy} - Hy\right)e^{-y^2/2}(-y)$$

$$= \left[\frac{dH^2}{dy^2} - 2y\frac{dH}{dy} - H + Hy^2\right]e^{-y^2/2}$$

Now substituting the expression for $\frac{d^2\Psi}{dy^2}$ and the expression for Ψ in equation (7), we get

$$\left[\frac{d^2H}{dy^2} - 2y\frac{dH}{dy} - H + Hy^2\right]e^{-y^2/2} + (\lambda - y^2)He^{-y^2/2} = 0$$

or

$$\frac{d^2H}{dy^2} - 2y\frac{dH}{dy} + (\lambda - 1)\ H = 0 \qquad ...(13)$$

Equation (13) is known as Hermit's differential equation. The solution of equation (13) can be under stood as follows.

Equation (13) is solved by the method of Frobenius According to it, the solution is a power series in positive powers of y. If the lowest power of y is zero, the solution will be of the form.

$$H(y) = A_0 + A_1 y + A_2 y^2 + A_3 y^3 + A_4 y^4 \qquad ...(14)$$

$$n = ¥$$

$$= S\, A_n y_n \qquad \text{...(15)}$$
$$n = 0$$

Differentiating equation (14) with respect to y, we get

$$\frac{dH}{dy}\; A_1 + 2\,A_2\,y + 3A_3y^2 + 4\,A_4\,y^3 \qquad \text{...(16)}$$

Multiplying this equation by (– 2y), we get

$$-\,2y\,\frac{dH}{dy} = -\,2\,(A_1y + 2\,A_2y^2 + 3A_3y + \ldots)$$

$$\begin{aligned} &n = \infty \\ &= \Sigma - Z_nA_ny^n \\ &n = 0 \end{aligned} \qquad \text{...(17)}$$

Differentiating equation (16) with respect to y, we get

$$\frac{d^2H}{dy^2} = 2A_2 + 3 \times 2A_3\;y + 4 \times A_4\;y^2 + \ldots$$

$$\begin{aligned} &n = \infty \\ &= \Sigma\,(n + 2)\,(n + 1)\,A_n + 2y^n \\ &= 0 \end{aligned} \qquad \text{...(18)}$$

Multiplying both sides of equation (15) by (l – 1), we get

$$\begin{aligned} &n = \infty \\ (l - 1)\,H\,(y) &= \Sigma\,(\lambda - 1)\,A_n\,y^n \\ &n = 0 \end{aligned} \qquad \text{...(19)}$$

Substituting the values of equations (17), 18, and (19) in equation (13), we get

$$= \sum_{n=0}^{n=\infty}\left[(n+2)(n+1)\,A_n + 2^{-2nA_n} + (\lambda - 1)\,A_n\right]y^n = 0$$

This equation must be true for all values of y. Therefore, the coefficient of each power of y must vanish separately. Thus,

$$(n + 2)\,(n + 1)\,A_n + 2^{-\,(2n + 1 - \lambda)\,A_n} = 0$$

$$\text{or}\quad A_{n+2} = \frac{(2n + 1 - \lambda)}{(n+2)(n+1)}\,A_n \qquad \text{...(20)}$$

where n = 0, 1, 2, 3, ...

If we now consider the second energy state as shown in Fig,. 2, it is seen that there are three sets (112), (121) and (211) of the quantum numbers, n_x, n_y, and n_z that give the same energy level, $E = 3h^2/4ma^2$. Such level is said to degenerate, and in this particular case tally all the energy levels are degenerate to some degree.

SIMPLE HARMONIC OSCILLATOR

$$\Rightarrow \frac{d^2\Psi}{dx^2}+\left(\frac{2mE}{h^2}-\frac{m^2\omega^2}{h^2}x^2\right)\Psi = 0 \qquad ...(1)$$

This is the Schrodinger wave equation for the oscillator.

Simplification of the Wave Equation

In order to simplify Eq. (1) let us introduce a dimensionless independent variable y which is related to x by the equation.

$$y = \sqrt{\frac{m\omega}{h}}\, x \qquad ...(2)$$

so that $x = \sqrt{\frac{h}{m\omega}}\, y$

Now we have

$$\frac{d\Psi}{dx} = \frac{d\Psi}{dy}\frac{dy}{dx} = \frac{d\Psi}{dy}\sqrt{\frac{m\omega}{h}}$$

$$\frac{d^2\Psi}{dx^2} = \frac{d^2\Psi}{dy^2}\cdot\frac{dy}{dx}\sqrt{\frac{m\omega}{h}} = \frac{d^2\Psi}{dy^2}\frac{m\omega}{h} = \frac{m\omega}{h}\frac{d^2\Psi}{dy^2} \qquad ...(3)$$

Substituting the values of $\frac{d^2\Psi}{dx^2}$ ans x^2 in Eq. (1) we obtain

$$\frac{m\omega}{h}\frac{d^2\Psi}{dy^2}+\left(\frac{2mE}{h^2}-\frac{m^2\omega^2}{h^2}\cdot\frac{h}{m\omega}y^2\right)\Psi = 0$$

$$\Rightarrow \frac{m\omega}{h}\frac{d^2\Psi}{dy^2}+\left(\frac{2mE}{h^2}-\frac{m\omega}{h}y^2\right)\Psi = 0$$

$$\Rightarrow \frac{m\omega}{h}\frac{d^2\Psi}{dy^2}+\frac{m\omega}{h}\left(\frac{h}{m\omega}\cdot\frac{2mE}{h^2}-y^2\right)\Psi = 0$$

$$\Rightarrow \frac{d^2\Psi}{dy^2}+\left(\frac{2E}{h\omega}-y^2\right)\Psi = 0 \qquad ...(4)$$

$$\Rightarrow \quad \frac{d^2\Psi}{dy^2} + (\lambda - y^2)\Psi = 0 \text{ where } l = \frac{2E}{h\omega} \quad ...(5)$$

Though Eq. (4) is in a simplified form, it is not easy to solve it. For large values of y, such that $y^2 >> l$, we may neglect λ. Therefore, Eq. (4) transformed to the form

$$\frac{d^2\Psi}{dy^2} - y^2\Psi = 0 \quad ...(6)$$

For large values of y, the approximate solution of this equation is

$$\Psi = e^{-y^2/2} \quad ...(7)$$

(Since $\Psi \to 0$, when $y \to \infty$, we reject the from $e^{+y^2/2}$)

If we substitute this equation into Eq. (6), we get

$$\frac{d^2\Psi}{dy^2} - (y^2 - 1)\Psi = 0 \quad ...(8)$$

For large values of y this equation is reduced to Eq. (6). This suggests that an accurate solution of Eq. (5) must be of the form

$$\Psi = e^{-y^2/2} H(y) \quad ...(9)$$

where H (y) is a finite polynomial in y.

To change the differential equation (5) into a differential equation for the dependent variable H (y) we differential equation (9) with respect to y, we obtain

$$\frac{d\Psi}{dy} = e^{-y^2/2}\frac{dH}{dy} + He^{-y^2/2}(-y)$$

$$= \left(\frac{dH}{dy} - Hy\right)e^{-y^2/2}$$

Now differentiating this equation with respect to y, we obtain.

$$\frac{d^2\Psi}{dy^2} = \left(\frac{d^2H}{dy^2} - y\frac{dH}{dy} - H\right)e^{-y^2/2} + \left(\frac{dH}{dy} - Hy\right)e^{-y^2/2}(-y)$$

$$= \left[\frac{d^2H}{dy^2} - 2y\frac{dH}{dy} - H + Hy^2\right]e^{-y^2/2}$$

Now substituting the expression for $\frac{d^2\Psi}{dy^2}$ and the expression for Ψ in Eq. (5) we get

$$\left[\frac{d^2H}{dy^2} - 2y\frac{dH}{dy} - H + Hy^2\right]e^{-y^2/2} + (\lambda - y^2)\,He^{-y^2/2} = 0$$

$$\Rightarrow \quad \frac{d^2H}{dy^2} - 2y\frac{dH}{dy} + (\lambda - 1)\;H = 0 \qquad ...(10)$$

This is a well known Hermit's differential equation.

Solution of Hermit's Differential Equation

This equation is solved by the method of Frobenius, according to which is the solution is a power series in positive powers of y. If lowest power of y is zero, the solution will be of the form

$$H\;(y) = A_0 + A_1y + A_2y^2 + A_3y^3 + A_4y^4 \qquad ...(11)$$

$$\sum_{n=0}^{n=\infty} A_n y^n \qquad ...(12)$$

On differentiating Eq. (11) with respect to y, we get

$$\frac{dH}{dy} = A_1 + 2A_2y + 3A_3y^3 + 4A_4y^3 \qquad ...(13)$$

Multiplying this equation by – 2y

$$-2y\;\frac{dH}{dy} = -2\;(A_1y + 2A_2y^3 + 3A_2y^3 + ...)$$

$$= \sum_{n=0}^{n=\infty} -2nA_n y^n \qquad ...(14)$$

On differentiating Eq. (11) with respect to y, we get

$$\frac{d^2H}{dy^2} = 2A_2 + 3.2A_2y + 4.3A_4y^2 + ...$$

$$= \sum_{n=0}^{n=\infty} (n+2)(n+1)\,A_{n+\lambda}y^n \qquad ...(15)$$

Multiplying both the sides of Eq. (12) by (λ – 1), we get

$$(1 - 1)\;H\;(y) = \sum_{n=0}^{n=\infty} (\lambda - 1)\,A_n y^n \qquad ...(16)$$

Substituting Eqs. (14), (15) and (16) in Eq. (10), we get

$$\sum_{n=0}^{n=\infty} [(n+2)(n+1)A_{n+2} - 2nA_n + (\lambda - 1)A_n]y^n = 0$$

$$\Rightarrow \quad \sum_{n=0}^{n=\infty} [(n+2)(n+1)A_{n+2} - (2n+1-\lambda)A_n]y^n = 0$$

This equation must be true for all values of y and, therefore, the coefficient of each power of y must vanish separately.

Hence we have

$$(n + 2)(n + 1)A_{n+2} - (2n + 1 - l)A_n = 0$$

or
$$A_{n+2} = \frac{2n+1-\lambda}{(n+1)(n+2)} A_n \qquad ...(17)$$

where n = 0, 1, 2, 3, ...

This equation is called the recursion formula connecting the coefficients

A_{n+2} and A_n.

In the solution of any second-order differential equation, there must be two arbitrary constants. So we consider A_0 A_1 as the two arbitrary constants and we determine all other coefficients in terms of A_0 and A_1.

On substituting n = 0, 2, 4, 6, .. successively in Eq. (17), we obtain

$$A_2 = \frac{(1-\lambda)}{2!} A_0$$

$$A_4 = \frac{(5-\lambda)}{3 \times 4} A_2 = \frac{(5-\lambda)}{3 \times 4} \frac{(1-\lambda)}{2!} A_0 = \frac{(1-\lambda)(5-\lambda)}{4!} A_2$$

$$A_6 = \frac{(9-y)}{5 \times 6} A_1 = \frac{(9-\lambda)}{5 \times 6} \frac{(1-\lambda)(5-\lambda)}{4!} A_5$$

$$= \frac{(1-\lambda)(5-\lambda)(9-\lambda)}{6!} A_0$$

Substituting n = 1, 3, 5, 7, ... successively in Eq. (17)

$$A_3 = \frac{(3-\lambda)}{2 \times 3} A_1 = \frac{(3-\lambda)}{3!} A_1$$

$$A_5 = \frac{(5-\lambda)}{4 \times 5} A_3 = \frac{(7-\lambda)}{4 \times 5} \frac{(3-\lambda)}{3!} A_1 = \frac{(3-\lambda)(7-\lambda)}{5!} A_1$$

Similarly

$$A_2 = \frac{(3-\lambda)(7-\lambda)(11-\lambda)}{7!} A_1$$

On substituting the value of these coefficients,

$$H(y) = A_0$$

$$\left[1 + \frac{(1-\lambda)}{2!}y^2 + \frac{(1-\lambda)(5-\lambda)}{4!}y^4 + \frac{(1-\lambda)(5-\lambda)(9-\lambda)}{6!}y^4 + ...\right] +$$

A_1

$$\left[y + \frac{(3-\lambda)}{3!}y^3 + \frac{(3-\lambda)(7-\lambda)}{5!}y^5 + \frac{(3-\lambda)(7-\lambda)(11-\lambda)}{7!}y^2 + ...\right] \qquad ...(18)$$

This equation shows that the polynomial H (y) is the sum of two infinite series. This means that if H (y) does not terminate for some value of n, the wave-function

$$\Psi = e^{-y^2/2} H(y)$$

will become infinite as y becomes infinite.

Conditions for Finite Value of the Wave-Function, and Eigen Values of Energy

The wave function will have a finite value at infinity *i.e.*, $\Psi(\infty) = 0$ if H (y) terminates at some value of n. For this to happen the following *two conditions* must be satisfied.

(1) For some value of n, $A_{n+2} = 0$

i.e., $$\frac{2n+1-\lambda}{(n+1)(n+2)} A_n = 0$$

or $$2n + 1 = 1 \qquad ...(19)$$

If in Eq. (19) n is an odd value, the second series in Eq. (18) will terminate at some value of n. But the first series will not terminate. *Hence when n is an odd value, A_0 must be zero.* For example, when n = 5, then 1 = 11 and the second series in Eq. (18) will terminate at y, but the first series will not terminate.

Substituting the value of l as given Eq. (19), we get the eigen-values of the total energy E_n..

$$\frac{2E}{h\omega} = (2n + 1)$$

or $$E_n = \frac{1}{2}(2n+1)\,h\omega$$

$$E_n = \left(n + \frac{1}{2}\right) h\omega \qquad \text{...(20)}$$

Substituting $h = \dfrac{2}{2\pi}$ and $\omega = 2\pi v$, this expression has the form

$$E_n = \left(n + \frac{1}{2}\right) h / v \qquad \text{...(21)}$$

where n = 0, 1, 2, ..., w is the angular frequency and v is the frequency of the classical harmonic oscillator, given by

$$E = \frac{\omega}{2\pi} = \frac{1}{2\pi k}\sqrt{\frac{k}{m}}$$

From Eq. (20) we get the following conclusion.

(1) The lowest energy of the oscillator is obtained by putting n =0 in Eq. (20) and it is

$$E_0 = \frac{1}{2} h\omega \qquad \text{...(22)}$$

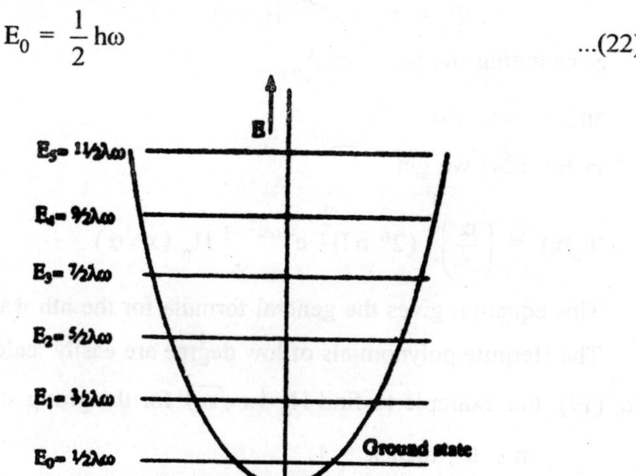

Fig. 5.13

This is called the ground state energy or the *zero point* vibrational energy of the harmonic oscillator. The zero-point energy is the characteristic result of quantum mechanics. The values of E_n in terms of E_0 are given by

$$En = (2n + 1) En \qquad ...(23)$$

where $n = 0, 1, 2, 3,...$

(2) The eigen values of the total energy depend only on one quantum number n. Therefore all the energy-levels of the oscillator are non-degenerate.

(3) The successive energy levels are equally spaced, the separation between two adjacent energy levels being hω. The energy level diagram for the harmonic oscillator is shown in Fig. 5.13.

Wave Functions of the Harmonic Oscillator

When the parameter l is given by

$$l = 2n + 1$$

H (y) is a polynomial of degree n with a constant multiplier. *If multiplier is chosen such that the coefficient of* y^n *is* 2^n, *the polynomial is denoted by* H^n *(y) and is called Hermite's Polynomial.*

The wave-function for the state E_n is given by

$$\Psi_n = A_n\, e^{-y^2/2}\, H_n(y) \qquad ...(24)$$

Substituting the value of A_n

and $y = x\sqrt{\alpha}$

in Eq. (24) we get

$$\Psi_n(x) = \left(\frac{\alpha}{\pi}\right)^{\frac{1}{4}} (2^n\, n!)^{\frac{1}{2}}\, e^{-\alpha x^2/2}\, H_n(x\sqrt{\alpha}) \qquad ...(25)$$

This equation gives the general formula for the nth wave-function.

The Hermite polynomials of low degree are easily' calculated from Eq. (19). For example to find $H_0\left(x\sqrt{\alpha}\right)$ for the group state we have

$$n = 0, \text{ then } \lambda = 1.$$

Now with the even value of n, the coefficient A_1 in Eq.(19) must be zero. Therefore, from Eq. (19)

$$H_0(y) = A_0$$

$$\Rightarrow \; H_0\left(x\sqrt{\alpha}\right) = 2^n = 2^0 = 1$$

Now substituting $n = 0$

and $\qquad H_0\left(x\sqrt{\alpha}\right) = 1$

in Eqn. (29) we get the wave-function for the ground state of the harmonic oscillator,

$$\Psi_0(x) = \left(\frac{\alpha}{\pi}\right)^{\frac{1}{4}} e^{-\alpha x^2/2} \qquad ...(26)$$

Table 5 : The first six Hermite polynomials are given in the following.

n	$\lambda = 2n + 1$	E_n	$H_n(y)$
0	1	$\frac{1}{2}h\omega$	$H_0(y) = 1$
1	3	$\frac{3}{2}h\omega$	$H_1(y) = 2y$
2	5	$\frac{5}{2}h\omega$	$H_2(y) = 4y^2 - 1$
3	7	$\frac{7}{2}h\omega$	$H_3(y) = 8y^2 - 12y$
4	9	$\frac{9}{2}h\omega$	$H_4(y) = 16y^4 - 48y^2 + 12$
5	11	$\frac{11}{2}h\omega$	$H_5(y) = 32y^5 - 160y^3 + 120y$

The first six wave-functions are shown in Fig. 5.14.

Comparison of the classical probability density and quantum mechanical probability density

Acc rding to classical mechanics the probability-density of a linear simple harmonic oscillator, having amplitude A and energy $E_n = \left(n + \frac{1}{2}\right)$ hw, is given by.

$$P(x) = \frac{1}{\pi\sqrt{A^2 - x^2}} = \frac{1}{\pi\sqrt{\frac{2E_n}{m\omega^2} - x^2}}$$

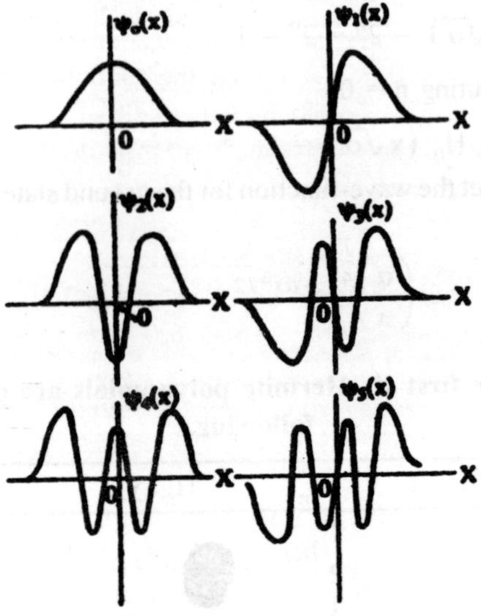

Fig. 5.14

This equation shows that the probability of finding the particle is maximum at $x = \pm A$, *i.e.*, at the extremities of the path, and it is minimum at $x = 0$, *i.e.*, at the equilibrium position. The classical probability – density for $n = 0$ and $n = 10$ is shown by dotted curves in Fig. 5.15(a) and Fig. 5.15(b).

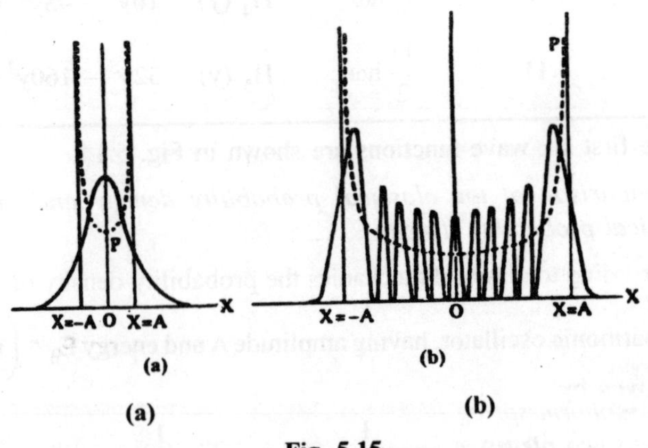

(a) (b)

Fig. 5.15

According to quantum mechanics the probability-density of the particle in the nth energy, state is given by

$$P(x) = \Psi_n(x)\,\Psi^*(x)$$

Calculations show that for n = 0, $|\Psi_0|$ is maximum at x = 0 and decreases on either side of x = 0 along the continuous curve (Fig. 5.1a). For n = 10, $|\Psi_{10}|^2$ has 5 peaks on either side of x = 0 (Fig. 5.1b). This means that the disagreement between two probability densities becomes less and less when n increases.

SOLVED EXAMPLES

Example 1:

The scale of a spring balance reading from 0 to 32 lb is 4.0 in long. A package suspended from the balance is found to oscillate vertically with a frequency of 2.0 oscillations per second. How much does the package weigh?

Solution:

Here, clearly, a mass of 32 *l*b suspended from the spring balance extends the spring by 4 inches or $\frac{4}{12} = \frac{1}{3}$ ft.

∴ *force constant of the spring, i.e.,*

$$C = \frac{\text{force applied}}{\text{extension produced}} = \frac{32\times 32}{1/3} = 32 \times 32 \times 3 \text{ poundals/ft.}$$

Now, *time-period of oscillation of the spring,* $T = 2\pi\sqrt{\frac{m}{C}}$.

$$\therefore\ T^2 = \frac{4\pi^2 m}{C}, \text{ whence, } m = \frac{CT^2}{4\pi^2}.$$

Since C = 32 × 32 × 3 *poundals/ft* and $T = \frac{1}{2}$ sec, we have

$$m = \frac{32\times 32\times 3\times 1/4}{4\pi^2} = \frac{192}{\pi^2} = 19.45\ lb.$$

The package thus weighs 19.45 *l*b.

Example 2:

A uniform spring of normal length l has a force constant C. It is cut into two pieces of lengths l_1 and l_2 such that $l_1 = nl_2$ where an is an integer. What are the force constants C_1 and C_2 of the two pieces respectively in terms of n and C?

Solution:

Here, clearly, $l = l_1 + l_2$.

Or, since $l_1 = nl_2$ and, therefore, $l_2 = l_1/n$, we have

$$l = l_1 + l_1/n = l_1(1 + l/n) = l_1(n + l)/n \qquad ....(i)$$

Also, $\quad l = nl_2 + l_2 = l_2(n + l) \qquad ...(ii)$

Now, for a given spring, its force constant is inversely proportional to its length, *i.e.*, $C \propto 1/l$ or Cl = a constant. We, therefore, have

$$C_1 l_1 = Cl = Cl_1\,(n + 1)/n,$$

whence, $\quad C_1 = C(n + 1)/n,$

Similarly, $\quad C_2 l_2 = Cl = Cl_2(n + 1),$

whence, $\quad C_2 = C(n + 1).$

Thus, the force constants of the two pieces of the spring are respectively $C_1 = C\,(n + 1)/n$ and $C_2 = C(n + 1)$.

Example 3(a):

A spherical Helmholtz resonator of capacity 10 litres has a narrow neck of length 1 cm and radios 1 cm. If the velocity of sound in air be 340 m/sec, calculate the frequency and the wavelength of the note to which it will sharply respond.

Solution:

We know that the frequency of the note to which a spherical resonator responds is given by $n = \left(\frac{v}{2\pi}\right)\sqrt{\frac{\alpha}{Vl}}$, where v is the *velocity of sound in air, l, the length and* α, *the area of cross-section of the neck* and V, the volume (or *capacity*) of the resonator.

Here, $\quad v = 340$ m/sec = 34000 cm/sec,

$$\alpha = \pi\,(1)^2 = \pi \text{ sq cm}, \; l = 1 \text{ cm}$$

and V = 10 litres = 10000 c.c.

$\therefore$ *frequency of the note,* $n = \frac{34000}{2\pi}\sqrt{\frac{\pi}{1000 \times 1}} = \frac{170}{\sqrt{\pi}} = 95.92/\text{sec}.$

And, *wavelength of the note,* $\lambda = \frac{v}{n} = \frac{34000}{95.92} = 345.5$ cm.

Example 3(b):

Calculate the frequency of an L-C circuit in which the inductance and capacitance are respectively 5 millihenry and 2μF. If the maximum potential difference across the capacitor be 10 volts, what is the energy of the system?

Solution:

For the frequency of the oscillations in an L-C circuit, we have the relation

$$n = \frac{1}{2\pi\sqrt{LC}},$$

where L and C are the *inductance* and *capacitance* in the circuit respectively.

Here, L = 5 *milli henry* = 5×10^{-3} henry

and C = 2μF = 2×10^{-6} farad.

∴ *frequency of the L.C circuit,*

$$n = \frac{1}{2\pi\sqrt{5\times10^{-3}\times2\times10^{-6}}} = \frac{10^4}{2\pi} = 1592/\text{sec.}$$

And, with V = 10 volts, *energy of the system*

$$= \frac{1}{2}CV^2 = \frac{1}{2} \times 2 \times 10^{-6} \times (10)^2 = 10^{-4} \text{ joule.}$$

Example 3(c):

(a) What is the reduced mass of each of the following diatomic molecules: O_2, HCl, CO? Express your answers in atomic mass units, the mass of a hydrogen atom being approximately 1.00 amu.

(b) An HCl molecule is known to vibrate at a fundamental frequency of 8.7×10^{13} cycles/sec. What is the effective 'force constant' C for the coupling forces between the atoms?

Solution:

(a) We know that a diatomic molecule behaves like a system of two masses connected by a small massless spring. So that, its reduced mass is given by

$$\mu = \frac{m_1 \times m_2}{m_1 + m_2},$$

where MI and mn are the masses of the two atoms respectively in amu.

Now, in the case of O_2, $m_1 = 16$ amu

and $m_2 = 16$ amu

$$\therefore \qquad \mu = \frac{16\times16}{16+16} = \frac{256}{32} = 8 \text{ amu.}$$

In the case of HCl, $m_1 = 1$ amu

and $m_2 = 35.5$ amu

$$\therefore \qquad \mu = \frac{1\times35.5}{1+35.5} = \frac{35.5}{36.5} = 0.97 \text{ amu.}$$

And, *in the case of* CO, $m_1 = 12$ amu and $m_2 = 16$ amu

$$\therefore \qquad \mu = \frac{12\times16}{12+16} = \frac{192}{28} = 6.857 \text{ amu.}$$

(b) The frequency of a two-body harmonic oscillator, as we know, is given by

$$n = \frac{1}{2\pi}\sqrt{\frac{C}{\mu}},$$

where C is the '*force constant*' for the coupling forces and μ, the *reduced mass of the system.*

Here, the reduced mass μ for the HCl molecule = 0.97 amu [See above under (a)], *i.e.*, equal to 0.97 × *mass of a hydrogen atom*

$$n = 8.7 \times 10^{23} \text{ cycles/sec}$$

So that, $$8.7\times10^{23} = \frac{1}{2\pi}\sqrt{\frac{C}{0.97\times1.67\times10^{-24}}},$$

whence, $C = 4\pi^2 (8.7 \times 10^{13})^2 \times 0.97 \times 1.67 \times 10^{-24}$

$= 485200$ dyne/cm $= 485$ N/m.

Example 4:

If a particle moves in a potential energy field $U = U_0 - ax + bx^2$, where a and b are positive constants, obtain an expression for the force acting on it as a function of position. At what point does the force vanish? Is this a point of stable equilibrium?

Calculate the force constant, time-period and frequency of the particle.

Solution:

(i) As we know, *force acting on the particle* is given by

$$F = -\frac{dU}{dx} = -\frac{d}{dx}\left(U_0 - ax + bx^2\right),$$

whence, $F = a - 2bx$.

(ii) The force vanishes at the point where

$$\frac{dU}{dx} = 0, \text{ i.e.,}$$

where $a - 2bx = 0$. So that, $2bx = a$

Or, $$x = \frac{a}{2b},$$

which gives the position of the point where the force vanishes.

(iii) We have $\frac{d^2U}{dx^2} = 2b$.

So that, b being *positive,* the point , $x = \frac{a}{2b}$ represents the point of *minimum potential energy* on the energy curve of the particle. *It is, therefore, a point of stable equilibrium.*

(iv) From the expression for force F in (i) above, it is clear that it is a *linear restoring force* and *the force constant C* is therefore equal to 2b.

(v) The *time-period of the particle,*

$$T = 2\pi\sqrt{\frac{1}{\mu}},$$

where, as know, $\mu = \frac{C}{m} = \frac{2b}{m}$.

So that, $$T = 2\pi\sqrt{\frac{1}{\frac{2b}{m}}} = 2\pi\sqrt{\frac{m}{2b}}.$$

And, therefore, *frequency of the particle,*

$$n = \frac{1}{T} = \frac{1}{2\pi}\sqrt{\frac{2b}{m}}.$$

Example 5:

(a) The potential energy of a harmonic oscillator of mass 2kg in its resting position is 5 joules, its total energy is 9 joules and its amplitude, 1cm. Calculate its time-period.

(b) A particle of mass 5gm lies in a potential field $V = (8x^2 + 200)$ ergs/gm. Calculate its time-period.

Solution:

(a) Clearly, the *energy gained by the oscillator in moving from its resting position to its extreme position, i.e., in moving through a distance equal to its amplitude a, is equal to (9 – 5) = 4 joules.*

Since, *energy gained = 1/2 Cx²*, where x is the *displacement* of the particle and C, the *force constant,* we have $\frac{1}{2}C(1)^2 = 4$,

whence, C = 4 × 2 = 8 joules/gm. [∵ x = a = 1 cm].

∴ *time period of the oscillator,*

$$T = 2\pi\sqrt{\frac{m}{C}} = 2\pi\sqrt{\frac{2}{8}} = 2\pi\sqrt{\frac{1}{4}}$$

$$= 2\pi \times \frac{1}{2} = \pi = 3.142 \text{ sec.}$$

(b) Here, P.E. *of the particle, i.e.,* U = mV = $5(8x^2 + 200)$ ergs.

∴ *force acting on the particle,*

$$F = -\frac{dU}{dx} = -5 \times 16x = 80x \text{ dynes.}$$

So that, its *equation of motion* is

$$\frac{md^2x}{dt^2} = -80x,$$

whence, $\frac{d^2x}{dt^2} = -\frac{80x}{m} = -\frac{80x}{5} = -16x = -\mu x.$

Or, $\frac{d^2x}{dt^2} \propto x.$

The particle thus executes a simple harmonic motion and its *time-period* is given by

$$T = 2\pi\sqrt{\frac{1}{\mu}} = 2\pi\sqrt{\frac{1}{16}} = 2\pi \times \frac{1}{4} = \frac{\pi}{2} = 1.571 \text{ sec.}$$

Example 6(a):

(a) Two discs D_1 and D_2 of masses 1 kg and 2 kg respectively are connected to the two ends of a massless spring and placed with the spring vertical and the heavier disc (D_2) resting on the table. Calculate the frequency of oscillation of the system, if the force constant of the spring be 10^6 dynes/cm.

(b) After the system has come to rest, the table is suddenly removed from under disc D_2. Calculate :

(i) the frequency of the system,

(ii) the instantaneous acceleration of the two discs and of the centre of mass of the system.

Solution:

(a) Obviously, here, the system functions as an ordinary *one-body harmonic oscillator*, with the lower end of the spring fixed and the upper one carrying disc D_1 of mass m_1 = 1kg.

If, therefore, n_1 be the frequency of the system, we have

$$n_1 = \frac{1}{2\pi}\sqrt{\frac{C}{m_1}} = \frac{1}{2\pi}\sqrt{\frac{10^6}{1000}} = 5.025 \text{ or } 5.0/\text{sec.}$$

(b) (i) When the table is suddenly removed from under disc D_2, the system oscillates as a two-body harmonic oscillator, with both the discs oscillating simultaneously.

If n_2 be the frequency of the system, we have

$$n_2 = \frac{1}{2\pi}\sqrt{\frac{C}{\mu}},$$

where μ is the *reduced mass of the system*, equal to

$$\frac{m_1 m_2}{m_1 + m_2} = \frac{1\times 2}{1+2} = \frac{2}{3} \text{ kg.}$$

$$\therefore \quad n_2 = \frac{1}{2\pi}\sqrt{\frac{10^6}{\left(\frac{2}{3}\right)\times 1000}} = \frac{1}{2\pi}\sqrt{\frac{3\times 10^6}{2000}} = \frac{1}{2\pi}\sqrt{\frac{3\times 10^3}{2}}$$

$$= 6.165/\text{sec.}$$

(ii) Clearly, with the sudden removal of the table from under disc D_2, the latter is subjected to an instantaneous force equal to

(1 kg + 2 kg) = 3kg ωt = $3 \times 10^3 \times g$ dynes.

If its acceleration be a, the force on it is clearly

$$m_2\, a = 2 \times 10^3\, a = 3 \times 10^3 \times g,$$

whence, $\quad a = 3g/2.$

Thus, the *instantaneous acceleration of disc* D_2 = 3/2 g cm/sec^2.

As to acceleration of disc D_1, we note that the system was at rest before removal of the table from under disc D_2. Since disc D_1 was at rest on top of the spring, its weight was being supported by the elastic reaction of the spring and it continues to be so supported at the instant the table is just removed. So that, *the instantaneous acceleration of disc* D_1 *is zero.*

The acceleration of the centre of mass of the system will naturally be the same as that of a body falling freely under the action of gravity, *viz.*, g. This may be shown as follows:

If y be the instantaneous displacement of the centre of mass (taking y-direction along the vertical) and y_1 and y_2, the corresponding displacements of the two discs of masses m_1 and m_2 respectively, we have $(m_1 + m_2)y = m_1y_1 + m_2y_2$. Differentiating with respect to t, we have

$$\frac{(m_1 + m_2)d^2y}{dt^2} = \frac{m_1d^2y_1}{dt^2} + \frac{m_2d^2y_2}{dt^2},$$

where, as we know, $\frac{d^2y_1}{dt^2} = 0$ and $\frac{d^2y_2}{dt^2} = \frac{3g}{2}$.

If, therefore, the acceleration of the centre of mass, *i.e.*,

$\frac{d^2y}{dt^2}$ be a′ cm/sec^2, we have

$$(1 + 2)\,a' = 0 + 2 \times \frac{3}{2}g.$$

Or, 3a′ = 3g, whence, a′ = g.

i.e., the acceleration of the centre of mass is g.

Example 6(b):

Two springs S_1 *and* S_2 *are connected to a mass M with the mass lying on a frictionless surface. If the respective force constants* of the spring be C_1 and C_2, obtain expressions for the time-periods of the mass in the two cases. What are their electrical analogues?

Solution:

(a) If x_1 and x_2 be the extensions produced in the two springs respectively as the mass (M) is displaced outwards, the *total extension* $x = x_1 + x_2$.

Since the restoring force due to each spring is the same, F, say, we have

$$F = -C_1x_1 = -C_2x_2;$$

so that, $\quad x_1 = -\dfrac{F}{C_1}$ and $x_2 = -\dfrac{F}{C_2}$.

$$\therefore \quad x = x_1 + x_2 = -F\left(\frac{1}{C_1} + \frac{1}{C_2}\right) = -F\left(\frac{C_1 + C_2}{C_1C_2}\right),$$

whence, $\quad F = -\left(\dfrac{C_1C_2}{C_1 + C_2}\right)x,$

indicating that the effective force constant of the combination of the two springs, say,

$$C = \left(\frac{C_1C_2}{C_1 + C_2}\right).$$

$\therefore$ *time-period of the oscillating mass,*

$$T = 2\pi\sqrt{\frac{M}{C}} = 2\pi\sqrt{\frac{M}{\left(\dfrac{C_1C_2}{(C_1 + C_2)}\right)}}$$

$$= 2\pi\sqrt{\frac{(C_1 + C_2)M}{C_1C_2}}.$$

(b) In this ease, if the mass (M) be; displaced to one side or the other, through a distance x, say, one spring gets extended and the other compressed, both exerting a restoring force on the mass *in the same direction*, tending to bring it back to its original position. Thus, if F_1 and F_2 be the restoring forces due to the two springs respectively, we have *resulting restoring force* $F = F_1 + F_2 = -C_1x - C_2x = -(C_1 + C_2)x = Cx$, indicating that the *effective force constant* C of the combination of the two springs is now equal to $(C_1 + C_2)$, the sum of their individual force constants.

The time-period of the oscillating mass is, therefore,

$$T = 2\pi\sqrt{\frac{M}{C}}$$

$$= 2\pi\sqrt{\frac{M}{(C_1 + C_2)}}.$$

The electrical analogues of the two systems are a parallel and a series arrangement of two capacitors respectively.

Example 7:

(a) Snow that the maximum tension in the string of a simple pendulum, when the amplitude θ_m, is small, is mg $(1 + \theta_m^2)$. At what position of the pendulum is the tension a maximum?

(b) Using conservation of energy, show that the angular speed of a simple pendulum is given by $\frac{d\theta}{dt} = \left[\frac{2}{ml^2}\{E - mgl(1-\cos\theta)\}\right]^{1/2}$ *where E is the total energy of oscillations, l and m are length and mass of the pendulum and θ is angular displacement from the vertical.*

Solution:

(a) Let SO be a simple pendulum of length l, with S, as its point of suspension and O, its mean or equilibrium position. If its angular amplitude be θ_m so that it is in the position SP, the whole of its energy is present in the potential form and is equal to mg × OP', where OP' is the vertical distance through which the bob rises with respect to its mean position O.

Or, since $OP' = SO - SP' = l - l\cos\theta_m$

$= l(1 - \cos\theta_m)$, we have

total energy of the pendulum $= mgl(1 - \cos\theta_m)$

Or, because θ_m is small,

we may take $\cos\theta_m = 1 - \frac{1}{2}\theta_m^2$.

So that, total energy of the pendulum

$$= mgl\left[1 - \left(1 - \frac{1}{2}\theta_{m2}\right)\right] = \frac{1}{2}mgl\theta_m^2$$

When the bob arrives at O, the whole of the energy of the pendulum is converted into the kinetic form, so that if v be its velocity at O tangential to the arc PO, its

$$K.E. = \frac{1}{2}mv^2.$$

Thus, $\frac{1}{2}mv^2 = \frac{1}{2}mgl\theta_m^2,$

whence, $v^2 = gl\theta_m^2.$

Or, $v = (gl\theta_m^{\ 2})^{1/2}$.

Now, the *centripetal force* necessary to keep the bob moving in an arc of radius $l = mv^2/l$ towards S and it must obviously be equal to T – mg, where T is the tension in the string along OS and mg, the weight of the bob, *i.e.*, $T - mg = mv^2/l$

Or, substituting the value of v from above, we have

$$T - mg = mgl\theta_m^{\ 2}/l = mg\theta_m^{\ 2},$$

whence, $T = mg + mg\theta_m^{\ 2} = mg\ (1 + \theta_m^{\ 2})$.

The tension is clearly the maximum when the bob passes through its mean or equilibrium position, because only in this position is the velocity of the bob, and hence the centripetal force acting upon it, the maximum, as also the entire weight of the bob effective.

(b) If the angular displacement of the pendulum bob be θ, *its potential energy in its displaced position* = $mgl\ (1 - \cos\theta)$, where, as we have just seen in part (a) above, $l(1 - \cos\theta)$ is the vertical distance through which the bob rises relative to its mean position.

Since θ is less than the amplitude of the pendulum, the bob in this position possesses also kinetic energy = $\frac{1}{2}mv^2$, where v is the velocity of the bob in the displaced position. Therefore,

total energy of the pendulum,

$$E = \frac{1}{2}mv^2 + mgl(1 - \cos\theta)$$

which throughout remains constant in accordance with the *law of conservation of energy.*

We thus have $\frac{1}{2}mv^2 = E - mgl\ (1 - \cos\theta)$.

Or, $mv^2 = 2E - 2mgl\ (1 - \cos\theta)$.

Now, $v = l\omega = l d\theta/dt$, where $\omega = d\theta/dt$ is the angular velocity of the bob.

So that, $$ml^2\left(\frac{d\theta}{dt}\right)^2 = 2E - 2mgl\ (1 - \cos\theta).$$

Or, $$\left(\frac{d\theta}{4t}\right)^2 = \left(\frac{2}{ml^2}\right)[E - mgl(1 - \cos\theta)].$$

And, therefore, $\frac{d\theta}{dt} = \left[\frac{2}{ml^2}\{E - mgl(1-\cos\theta)\}\right]^{1/2}$

Example 8:

An inductance coil of inductance 0.05 henry and resistance 12.4 ohms is connected to an A.C. supply at 200 volts and 50 cycles. Find (a) power factor, and (b) angle of lag. (c) Show that the power consumed in just equal to the rate of production of heat.

Solution:

(a) We know that the power factor,

$$\cos\phi = \frac{R}{Z} = \frac{R}{\sqrt{R^2 + (Lp)^2}}.$$

Here, R = 12.4 ohms, L = 0.05 henry

and $\pi = 2\pi n = 2\pi \times 50 = 100\pi$.

$\therefore$ *power factor*, $\cos\phi = \frac{12.4}{\sqrt{(12.4)^2 + (0.05 \times 100\pi)^2}}$

$$= \frac{12.4}{\sqrt{153.8 + 246.7}} = \frac{12.4}{20} = 0.62.$$

(b) And, therefore, *angle of lag*, f = $\cos^{-1}$ (0.62) = 51° 41'.

(c) Clearly, *power consumed*, $E_v I_v \cos\phi$.

Now, E_v = 100 volts,

$$I_v = \frac{E_v}{Z} = \frac{100}{\sqrt{(12.4)^2 + (0.05 + 100\pi)^2}}$$

$$= \frac{100}{20} = 5 \text{ ampere and } \cos\phi = 0.62.$$

$\therefore$ *power consumed* = 100 × 5 × 0.62 = 310 watts.

And, *rate of production of heat* = I^2R = 25 × 12.4 = 310 watts.

Thus, *power consumed* = *rate of production of heat.*

Example 9:

Write down the equation of motion of the damped harmonic oscillator of example 4 above and calculate the time in which (i) its amplitude (ii) its energy falls to l/eth of its undamped value.

If the mass of the oscillator be 1.127 gm, calculate its average rate of loss of energy.

Solution:

The differentia) equation of a damped harmonic oscillator, as we know (See 2.2), is

$$\frac{d^2x}{dt^2}+2k\frac{dx}{dt}+\omega_0^2x=0.$$

which, when the oscillator is underdamped, *i.e.*, when $k << \omega_0$, gives

$$x = a_0e^{-kt}\sin(\omega t + \phi).$$

So that its amplitude is given by $a = a_0e^{-kt} = a_0e^{-t/2\tau}$, where a_0 is its amplitude the absence of damping.

(i) Now, as we have seen under example 2 above, the damping constant k for the given oscillator works out to 0.02. If, therefore, its amplitude is reduced to $1/e^{th}$ of its undamped value, we have

$$a=\frac{a_0}{e}=a_0e^{-1}$$

and, therefore, $a_0e^{-1} = a_0e^{-kt}$. Or, $kt = 1$,

whence, $$t=\frac{1}{k}=\frac{1}{0.02}=50\text{ sec.}$$

Thus, *the amplitude of the oscillator falls to $1/e^{th}$ of its undamped value in 50 sec.*

(ii) Again, as we have seen in 2.3, the energy of a damped harmonic oscillator at a given instant t is given by $E = E_0e^{-2kt}$. So that, when its energy falls to $1/e^{th}$ of its value E_0 in the absence of damping, we have

$$E=\frac{E_0}{e}=E_0e^{-1}.$$

And, therefore, $E_0e^{-1} = E_0e^{-2kt}$. Or, $2kt = 1$,

whence, $$t=\frac{1}{2k}=\frac{1}{2}\times 0.02=25\text{ sec.}$$

Thus, *the energy of the oscillator falls to $1/e^{th}$ of its value in the absence of damping in 25 sec.*

The *rate of loss of energy* $=\frac{E}{\tau}=2kE.$

Now, $$E=\frac{1}{2}ma_0^2\omega_0^2e^{-2kt},$$

where m = 1.127 gm, a_0 = 2.011 cm,

$$\omega_0 = \frac{2\pi}{T} = \frac{2\pi}{1.15}, \; k = 0.02$$

and t = 25 sec. So that,

rate of loss energy

$$= 2 \times 0.02 \times \frac{1}{2} \times 1.127 \times (2.011)^2 \times (2\pi/1.15)^2 \, e^{-2(0.02 \times 25)}$$

$$\frac{0.02 \times 1.127 \times (2.011)^2 \times 4\pi^2}{(1.15)^2 \times e} = \frac{0.02 \times 1.127 \times (2.011)^2 \times 4\pi^2}{(1.15)^2 \times 2.7183}$$

$$= 1 \text{ erg/sec.}$$

Example 10:

If the amplitude of a seconds pendulum, with a bob of mass 200gm, is reduced to half its undamped value in 200 seconds, what fa its quality factor Q? What should be the mass of the bob, its size and shape remaining unchanged, in order that the damping of the pendulum may become critical? (Take $\log_e$ 10 = 2.30).

Solution:

We know that the amplitude of the pendulum (a damped harmonic oscillator) at a given instant t is given by $a = a_0 e^{-kt} = a_0 e^{-t/2\tau}$.

Here, $a = \dfrac{a_0}{2}$ and t = 200 sec.

So that, $\dfrac{a_0}{2} = a_0 e^{-200/2\tau} = a_0 e^{-100/\tau}$.

Or, $\dfrac{100}{\tau} = \log_e 2 = 2.30 \log_{10} 2 = 2.30 \times 0.3010$

whence, $\tau = \dfrac{100}{2.30} \times 0.3010 = \dfrac{100}{0.6923}$.

And, since $\omega_0 = \dfrac{2\pi}{T} = \dfrac{2\pi}{2} = \pi$ ($\because$ T = 2 sec), we have

Quality factor of the pendulum, $Q = \omega_0 \tau = \pi \dfrac{100}{0.6923} = 453.7$.

For *critical damping*, as we know, the condition is that $k = \omega_0$. So that, τ' be the value of the *relaxation time* in this case, we have $1/2\tau' = \omega_0$

Or, since $\omega_0 = \pi$, we have $1/2\tau' = \pi$ and, therefore, $\tau' = \dfrac{1}{2\pi}$.

If m and m' be the masses of the bob in the two cases respectively, we have $\gamma/m = 1/\tau$ and $\gamma/m' = 1/\tau'$ ($\because$ γ remains the same in either case).

$$\therefore \quad \frac{\tau'}{\tau} = \frac{\left(\frac{\gamma}{m}\right)}{\left(\frac{\gamma}{m'}\right)} = \frac{m'}{m}, \text{ whence, } m' = m \times \frac{\tau'}{\tau}.$$

Substituting the values of m, τ' and τ, therefore, we have

$$m' = 200 \times \frac{1}{2\pi} \times \frac{0.6923}{100} = \frac{0.6923}{\pi} = 0.2204 \text{ gm.}$$

Thus, *for the damping to be critical, the mass of the bob must be* 0.2204 gm.

Example 11:

A massless spring, suspended from a rigid support, carries a mass of 500 gm at its lower end and the system oscillates with a frequency of S/sec, with the amplitude reduced to half its undamped value in 20 sec. Calculate (i) the force constant of the spring. (ii) the relaxation time of the system and (iii) its quality factor.

(iv) What would be the quality factor of the system if the suspended mass be reduced to 150 gm?

Solution:

(i) We have the relation $\omega_0 = \sqrt{C/m}$ or $C = m\omega_0^2$, where C is the *force constant* of the spring and m, the *mass* suspended from it.

Here, $\omega_0 = 2\pi n \times 2\pi \times 5 = 10\pi$ radian/sec and m = 500 gm.

So that, *force constant of the spring,* $C = 500 \times (10\pi)^2 = 5\pi^2 \times 10^4$

$= 4.934 \times 10^5$ dynes/cm.

(ii) The amplitude at an instant t is given by $a = a_0 e^{-kt}$.

Since $a = a_0/2$ and t = 20 sec, we have $a_0/2 = a_0 e^{-20k}$.

Or, $\log_e 2 = 20k$.

Or, $2.30 \log_{10} 2 = 20k$.

Or, $2.30 \times 0.3010 = 20k$,

whence, $k = 2.30 \times 0.3010/20$.

∴ *relaxation time of the system,*

$$\tau = \frac{1}{2k} = \frac{1\times 20}{2\times 2.30\times 0.3010} = 14.44 \text{ sec.}$$

(iii) The *quality factor of the system,* $Q = \omega_0\tau = 10\pi \times 14.44 = 453.7$.

(iv) Since the force constant (C) of the spring remains the same irrespective of the mass suspended from it, and so does the damping factor $\gamma = 2km = 2 \times 2.30 \times 0.010 \times 500/20 = 115 \times 0.3010$, we have

$$\omega'_0 = \sqrt{\frac{C}{m'}} = \sqrt{5\pi^2 \times \frac{10^4}{150}} = 18.26\ \pi, \text{ since } m' = 150 \text{ gm.}$$

∴ τ', the relaxation time now $= 1/2k = m'/g = \frac{150}{115} \times 0.3010$.

And, therefore, the *quality factor now,* $Q' = \omega_0'\tau'$

$$= 18.26\pi \times \frac{150}{115} \times 0.3010 = 248.5.$$

Example 12:

The restoring torque exerted by the suspension fibre of a ballistic galvanometer is 103 dyne-cm/radian. If the period of vibration of the coil be 2π seconds and its amplitude be reduced to $1/e^{th}$ of its undamped value in 100 seconds, calculate the values of :

(i) the moment of inertia of the suspended system,

(ii) the damping factor γ, and

(iii) the quality factor Q of the system.

Solution:

(i) The oscillating system in a ballistic galvanometer, as we know, is an underdamped one (2.5I (ii)) and, therefore, the angular frequency of the system is given by $\omega = \sqrt{\omega_0^2 - k^2}$, where ω_0 is the undamped angular frequency of the system and k, the damping constant.

∴ *time-period of the suspended system or the coil,*

$$T = \frac{2\pi}{\sqrt{\omega_0^2 - k^2}}.$$

Or, since $T = 2\pi$ sec, we have

$$2\pi = \frac{2\pi}{\sqrt{\omega_0^2 - k^2}},$$

whence, $\sqrt{\omega_0^2 - k^2} = 1.$

Or, $\omega_0^2 - k^2 = 1$.

Now, the amplitude of the coil at an instant t is given by $\theta = \theta_0 e^{-kt}$, where θ_0 is its undamped amplitude. And, since the amplitude is reduced to $\theta_0/e = \theta_0 e^{-1}$ in time t = 100 sec, we have

$$\theta_0 e^{-1} = \theta_0 e^{-100k}$$

or, $100\ k = 1$

or $k = \frac{1}{100} = 0.01.$

$\therefore$ $\omega_0^2 - k^2 = \omega_0^2 - (0.01)^2 = 1.$

Or, $\omega_0^2 \approx 1$ and $\therefore\ \omega_0 \approx 1$ rad/sec.

Hence $C/I = \omega_0^2 = 1$, where C is the torque exerted by the suspension fibre per radian twist of the fibre.

Since C = 10^3 *dyne-cm/radian* (given), we have $\frac{10^3}{I} = 1.$

Or, $I = 10^3$ dyne-cm^2.

Thus, *the moment of Inertia of the suspended system about the suspension fibre, i.e.*

$$I = 10^3$$

or 1000 dyne-cm^2.

(ii) Now, $(\gamma \times \wedge/R)/I = 2k$. Or, $\gamma/I = 2k$ (the electromagnetic damping being obviously ignored in the problem). So that,

damping factor $\gamma = I \times 2k = 1000 \times 2 \times 0.1 = 20$ dyne-cm/sec.

(iii) *The quality factor of the system,*

$$Q = \omega_0 \tau = \frac{\omega_0}{2k} = \frac{1}{2} \times 0.01 = 50.$$

Example 13:

In an experiment on forced oscillations, the frequency of a sinusoidal during force is changed while its amplitude is kept constant. It is found that the amplitude of vibrations is 0.01 mm at very low frequency of the *driver and-goes upto a maximum of 5.0 mm at driving frequency 200 sec^{-1}. Calculate (i) Q of the system, (ii) relaxation time T and (iii) half width of resonance curve.*

Solution:

We know that if the driving frequency be negligibly small and the damping be low, the *amplitude of the driven oscillator*, $A = f_0\omega_0^2$ and that at low damping, the maximum amplitude (or amplitude at resonance) is $A_{max} = f_0\tau/\omega_0$, where ω_0 is the *natural frequency of the oscillator* (or the system) and T, its relaxation time.

(i) We, therefore, have $\dfrac{A_{max}}{A} = \dfrac{\dfrac{f_0\tau}{\omega_0}}{\dfrac{f_0}{\omega_0^2}} = \omega_0\tau = Q.$

Since A_{max} ∴ 5.0 m and A = 0.01 mm, we have

quality factor of the system, $Q = \dfrac{5}{0.01} = 500.$

(ii) Since $Q = w_0 t$, we have $\tau = \dfrac{Q}{\omega_0}$.

Here, Q = 500 and $\omega_0 = 2\pi n = 2\pi \times 200 = 400\pi$

∴ *relaxation time of the system,* $\tau = \dfrac{500}{400\pi} = 0.3971 \approx 0.40$ sec.

(iii) For *low damping, half width of the resonance curve,* $\Delta p = \sqrt{3}$.

Since $k = \dfrac{1}{2\tau} = \dfrac{1}{2} \times 0.40 = 1.25$, we have *half width of the resonance curve,*

$\Delta p = \sqrt{3} \times 1.25 = 2.165$ radian/sec

$= \dfrac{2.165}{2\pi} = 0.3445$ c.p.s.

Example 14:

A harmonic oscillator of quality factor 10 is subjected to a sinusoidal applied force of frequency one and a half times the natural frequency of the oscillator. If the damping be small, obtain (i) the amplitude of the forced oscillation in tarns of its maximum amplitude and (ii) the angle θ by which it will be out of phase with the driving force.

Solution:

As we know, the *amplitude of the forced oscillation* is given by

$$A = \frac{f_0}{\sqrt{(\omega_0^2 - p^2)^2 + 4k^2p^2}} = \frac{f_0}{\omega_0^2\sqrt{\left(1-\frac{p^2}{\omega_0^2}\right)^2 + 4\frac{k^2p^2}{\omega_0^4}}},$$

where ω_0 is the *natural angular frequency of the oscillator and p, the angular frequency of the driving force.*

Now, the quality factor $Q = \omega_0\tau = \frac{\omega_0}{2k} = 10$.

So that, $2k = \frac{\omega_0}{10}$ and $\frac{p}{\omega_0} = \frac{3}{2}$ (given). We, therefore, have

$$A = \frac{f_0}{\omega_0^2\sqrt{\left(1-\frac{p^2}{\omega_0^2}\right)^2 + \frac{1}{100}\times\frac{p^2}{\omega^2}}} = \frac{f_0}{\omega_0^2\left(\frac{25}{16}+\frac{1}{100}\times\frac{9}{4}\right)^{1/2}}$$

$$= \frac{f_0}{\omega_0^2\left(\frac{634}{400}\right)^{1/2}} = \frac{20f_0}{\omega_0^2\sqrt{634}}.$$

And, *for low damping,* amplitude is maximum when $p = \omega_0$. So that,

$$A_{max} = \frac{f_0}{\omega_0^2\sqrt{(1-1)^2 + \frac{1}{100}\times 1}} = \frac{f_0}{\omega_0^2\left(\frac{1}{100}\right)^{1/2}}$$

$$= \frac{f_0}{\omega_0^2\left(\frac{1}{10}\right)} = \frac{10f_0}{\omega_0^2}.$$

$$\therefore \frac{A}{A_{max}} = \frac{20f_0}{\omega_0^2\sqrt{634}}\times\frac{\omega_0^2}{10f_0} = \frac{2}{\sqrt{634}} = 0.07943 \approx 0.08.$$

Or, $A = 0.08\, A_{max}$.

Thus, *the amplitude of the forced oscillation will be 0.08 of the maximum amplitude.*

$$\text{Now, } \tan\theta = -\frac{2kp}{\omega_0^2 - p^2} = \frac{p\omega_0/10}{\omega_0^2 - p^2} = \frac{\left(\frac{3\omega_0}{2}\right)\left(\frac{\omega_0}{10}\right)}{\omega_0^2 - \frac{9\omega_0^2}{4}}$$

$$= \frac{3\omega_0^2}{20} \times \frac{4}{5\omega_0^2} = \frac{3}{25} = 0.12.$$

Or, $\theta = \tan^{-1} 0.12 = 6°51'$

Therefore, *the forced oscillation is* (180° – 6°51') = 173°9' *out of phase with the driving force.*

Example 15(a):

Give the theory of oscillations in a series LCR circuit with small damping. Deduce the frequency and quality factor for a circuit with L = 2mH, C = 5mF and R = 0.2 ohm.

Solution:

For answer to the first part of the question. The circuit being equivalent to an *underdamped harmonic oscillator*, we have

$$\omega = \sqrt{\omega_0^2 - k^2}$$

and, therefore, *frequency of the oscillations*,

$$n = \frac{\omega}{2\pi} = \sqrt{\omega_0^2 - \frac{k^2}{2\pi}}.$$

$$\text{Or, } n = \frac{1}{2\pi}\sqrt{\frac{1}{LC} - \frac{R^2}{4L^2}}.$$

Here, L = 2mH = 0.002 henry,

C = 5μF = 5 × 10^{-6} farad

and R = 0.02 ohm. We, therefore, *have frequency of the circuit,*

$$n = \frac{1}{2\pi}\sqrt{\frac{1}{0.002 \times 5 \times 10^{-6}} - \frac{(0.02)^2}{4(0.002)^2}} = \frac{1}{2\pi}\sqrt{\frac{4\times 10^8 - 10^4}{4}}$$

$$= \frac{10^2}{2\pi}\sqrt{10^4} - 1 = \frac{10^4}{2\pi} = 1.592 \times 10^3 \text{ cps.}$$

Now, the resistance being small, *quality factor*

$$Q = \frac{L\omega_0}{R} = \frac{L\omega}{R},$$

where $\omega = 2\pi n = 2\pi \times 10^4/2\pi = 10^4$ rad/sec.

$\therefore$ *quality factor for the circuit,*

$$Q = 0.002 \times \frac{10^4}{0.2} = 0.01 \times 10^4 = 100.$$

Example 15(b):

A torsion pendulum of moment of inertia 10^5 gm-cm^2 suffers an angular displacement of 4° under a constant torque of 2000 dynes-cm. What should be the frequency of the periodic torque that would set the pendulum in resonant vibration?

Solution:

Here, *angular di placement of the pendulum* $= 4° = 4 \times \frac{\pi}{180}$ radian and the torque applied = 2000 dyne-cm.

$\therefore$ *torque per unit angular displacement* (C) $= \frac{\text{torque applied}}{\text{angular displacement}}$

$$= \frac{2000}{4\pi/180} = \frac{2000 \times 180}{4\pi} \text{ dyne-cm/rad.}$$

Now, *for setting the pendulum into resonant vibration, the frequency of the applied torque must be the same, or very nearly the same, us the natural frequency of the pendulum,*

$$n = \frac{1}{2\pi}\sqrt{\frac{C}{I}}.$$

Substituting the values of C and I, therefore, we have

required frequency of the periodic torque,

$$n = \frac{1}{2\pi}\sqrt{\frac{2000 \times 180}{4\pi \times 10^5}}$$

$$= 0.1704 \text{ or } 0.17 \text{ sec}^{-1}.$$

Example 15(c):

A particle of mass 50 gm, moving with an initial velocity of 100 cm/ sec, is acted upon by a damping force which brings it to rest in a distance of. 10 metres. Assuming the damping force to be proportional to velocity, *calculate (i) its relaxation time, (ii) the time in which (a) its velocity is halved, (b) its kinetic energy is halved, (iii) the damping force on it when its velocity is 20 cm/sec. (Take $\log_e 10 = 2.30$).*

Solution:

We have the relation $x = v_0\tau\,(1 - e^{t/\tau})$ for the distance covered by a particle in time t, where v_0 is its *initial velocity* and τ, its relaxation time.

(i) Clearly, x will have its maximum value as $\tau \rightarrow \infty$ and, therefore, the quantity $e^{-t}/\tau \rightarrow 0$.

So that, $x_{max} = v_0\tau$

and ∴ $\tau = x_{max}/v_0$.

Here, $x_{max} = 10\text{ m} = 1000\text{ cm}$ and v_0 100 cm/sec. We, therefore, have

relaxation time $\tau = \dfrac{1000}{100} = 10$ sec.

(ii) (a) The velocity of the particle after time t from the start is given by the relation $v = v_0 e^{-t/\tau}$,

whence, $\log_e (v/v_0) = -t/\tau$.

Or, $t = \tau \log_e(v_0/v)$.

∴ the velocity will be reduced to half, *i.e.*, 50 cm/sec in time t given by

$$t = 10 \times 2.30 \log_{10}\left(\frac{100}{50}\right)$$

$$= 10 \times 2.30 \times \log_{10} 2$$

$$= 10 \times 2.30 \times 0.3010 = 6.92 \text{ sec.}$$

(b) We know that if T be the kinetic energy of the particle after time t and T_0, its initial kinetic energy, we have $T = T_0\, e^{-2t/\tau}$.

∴ $$\log_e\left(\frac{T}{T_0}\right) = -\frac{2t}{\tau},$$

whence, $$2t = \tau \log_e\left(\frac{T_0}{T}\right).$$

Or, $$2\tau = 10 \times 2.30 \log_{10}\left(\frac{2}{1}\right).$$

Or, $t = 5 \times 2.30 \log_{10} 2$

$= 5 \times 2.30 \times 0.3010 = 3.46$ sec.

Or, we could have obtained this result directly from (a) above, since, as we know, relaxation time for K.E. is half that for velocity.

Thus, *the K.E. of the particle will be reduced to half its initial value in 3.46 sec.*

(iii) The *damping force* = γv, where, $\gamma = \frac{m}{\tau}$.

$\therefore$ *damping force at velocity* 20 cm/sec

$$= mv/\tau = 50 \times \frac{20}{10} = 100 \text{ dynes.}$$

Example 15(d):

A charged oil drop falls through air under the combined action of an electric field E and gravity g, both acting along the same direction. If m be the mass of the drop and q, the charge on it, show that the terminal velocity attained by it (i.e., when $t \to \infty$) is given by $v_{terminal} = (qE/m)\tau + g\tau$,, assuming a solution of the form $v = A + Be^{-\alpha\tau}$. (Assume damping force to be proportional to velocity).

Solution:

Since the electric field E and g are both similarly directed, they both act downwards. So that, the force acting on the drop is given by

$$F = m\left(\frac{dv}{dt}\right) = mg + qE - \gamma v,$$

where γv is the damping force.

Or, $$m\left(\frac{dv}{dt}\right) + \gamma v = mg + qE.$$

Or, $$\left(\frac{dv}{dt}\right) + \left(\frac{\gamma}{m}\right)v = g + \left(\frac{q}{m}\right)E.$$

Or, since $\frac{\gamma}{m} = \frac{1}{\tau}$, where τ is the *relaxation time*, the *equation of motion* of the drop is

$$\frac{dv}{dt} + \frac{v}{\tau} = g + \frac{qE}{m}. \quad ...(I)$$

Now, let us assume a solution of this equation to be $v = A + Be^{-\alpha t}$ where A and B are *arbitrary constants* to be determined from the initial conditions.

Since at t = 0, *i.e., when the drop Just starts from rest*, $v_0 = 0$, we have 0 = A + B, whence, B = – A. And, therefore,

$$v = A + Ae^{-\alpha t} = A(1 - e^{-\alpha t}). \qquad \text{...(II)}$$

Hence, $$\frac{dv}{dt} = A\alpha e^{-\alpha t}.$$

Substituting this value of dv/dt in the equation of motion (I) above, we have

$$A\alpha e^{-\alpha t} + \frac{1}{\tau}A\left(1 - e^{-\alpha t}\right) = g + \frac{qE}{m}.$$

Clearly, at $t \to \infty$, we have $\frac{A}{\tau} = g + \frac{qE}{m}$.

Or, $$A = \tau\left(g + \frac{qE}{m}\right) \qquad \text{...(III)}$$

And, at t = 0, we have $$A\alpha = g + \frac{qE}{m}. \qquad \text{...(IV)}$$

.From relations III and IV, we thus have $\alpha = \frac{1}{\tau}$.

Hence from equation II, we have

$$v = \tau\left(g + \frac{qE}{m}\right)\left(1 - e^{-\alpha t}\right) = \tau\left(g + \frac{qE}{m}\right)\left(1 - e^{-t/\tau}\right) \qquad \text{...(V)}$$

Clearly, therefore, when $t \to \infty$, we have

$$v_{terminal} = \tau\left(g + \frac{qE}{m}\right) = \left(\frac{qE}{m}\right)\tau + g\tau.$$

Example 16:

A small ball bearing is falling freely through a viscous medium, of coefficient of viscosity η, under the action of gravity alone. The viscous drag on it due to the medium is $-6\pi\eta av$, where a is its radius and v, its velocity. Obtain (i) the relaxation time, (ii) the terminal velocity of the ball bearing.

(iii) If the ball bearing has an initial horizontal velocity u, what will be the maximum horizontal distance covered by it?

(iv) If the radios of the ball bearing be doubled, how will it affect its terminal velocity?

Solution:

(i) Here, obviously, the *damping force* = $6\pi\eta av = \gamma v$, where $\gamma = 6\pi\eta a$ and is thus *proportional to velocity.*

$$\textit{relaxation time } \tau = \frac{m}{\gamma} = \frac{m}{6\pi\eta a}.$$

(ii) Proceeding as in worked example 2 above and remembering that, here, the ball bearing (a tiny sphere) is falling only under the action of gravity, we have terminal velocity of the ball bearing given by

$$v_{terminal} = gt = g\left(\frac{m}{6\pi\eta a}\right) = \frac{mg}{6\pi\eta a}.$$

(iii) The *maximum horizontal distance covered by the ball bearing* is, as we know, equal to its *initial velocity* (u) × relaxation time (τ) = $u(m/6\pi\eta a) = mu/6\pi\eta a$.

(iv) We have seen under (u) above that the *terminal velocity of the ball bearing*

$$= \frac{mg}{6\pi\eta a} = \frac{4}{3}\frac{\pi a^3 \rho g}{6\pi\eta a} = \frac{2a^2\rho g}{9\rho}$$

where $\frac{4}{3}\pi a^2$ is the volume and,

therefore, $\frac{4}{3}\pi a^3\rho$, the mass (m) of the ball bearing.

Thus, the *terminal velocity* α a^2, (ρ and g being constants).

Hence, *if the radius of the ball bearing be doubled, its terminal velocity will be four times its previous value, i.e.*, equal to

$$\frac{4mg}{6\pi\eta a} = \frac{2mg}{3\pi\eta a}.$$

Example 17:

An underdamped harmonic oscillator has its amplitude reduced to 1/10th of its initial value after 100 oscillations. If its time-period be 1.15 sec, calculate (i) the damping constant, (ii) the relaxation time.

If the observed value of the first amplitude of the oscillator be 2 cm. what would be its value in the absence of damping?

Solution:

(i) The amplitude (a) of an underdamped oscillator is, as we know, given by the relation $a = a_0e^{-kt}$ where a_0 is its amplitude in the absence of damping and k, the *damping constant.*

Now, for the first amplitude (a_1), obviously, $t = \frac{T}{4}$ where T is the *time-period* of the oscillator. So that, $a_1 = a_0e^{-kT/A}$.

Since successive amplitudes occur at intervals of T/2 sec each, the 201th amplitude after 100 oscillations occurs after (200T/2 + T/4) = (100T + T/4) sec. And, therefore,

$$a_{201} = a_0e^{-k(100T + T/4)}$$

Hence $$\frac{a_{201}}{a_1} = \frac{a_0e^{-k}(100T + T/4)}{a_0e^{-kT/4}} = e^{-100kT}$$

But we are given that $\frac{a_{201}}{a_1} = \frac{1}{10}$

or 10^{-1} and T = 1.15 sec.

$\therefore$ $10^{-1} = e^{-100k(1.15)} = e^{-115k}$.

Or, $\log_e 10 = 115$ k.

Or, $230 \log_{10} 10 = 115k$,

i.e., 2.30 = 115*l*,

whence, $k = \frac{2.30}{115} = 0.02$.

Thus, the *damping constant* = 0.02.

(ii) And, therefore, *relaxation time* $\tau = \frac{1}{2k} = \frac{1}{2} \times 0.02 = 25$ sec.

Now, since *the first amplitude* $a_1 = a_0e^{-kT/4} = 2$ cm, we have

$$2 = e_0e^{-(0.02 \times 1.15/4)} \text{ or, } a_0 = 2e^{0.023/4}$$

or, $\log_{10} a_0 = \log_{10} 2 + \frac{0.023}{4} \times \frac{1}{2.30}$

Or, $\log_{10} a_0 = 0.3010 + 0.0025 = 0.3025$,

whence, $a_0 = 2.011$ cm.

Thus, *the undamped amplitude of the oscillator would be* 2.011 cm.

Example 18:

If in problem 14 above, the rate of dissipation of energy is desired to be halved, what should be (i) the amplitude of the applied periodic force, its frequency remaining unchanged, (ii) the frequency of the applied periodic force, its amplitude remaining unchanged?

Solution:

(i) We know that the *rate of dissipation of energy* α *(amplitude of the force)*^{2}.

Thus, if R_1 and R_2 be the rates of dissipation of energy and f_{01} and f_{02} the amplitudes of the force in the two cases respectively, *the frequency of the force remaining the same in either case,* we have

$$\frac{R_1}{R_2} = \left(\frac{f_{01}}{f_{02}}\right)^2.$$

We know that the *effective current in an LCR circuit*

$$= \frac{\text{effective emf}}{\text{impedance of}} = \frac{E}{Z}.$$

Here, *effective* emf E = 500 *volts* and *impedance*

$$Z = \sqrt{R^2 + L^2 p^2}$$
$$= \sqrt{R^2 + L^2 (2\pi n)^2},$$

$$\text{Or, } Z = \sqrt{(4)^2 + (0.68)^2 (2\pi \times 120)^2} = \sqrt{16 + 262800} = 512.6 \text{ ohms.}$$

$$\therefore \textit{ effective current in the circuit, } I = \frac{500}{512.6} = 0.9754 \text{ amp.}$$

$$\text{And, } \textit{power factor of the circuit} = \frac{R}{Z} = \frac{4}{512.6} = 0.0078.$$

Example 19(a):

An electric lamp which runs at 40 volts D.C. and consumes 10 amperes current is connected to A.C. mains at 100 volts, 50 cycles. Calculate the inductance of the choke.

Solution:

$$\text{Clearly, } \textit{resistance of the lamp, } R = \frac{\text{voltage}}{\text{current}} = \frac{40}{10} = 4 \text{ ohms.}$$

And, *impedance of A.C. circuit,* $Z = \sqrt{R^2 + (Lp)^2}$, where L is the *inductance of the choke.*

Here, R = 4 ohms, π = 2πn × 50 = 100π and I = 10.

$$\therefore \quad I = \frac{E}{Z} = \frac{E}{\sqrt{R^2 + (Lp)^2}} = \frac{100}{\sqrt{4^2 + (L + 100\pi)^2}} = 10$$

So that, $\dfrac{10^4}{16 + 10^4 \pi^2 L^2} = 100.$

Or, $10^6\pi^2L^2 = 10^4 - 1600 = 8400.$

So that, $L^2 = \dfrac{8400}{10^6 \pi^2}.$

Or, $L = \sqrt{\dfrac{84}{10^4 \pi^2}} = \dfrac{\sqrt{84}}{100\pi} = 0.02917.$

Thus, the *inductance of the choke* = 0.02917 henry.

Example 19(b):

An inductance coil of inductance 0.05 henry and resistance 12.4 ohms is connected to an A.C. supply at 200 volts and 50 cycles. Find (a) power factor, and (b) angle of lag. (c) Show that the power consumed in just equal to the rate of production of heat.

Solution:

(a) We know that the power factor,

$$\cos\phi = \frac{R}{Z} = \frac{R}{\sqrt{R^2 + (Lp)^2}}.$$

Here, R = 12.4 ohms, L = 0.05 henry

and π = 2πn = 2π × 50 = 100π.

$$\therefore \textit{ power factor, } \cos\phi = \frac{12.4}{\sqrt{(12.4)^2 + (0.05 \times 100\pi)^2}}$$

$$= \frac{12.4}{\sqrt{153.8 + 246.7}} = \frac{12.4}{20} = 0.62.$$

(b) And, therefore, *angle of lag*, f = $\cos^{-1}$ (0.62) = 51° 41'.

(c) Clearly, *power consumed*, $E_v I_v \cos\phi$.

Now, E_v = 100 volts,

$$I_v = \frac{E_v}{Z} = \frac{100}{\sqrt{(12.4)^2 + (0.05 + 100\pi)^2}}$$

$$= \frac{100}{20} = 5 \text{ ampere and } \cos\phi = 0.62.$$

∴ *power consumed* = 100 × 5 × 0.62 = 310 watts.

And, *rate of production of heat* = I^2R = 25 × 12.4 = 310 watts.

Thus, *power consumed = rate of production of heat.*

Example 19(c):

A harmonic oscillator consisting of a 60 gm mass attached to a massless spring has a quality factor 200. If it oscillates with an amplitude of 2 cm in resonance with a periodic force of frequency 20 c.p.s., calculate :

(i) the average energy stored in it,

(ii) the rate of dissipation of energy.

Solution:

In the case of resonant vibration, the average energy stored in the oscillator is equal to its maximum potential energy $= \frac{1}{2}Cx_0^2 = \frac{1}{2}m\omega_0^2 x_0^2$.

$[\because C = m\omega_0^2]$.

Here, $m = 50$ gm, $\omega_0 = 2\pi n = 2\pi \times 20 = 40\pi$

and x_0 = 2cm. So that,

average energy absorbed or stored up in the oscillator

$$= \frac{1}{2} \times 50 \times (40\pi)^2 \times (2)^2 = 16 \times 10\pi^2 = 1.58 \times 10^6 \text{ ergs.}$$

Now, by its very definition, the quality factor

$$Q = 2\pi \frac{\text{everage energy stored}}{\text{energy dissipated per cycle}}$$

$$\therefore \textit{energy dissipated per cycle} = 2\pi \frac{\text{average energy stored}}{Q}$$

$$= 2\pi \frac{1.58 \times 10^6}{200} = \pi\ 1.58 \times 10^4 \text{ ergs.}$$

Since there are 20 cycles to the second, we have

energy dissipated per second or *rate of diss:pation of energy*

$$= 20\pi \times 1.58 \times 10^1 = 9.926 \times 10^5 \approx 10^6 \text{ erg/sec}$$

Example 20:

Show that if the displacement of a moving point at any time is given by an equation of the form $x = a \cos \omega t + b \sin \omega t$, the motion is simple harmonic. If $a = 3$, $b = 4$ and $\omega = 2$, determine the period, amplitude, maximum velocity and maximum acceleration of the motion.

Solution:

Since the *displacement of the particle or the point* is given by $x = a \cos \omega t + b \sin \omega t$, its *velocity,*

$$v = \frac{dx}{dt} = -a\,\omega \sin \omega t - b\,\omega \cos \omega t$$

and its *acceleration,*

$$a = \frac{d^2x}{dt^2} = -\omega^2(a \cos \omega t + b \sin \omega t) = -\omega^2 x.$$

The particle, therefore, executes a *simple harmonic motion* of *amplitude* $\sqrt{a^2 + b^2} = \sqrt{3^2 + 4^2} = 5\text{cm}.$

The *time-period of the particle,* $T = \frac{2\pi}{\omega} = \frac{2\pi}{2} = \pi = 3.142$ sec.,

maximum velocity of the particle $= \omega a = 2 \times 5 = 10$ cm/sec.

and its *maximum acceleration* $= \omega^2 a = (2)^2 \times 5 = 20$ cm/sec^2.

Example 21:

The angular vibrational frequency of the carbon monoxide molecule (CO) is 0.6×10^{15}/sec. Calculate (i) the reduced mass of the molecule in gms, (ii) the force constant for stretching the molecule and (iii) the work done in stretching die molecule by 0.5 Angstrom unit.

Solution:

(i) Clearly, *reduced mass of the CO molecule, i.e.,*

$$\mu = \frac{m_1 m_2}{m_1 + m_2} = \frac{12 \times 16}{12 + 16}$$

$$= \frac{192}{28} \text{amu} = \frac{192}{28} \times \textit{mass of hydrogen atom}$$

$$= \frac{192}{28} \times 1.67 \times 10^{-24}$$

$$= 1.145 \times 10^{-23} \text{ gm.}$$

(ii) As we know, the vibrational frequency of the molecule is given by

$$n = \left(\frac{1}{2\pi}\right)\sqrt{\frac{C}{\mu}}.$$

Or, $$2\pi n \sqrt{\frac{C}{\mu}}.$$

Now, $$\sqrt{\frac{C}{\mu}} = \omega, \text{ whence, } C = \mu\omega^2,$$

where, $\mu = 1.145 \times 10^{-23}$ gm

and $\omega = 0.6 \times 10^{15}$ (given).

∴ *force constant* $C = (1.145 \times 10^{-23})\ (0.6 \times 10^{15})^2$

$= 1.145 \times 0.36 \times 10^{7}$

$= 4.122 \times 10^{6}$ dynes/cm.

(iii) Work done in stretching the molecule by $(r - r_0)$ is given by $\frac{1}{2}C\ (r - r_0)^2$.

Here, $C = 4.122 \times 10^{6}$ dynes/cm and

$(r - r_0) = 0.5 \text{ A} = 0.5 \times 10^{-8}$ cm.

∴ *work done in stretching the molecule through 0.5 A*

$$= \frac{1}{2} \times 4.122 \times 10^{5} \times (0.5 \times 10^{-8})^2$$

$$= 2.061 \times 10^{6} \times 0.25 \times 10^{-26}$$

$$= 5.15 \times 10^{-11} \text{ erg.}$$

Example 22:

HCl gas absorbs light of wave length λ, thereby causing the transition in the vibrational energy level from $n = 0$ to $n = l$. Calculate the force *constant of the HCl molecule and the wavelength of the light absorbed if the vibrational energy levels be separated by 0.36 eV.*

Solution:

Here, the transition in the energy level occurring from n = 0 to n = 1, *i.e.*, with n changing by 1, the *energy absorbed* = hv = hv_0, where h is the *Planck's Constant* = 6.6×10^{-27} erg-sec and v = frequency of the light absorbed.

Since $hv = 0.36 \text{ eV} = 0.36 \times 1.6 \times 10^{-12}$ ergs, we have

$$v = v_0 = 0.36 \times 1.6 \times 10^{-12}/6.6 \times 10^{-27}$$

$$= 8.728 \times 10^{13}/\text{sec.}$$

Now, $v = v_0 = \left(\frac{1}{2\pi}\right)\sqrt{\frac{C}{\mu}}$, where C is the *force constant* and μ, the *reduced mass of the molecule.*

Clearly, $\mu = \frac{1 \times 35.5}{1 + 35.5} \times 1.67 \times 10^{-24}$

$$= 0.97 \times 1.67 \times 10^{-24} \text{ gm.}$$

We, therefore, have

$$8.728 \times 10^{13} = \frac{1}{2\pi}\sqrt{\frac{C}{0.97 \times 1.67 \times 10^{-24}}},$$

whence, $C = (8.728 \times 10^{13})^2 \times 4\pi^2 \times 0.97 \times 1.67 \times 10^{-24}$.

Or, *force constant* $C = (8.728)^2 \times 0.97 \times 1.67 \times 4\pi^2 \times 10^2$

$$= 4.87 \times 10^5 \text{ dynes/cm.}$$

And, *wavelength of the light absorbed,*

$$\lambda = \frac{\text{velocity of light (c)}}{\text{frequency (v)}} = \frac{3 \times 10^{10}}{8.728 \times 10^{13}}$$

$$= 3.438 \times 10^{-4} \text{ cm} = 34380 \text{ A.}$$

Example 23:

(a) If the transition from one energy level to an adjacent one involves an energy change of 0.5 eV in the case of hydrogen, calculate the energy change involved in a similar transition in the case of HCl

(b) While comparing the potential energy curves of two diatomic

molecules A and B, it is observed that the potential well of the latter is narrower. What conclusions can be drawn from this? Can an estimate of the binding energies of the two molecules-be-made from these curves?

Solution:

(a) The energy change involved in transition between adjacent energy levels is, as we know, hv, where $hv = hv_0 = \left(\frac{h}{2\pi}\right)\sqrt{\frac{C}{\mu}}$, with C as the *force constant of the molecule* and μ, its *reduced mass.*

Now, in the case of hydrogen, hv = 0.5 eV, and, taking the mass of an atom of hydrogen to be m amu, μ = m × m/(m + m) = m/2 amu. And, therefore,

$$0.5 = \frac{h}{2\pi}\sqrt{\frac{C}{m/2}} = \frac{h}{2\pi}\sqrt{\frac{2C}{m}}. \quad \text{...(i)}$$

In the case of HCl, $\mu = \frac{m \times 35.5m}{m + 35.5m} = \frac{35.5m^2}{36.5m} = 0.97$ m amu.

If, therefore, the energy change involved in transition between adjacent energy levels be E, we have

$$E = \frac{h}{2\pi}\sqrt{\frac{C}{0.97m}}. \quad \text{...(ii)}$$

∴ dividing relation (ii) by (i), we have

$$\frac{E}{0.5} = \sqrt{\frac{C}{0.97m} \times \frac{m}{2C}} = \sqrt{\frac{1}{2 \times 0.97}}.$$

Or, $E = \sqrt{\frac{1}{2 \times 0.97}} \times 0.5 = 0.359 = 0.36$ eV.

Thus, *the energy change involved in transition between adjacent energy levels in the case of HCl is 0.36 electron volts.*

(b) A narrower potential well in the case of molecule B signifies a higher value of d^2U/dr^2 and, in consequence, *a larger value of the force constant* C. So that, for the same value of the reduced mass (μ) in the two cases, the vibrational frequency n [equal to $(1/2\pi)\sqrt{C/\mu}$] will be higher in the case of molecule B.

An estimate of the binding energies of the two molecules can be made by observing the minima of the potential energies in the two cases.

For, *if we ignore the zero-point energy, the numerical values of the minimum potential energy and the binding energy of a molecule are the same.*

Example 24:

The vibrational frequency of H^2Cl^{35} molecule is 8990×10^{10} cycles/sec. Deduce the vibrational frequency of H^2Cl^{35} molecule, assuming the force constant to be the same in the two cases.

Solution:

Since the vibrational frequency of a diatomic molecule is given by $(1/2\pi)\sqrt{C/\mu}$, it is clear that for the same value of the force constant (C), $n \propto \sqrt{1/\mu}$.

If, therefore, μ_1 and μ_2 be the reduced masses of the HCl molecule in the two cases, and n_1 and n_2, their vibrational frequencies respectively, we have

$$\frac{n_2}{n_1} = \sqrt{\frac{\mu_1}{\mu_2}}, \text{ whence, } n_2\sqrt{\frac{\mu_1}{\mu_2}}n_1 \quad \text{[C being the same for both]}$$

Now,

$$\mu_1 = 1 \times \frac{35}{(1+35)} = \frac{35}{36},$$

$$\mu_2 = 2 \times \frac{35}{(2+35)} = \frac{70}{37} \text{ and}$$

$$n_1 = 8990 \times 10^{10} \text{ c.p.s. (given).}$$

$$\therefore \quad n_2 = \sqrt{\frac{35}{36} \times \frac{37}{70}} \times 8990 \times 10^{10} = \sqrt{\frac{37}{72}} \times 8990 \times 10^{10}$$

$$= 6445 \times 10^{10} \text{ c.p.s.}$$

Thus, the vibrational frequency of the H^2Cl^{35} molecule

$$= 6445 \times 10^{10} \text{ cycles/sec.}$$

Example 25:

An oscillating particle of mass 1 gm has potential energy equal to $5000x^2 + 200x^4$ ergs what is its frequency for small oscillations? What will be its frequency if the amplitude of its fundamental component be 3 cm? Also calculate the frequency and the amplitude of the first overtone in the latter rase.

Solution:

We have here $U = 5000x^2 + 200x^4$ ergs.

$$\therefore \quad F = -\frac{dU}{dx} = -10{,}000x - 800x^3 \text{ dynes.}$$

For small oscillations, the second term in x^3 may be neglected. So that,

$$F = -10{,}000x \text{ dynes.}$$

$$\therefore \textit{ acceleration } \frac{d^2x}{dt^2} = -\frac{10{,}000x}{1} = -10{,}000x = -\left(\frac{C}{m}\right)x = -\omega_0^2 x,$$

i.e., $\omega_0^2 = 10{,}000$ and therefore, $\omega_0 \sqrt{10{,}000} = 100$.

$$\therefore \textit{ frequency for small oscillations, } n_0 = \frac{\omega_0}{2\pi} = \frac{100}{2\pi} = 15.92/\text{sec.}$$

Now, $\omega = \omega_0 + \frac{3}{8}\beta\frac{B_1^2}{\omega_0}$ (Equation VII, 1.9).

And, here, $\beta = 800$, $B_1 = 3$cm.

So that, $\omega = 100 + \frac{3}{8} \times 800 \times \frac{(3)^2}{100} = 127$.

$\therefore$ frequency, when amplitude of fundamental component is 3 cm is

$$n = \frac{\omega}{2\pi},$$

whence, $n = \frac{127}{2\pi} = 20.21/\text{sec.}$

Clearly, the *first overtone* here has thrice the frequency of the fundamental component, *i.e.*, a frequency n', say,

$$= \frac{3\omega}{2\pi} = 20.21 \times 3 = 60.63/\text{sec.}$$

and *its amplitude* is

$$\frac{\beta B_1^3}{32\omega_0^2} = \frac{800 \times (3)^3}{32 \times (100)^2} = \frac{27}{400} = 0.0675 \text{ cm.}$$

Example 26(a):

A vertical U-tube of uniform cross-section contains water to a height of 30 cm. Show that if the water on one side is depressed and then released, its motion up and down the two sides of the tube is simple harmonic, and calculate its time-period.

Solution:

Let AA′ be the initial level of water in the U-tube and let the column on the *left* be depressed through distance y to B. Then, obviously, the column on the right will rise up through the same distance y (liquids being incompressible) to the level C, so that the *difference of level between the two columns* = B′C = 2y, where B′ is in a level with B.

The *weight* of this column of water = 2y × a × ρ × g, where a is the internal area of cross-section of the tube (or of the water column), ρ, the density of water and g, the acceleration due to gravity at the place.

- This is then the force acting on the *total mass of water*, m, say, = 2haρ = 2 × 30 × a × ρ = 60 aρ gm in the two limbs of the U-tube.

∴ *acceleration of the mass of water = force*

mass of water, m, say, = 2haρ = 2 × 30 × a × ρ = negative sign indicating that it is directed *opposite to the direction of displacement.*

Thus, the *acceleration of the mass of water is proportional to its displacement and is directed opposite to it.* It, therefore, executes a S.H.M. and its *time-period* is given by

$$T = 2\pi\sqrt{\frac{1}{\mu}} = 2\pi\sqrt{\frac{1}{\frac{g}{30}}} = 2\pi\sqrt{\frac{30}{g}} = 2\pi\sqrt{\frac{30}{981}} = 1.098 \text{ sec.}$$

Example 26(b):

(a) A hyrdrogen atom has a mass of 1.68×10^{-24} gm. When attached to a certain massive molecule, it oscillates as a classical oscillator with a frequency of 10^{14} cycles per second and with an amplitude of 10^{-9} cm. Calculate the force acting on the hydrogen atom.

(b) A test tube of weight 6 gm and of external diameter 2 cm is floated vertically in water by placing 10 gm of mercury at the bottom of the tube. The tube is depressed by a small amount and then released. Find the time of oscillation.

Solution:

(a) We know that the frequency of a panicle executing S.H.M. is given by

$$n = \frac{\omega}{2\pi}. \text{ Here, } n = 10^{14}/\text{sec.}$$

So that, 10^{14}/sec. So that, $10^{14} = \frac{\omega}{2\pi}$,

whence, $\omega = 2\pi \times 10^{14}$.

Now, acceleration of the particle $= -\omega^2 x$, where x is its displacement from its mean or equilibrium position, the –ve sign indicating that it is directed towards that position, *i.e.* , opposite to displacement.

Here, since *displacement* = *amplitude* a, we have *acceleration of the particle, or the hydrogen atom*

$$= -\omega^2 a = -(2\pi \times 10^{14})^2 \times 10^{-9} \text{ cm/sec}^2$$

$\therefore$ *force acting on the hydrogen atom* = *mass* × *acceleration*

$$= 1.68 \times 10^{-24} \times (2\pi \times 10^{14})^2 \times 10^{-9}$$

$$= 6.63 \times 10^{-4} \text{ dynes.}$$

(b) Here *mass of the tube* + *mercury* = 6 + 10 = 16 gm, *external radius of the tube* = 2/2 = 1 cm and, therefore, its *area of cross section* $= \pi r^2 = \pi(1)^2 = \pi$ sq cm.

If the tube be depressed into water through a distance y cm, clearly, *volume of water displaced* = y × π c.c. Hence, *upthrust on the tube due to displaced water*

$$= y \times \pi \times \rho \times g = y\pi g \ (\because \rho \text{ for water} = 1 \text{ gm/c.c.}).$$

$\therefore$ *Acceleration of the tube* $= \frac{\text{force}}{\text{mass}} = -\frac{\pi g}{16} y = -\mu y$,

where $\frac{\pi g}{16} = \mu$.

The –ve sign, as usual, indicates that the acceleration is directed oppositely to displacement.

Thus, the acceleration ∝ displacement and is directed oppositely to it. The tube, therefore, executes a S.H.M. and its *time-period* is given by

$$T = 2\pi\sqrt{\frac{1}{\mu}} = 2\pi\sqrt{\frac{1}{\frac{\pi g}{16}}} = 2\pi\sqrt{\frac{16}{\pi g}} = \sqrt{4\pi \times \frac{16}{g}} = 0.4527 \text{ sec.}$$

Example 26(c):

A particle is moving with S.H.M. in a straight line. When the distance of the particle from the equilibrium position has the values x_1 and x_2, *the corresponding values of the velocity are u_1 and u_2. Show that the period is $2\pi[(x_2^2 - x_1^2)/(u_1^2 - u_2^2)]^{1/2}$.*

Find also the maximum velocity and amplitude of the particle.

Solution:

As we know, the velocity of a particle, executing S.H.M., at a distance x from its equilibrium position is given by $u = \omega\sqrt{a^2 - x^2}$, where ω is its *angular velocity* and a, its *amplitude*. So that, we have here

$$u_1 = \omega\sqrt{a^2 - x_1^2} \text{ and } u_2 = \omega\sqrt{a^2 - x_2^2}.$$

$$\therefore\ u_1^2 - u_2^2 = \omega^2\left(x_2^2 - x_1^2\right).$$

$$\text{Or, } \omega = \left[\frac{\left(u_1^2 - u_2^2\right)}{\left(x_2^2 - x_1^2\right)}\right]^{1/2} \qquad \text{...(i)}$$

And, therefore, *time-period of the particle,* $T = \dfrac{2\pi}{\omega}$.

$$= 2\pi\left[\frac{\left(x_2^2 - x_1^2\right)}{\left(u_1^2 - u_2^2\right)}\right]^{1/2}$$

Now, we have, from above, $u_1^2 = \omega^2\left(a^2 - x_1^2\right)$.

$$\text{Or, } \omega^2 = \frac{u_1^2}{\left(a^2 - x_1^2\right)} \qquad \text{...(ii)}$$

From relations (i) and (ii), therefore, we have

$$\frac{u_1^2 - u_2^2}{x_2^2 - x_1^2} = \frac{u_1^2}{\left(a^2 - x_1^2\right)}.$$

$$\text{Or, } a^2 - x_1^2 = u_1^2\left(\frac{x_2^2 - x_1^2}{u_1^2 - u_2^2}\right)$$

$$\therefore\ a^2 = \frac{u_1^2 x_2^2 - u_1^2 x_1^2}{u_1^2 - u_2^2} + x_1^2 = \frac{u_1^2 x_2^2 - u_1^2 x_1^2 + u_1^2 x_1^2 - u_2^2 x_1^2}{u_1^2 - u_2^2}$$

whence, *amplitude of the particle,*

$$a = \left(\frac{u_1^2 x_2^2 - u_2^2 x_1^2}{u_1^2 - u_2^2} \right)^{1/2}$$

Finally, we know that the maximum velocity of a particle executing S.H.M, is given by $v_{max} = a\omega$. So that here, *maximum velocity of the particle,*

$$v_{max} = \left(\frac{u_1^2 x_2^2 - u_2^2 x_1^2}{u_1^2 - u_2^2} \right)^{1/2} \left(\frac{u_1^2 - u_2^2}{x_2^2 - x_1^2} \right)^{1/2}$$

$$= \left(\frac{u_1^2 x_2^2 - u_2^2 x_1^2}{x_2^2 - x_1^2} \right)^{1/2}$$

Example 27:

The total energy of a particle executing a simple harmonic motion of period 2π sec is 10240 ergs. $\pi/4$ sec after the particle passes the mid-point of the swing, its displacement is $8\sqrt{2}$ cm. Calculate the amplitude of the motion and the mass of the particle.

Solution:

We know that *total energy of a particle executing S.H.M.* $= 2\pi^2 ma^2/T^2$, where the symbols have their usual meanings.

Here, $T = 2\pi$ sec and, therefore, *total energy of the particle* $= 2\pi^2 ma^2/T^2$, where the symbols have their usual meanings.

Here, $T = 2\pi$ sec and, therefore, *total energy of the particle* $= 2\pi^2 ma^2/(2\pi^2)^2 = 10240$,

whence, $\frac{1}{2} ma^2 = 10240.$

Or, $ma^2 = 20480.$ ...(i)

Since the time here is counted from the instant the particle passes through its mean or equilibrium position, we have $\phi = 0$. Hence its *displacement* is given by the relation $x = a \sin \omega t$.

Now, $T = \frac{2\pi}{\omega} = 2\pi$ (given).

So that, $\omega = 1$ radian/sec,

$$x = 8\sqrt{2} \text{ cm and } t = \frac{\pi}{4} \text{ sec.}$$

So that, $$8\sqrt{2} = a \sin\left(1 \times \frac{\pi}{4}\right) = \frac{a \sin \pi}{4} = a \times \frac{1}{\sqrt{2}},$$

whence, $$a = 8\sqrt{2} \times \sqrt{2} = 8 \times 2 = 16 \text{ cm.},$$

i.e., amplitude of the particle is 16 cm.

Substituting this value of a in expression (i) above, we have

$$m \times (16)^2 = 20480,$$

whence, $$m = 20480/16 \times 16 = 80 \text{ gm.}$$

Thus, the *mass of the particle* is 80 gm.

Example 28:

(a) What is the frequency of a simple pendulum 2.0 metres long?

(b) Assuming small amplitude, what would its frequency be in an elevator accelerating upward at a rate of 2.0 metres/sec^2?

(c) What would its frequency be in free fall?

Solution:

(a) We know that the *time-period* of a simple pendulum is given by $T = 2\pi\sqrt{\frac{l}{g}}$. Hence, its *frequency* $n = \frac{1}{T} = \frac{1}{2\pi}\sqrt{\frac{g}{l}}$.

Here, $l = 2.0$ metres. Therefore, taking $g = 9.8$ m/sec^2, we have *frequency of the pendulum,*

$$n = \frac{1}{2\pi}\sqrt{\frac{9.8}{2}} = \frac{1}{2\pi}\sqrt{4.9} = 0.3524/\text{sec.}$$

(b) In the elevator going up with an acceleration a, the *effective weight* of the pendulum is $mg\left(1 + \frac{a}{g}\right) = m(g + a)$. So that, here, the effective value of g is, say, $g' = (g + a) = (9.8 + 2.0) = 11.8$ m/sec^2.

∴ *frequency of the pendulum in the elevator,* say,

$$n' = \frac{1}{2\pi}\sqrt{\frac{g'}{l}} = \frac{1}{2\pi}\sqrt{\frac{11.8}{2}} = \frac{1}{2\pi}\sqrt{5.9} = 0.3867/\text{sec.}$$

(c) In *free fall* of the pendulum, the *effective weight* of the pendulum is $mg\left(1-\frac{g}{g}\right)$ = 0, *i.e.*, the effective value of g, *i.e.*, g' = 0.

∴ *time-period of the pendulum* = $2\pi\sqrt{\frac{l}{g'}}$ = *infinite, i.e., there is no oscillation of the pendulum at all. Its frequency (n) is, therefore, zero.*

Example 29:

A simple pendulum of length l and mass m is suspended in a car that is travelling with a constant speed v around a circle of radius R. If the pendulum undergoes small oscillations about Its equilibrium position, what will its frequency of oscillation be?

Solution:

Here, in addition to g, the acceleration due to gravity, the pendulum will also be subject to the centripetal acceleration v^2/R in a direction perpendicular to that of g on account of the car going round with speed v in a circle of radius R. Thus, *resultant acceleration to which the pendulum is subjected*

$$= \left[g^2 + \left(\frac{v^2}{R}\right)^2\right]^{1/2} = \left(g^2 + \frac{v^4}{R^2}\right)^{1/2}.$$

On a small angular displacement θ from its mean or equilibrium position, therefore, the *restoring couple acting on the pendulum*

$$= m\left(g^2 \times \frac{v^4}{R^2}\right)^{1/2} l \sin\theta$$

$$= m\left(g^2 + \frac{v^4}{R^2}\right)^{1/2} l\theta. \qquad [\because \text{ is small}].$$

If $d^2\theta/dt^2$ be the acceleration of the pendulum and I, its moment of inertia about the suspension thread, we have the couple on the pendulum also equal to $I.d^2\theta/dt^2$.

Or equal to $ml^2d^2\theta/dt^2$, because $I = ml^2$.

We, therefore, have $ml^2d^2\theta/dt^2 = -\, m\left(g^2 + \frac{v^4}{R^2}\right)^{1/2} l\theta$

whence, $$\frac{d^2\theta}{dt^2} = -\frac{\left(g^2 + \frac{v^4}{R^2}\right)^{1/2}\theta}{l} = -\mu\theta,$$

where $$\mu = \frac{\left(g^2 + \frac{v^4}{R^2}\right)^{1/2}}{l}.$$

The pendulum thus executes a S.H.M. and, therefore, *frequency of the pendulum*

$$= n = \frac{1}{T} = \frac{1}{2\pi}\sqrt{\mu} = \frac{1}{2\pi}\sqrt{\frac{\left(g^2 + \frac{v^4}{R^2}\right)^{1/2}}{l}}.$$

Example 30:

A. simple pendulum of length 100 cm has an energy equal to 2×10^6 ergs when its amplitude is 4 cm. Calculate its energy when (i) its length is doubled (ii) its amplitude is doubled.

Solution:

We have *energy of a simple pendulums* = its max. P.E. = its max.

$$\text{K.E.} = \frac{1}{2}mv^2 = \frac{1}{2}m\omega^2a^2.$$

Now, $\omega = \frac{2\pi}{T}$, where T is the *time-period* of the pendulum

$$= 2\pi\sqrt{\frac{l}{g}}. \text{ Or, } \omega = \sqrt{\frac{g}{l}}.$$

Initially, l = 100 cm.

$$\therefore \qquad \omega = \sqrt{\frac{g}{100}}.$$

So that, initially, with amplitude 4 cm, the energy of the pendulum

$$= E = \frac{1}{2}m\left(\frac{g}{100}\right)(4)^2$$

$$= 8mg/100 = 2 \times 10^6 \text{ ergs}.$$

(i) *If the length of the pendulum is doubled*, we have $l = 200$ cm and

$$\therefore \qquad \omega = \sqrt{\frac{g}{200}}.$$

Hence, *energy of the pendulum,*

$$E' = \frac{1}{2}m\left(\frac{g}{200}\right)(4)^2 = \frac{8mg}{200}.$$

So that,
$$\frac{E'}{E} = \frac{\frac{8mg}{200}}{\frac{8mg}{100}} = \frac{1}{2}.$$

Or,
$$E' = \frac{1}{2}E.$$

Since the *initial energy of the pendulum*, E = 2 × 10^6 ergs (given), we have *energy of the pendulum* ndw

$$= \frac{1}{2} \times 2 \times 10^6 = 10^6 \text{ ergs.}$$

(ii) *If the amplitude of the pendulum be doubled*, a = 8 cm, its length remaining 100 cm; we have energy of the pendulum,

$$E'' = \frac{1}{2}m\left(\frac{g}{100}\right)(8)^2 = 32\ mg/100.$$

$$\therefore \qquad \frac{E''}{E} = \frac{\frac{32mg}{100}}{\frac{8mg}{100}} = 4.$$

Or, E" = 4E.

Or, E" = 4 × 2 × 10^6 = 8 × 10^6 ergs.

Thus, *the energy of the pendulum is now* 8 × 10^6 ergs.

Example 31(a):

Calculate the percentage change in the time-period of a simple pendulum if the angular amplitude of the pendulum be (i) 30°, (ii) 60°.

Solution:

Let the length of the simple pendulum be *l*. Then, its *true time-period* (for infinitely small amplitudes) is

$$T_0 = 2\pi\sqrt{\frac{l}{g}}.$$

If its amplitude be θ_1, the time-period

$$T = T_0\left(1+\frac{\theta_1^2}{16}\right).$$

∴ *percentage change in the time-period*

$$= \left(\frac{T-T_0}{T}\right)\times 100 = \frac{\theta_1^2}{16}\times 100.$$

Now, in case (i) $\theta_1 = 30° = \frac{\pi}{6}$ radian. So that, we have

percentage change in the time-period

$$= \frac{(\pi/6)}{16}\times 100 = \frac{25\pi^2}{144} = 1.7.$$

And, in case (ii), $\theta_1 = 60° = \frac{\pi}{3}$ radian and we, therefore, have *percentage change in the time-period*

$$= \frac{(\pi/3)^2}{16}\times 100 = \frac{25\pi^2}{36} = 6.85.$$

Example 31(b):

A uniform circular disc of radius R oscillates in a vertical plane about a horizontal axis. Find the distance of the axis of rotation from the centre for which the period is minimum. What is the value of this period?

Solution:

The circular disc here oscillates as a compound or a physical pendulum of length l, whose time-period is given by

$$T = 2\pi\sqrt{\frac{\left(\frac{K^2}{l}\right)+l}{g}},$$

where k its *radius of gyration* about an axis through its *e.g.*, parallel to the axis of suspension.

Now, as we know, the *time-period of a compound pendulum is the minimum, when its length is equal to its radius of gyration about its e.g., i.e.*, when l = k.

So that, $$T_{min.} = 2\pi\sqrt{\frac{\left(\frac{k^2}{k}\right)+k}{g}} = 2\pi\sqrt{\frac{2k}{g}}.$$

Since the moment of inertia of a disc about an axis perpendicular to its plane and passing through its centre is equal to

$$I = Mk^2 = \frac{1}{2}MR^2,$$

where M is the *mass* of the disc and R, its *radius*, we have

$$k^2 = \frac{R^2}{2} \text{ and}$$

$$\therefore \quad k = \frac{R}{\sqrt{2}}.$$

Thus, *the disc will oscillate with the minimum time-period when the distance of the axis of rotation from the centre is* $\frac{R}{\sqrt{2}}$.

And the value of this minimum time-period will be

$$T_{min} = 2\pi\sqrt{\frac{\frac{2R}{\sqrt{2}}}{g}} = 2\pi\sqrt{\frac{\sqrt{2}R}{g}} = 2\pi\sqrt{\frac{1.414\,R}{g}}.$$

Example 32:

A solid sphere of radius 0.3 metre executes torsional oscillations of time-period $2\pi\sqrt{12}$ *sec at the end of a suspension wire whose upper end is fixed in a rigid support. If the torque constant of the wire be 6 × 10^{-3} N-m/radian, calculate the mass of the sphere.*

Solution:

We know that the M.I. of a solid sphere about any diameter, *i.e.*, about any axis passing through its centre is 2/5 MR^2, where M is its mass and R, its radius.

We, therefore, have

$I = \frac{2}{5}MR^2$; *torque constant or torque per unit twist of the suspension wire, i.e.,*

$C = 6 \times 10^{-2}$ N-m/radian and $T = 2\pi\sqrt{12}$ sec.

Substituting these values in the relation $T = 2\pi\sqrt{\frac{I}{C}}$ for the time-period of a torsional oscillation, we have

$$2\pi\sqrt{12} = 2\pi\sqrt{\frac{I}{C}},$$

whence, $\sqrt{12} = \sqrt{\frac{I}{C}}$.

Or, $\frac{I}{C} = 12.$

Or, $I = 12C,$

i.e., $\frac{2}{5}MR^2 = 12 \times 6 \times 10^{-3} = 72 \times 10^{-3}.$

Or, $MR^2 = 72 \times 10^{-3} \times \frac{5}{2}$

$= 180 \times 10^{-3} = 18 \times 10^{-2}.$

$$\therefore \quad M = \frac{18\times10^{-2}}{R^2} = \frac{18\times10^{-2}}{(0.3)^2} = \frac{18\times10^{-2}}{9\times10^{-2}} = 2\,\text{kg}.$$

Thus, *the mass of the sphere is 1 kg.*

Example 33:

The tune-period of a rod of mass 120 gm and length 10 cm, executing torsional vibration about a suspension wire passing through its centre and perpendicular to its length is 2 sec. The period of a flat plate of the shape of an equilateral triangle suspended from the suspension wire, passing through its centre of mass and perpendicular to its plane is found to be 4 sec. What is the moment of inertia of the triangular plate about the suspension wire as axis?

Solution:

Let I_1 be the M.I. of the rod about the suspension wire as axis and I_2, that of the triangular plate about the same axis. Then, if T_1 and T_2 be the respective time-periods of the two torsional oscillations, we have

$$T_1 = 2\pi\sqrt{\frac{I_1}{C}}$$

and $$T_2 = 2\pi\sqrt{\frac{I_2}{C}}.$$

So that, $$\frac{T_2}{T_1} = \sqrt{\frac{I_2}{I_1}}.$$

Or, $$\frac{I_2}{I_1} = \frac{T_2^2}{T_1^2},$$

whence, $$I_2\left(\frac{T_2^2}{T_1^2}\right)I_1 = \left(\frac{T_2}{T_1}\right)^2 I_1.$$

Now, *M.I. of the rod about the suspension wire, i.e.*

$$I_1 = \frac{Ml^2}{12} = 120 \times (10)^2/12 = 10^3 \text{ gm-cm}^2,$$

$$T_1 = 2 \text{ sec and } T_2 = 4 \text{ sec.}$$

∴ *M.I. of the triangular plate about the suspension wire,*

$$I_2 = \left(\frac{4}{2}\right)^2 \times 10^3 = 4 \times 10^3 \text{ gm-cm}^2.$$

Example 34:

The balance wheel of a watch vibrates with an angular amplitude of n radians and a period of 0.5 sec. Find (a) the maximum angular speed of the wheel, (b) the angular speed of the wheel when its displacement is π/2 radians and (c) the angular acceleration of the wheel when its displacement is π/4 radians.

Solution:

(a) Just as in *linear* S.H.M., so also, here in torsional vibration, we have

$$\theta = \theta_0 \sin(\omega t + \phi),$$

where θ is the angular displacement at the given instant and θ_0, the *maximum angular displacement or amplitude.*

∴ angular speed, $\frac{d\theta}{dt} = \theta_0\omega\cos(\omega t + \phi)$ the maximum value of which will clearly be θ_0, *i.e.*, when $\cos(\omega t + \phi) = 1$.

Here, $\theta_0 = \pi$ radians

and $\omega = 2\pi n = 2\pi \times \frac{1}{T} = 2\pi \times \frac{1}{0.5} = 4\pi$.

∴ *maximum angular speed of the balance wheel* $= \theta_0\omega = \pi \times 4\pi$ $= 4\pi^2 \approx 39.48$ *radians/sec.*

(b) Again, as in linear S.H.M., so also here, *angular speed at angular displacement* $\theta = \omega\sqrt{\theta_0^2 - \theta^2}$. So that, *angular speed of the wheel at displacement* $\pi/2$ *radian*

$$= 4\pi\sqrt{\pi^2 - \left(\frac{\pi}{2}\right)^2} = 4\pi\sqrt{\frac{3\pi^2}{4}}$$

$$= 2\pi^2\sqrt{3} = 34.18 \text{ radians/sec.}$$

(c) Again, acceleration of the balance wheel at displacement θ is given by

$$\frac{d^2\theta}{dt^2} = -\omega^2\theta.$$

Therefore, *Acceleration of the balance wheel at displacement* $\theta = \frac{\pi}{4}$ *radian* is clearly equal to $-(4\pi)^2\left(\frac{\pi}{4}\right) = -4\pi^3 = -124$ radians/sec^2.

Example 35(a):

If the quality factor of an underdamped harmonic oscillator of frequency 512 be 8 × 10^4, calculate the time in which its energy is reduced to 1/eth of its energy in the absence of damping. How many oscillations does the oscillator make in this time?

Calculate the percentage reduction in the frequency of the oscillator due to damping. What inference do you drew from it?

Solution:

The energy of a damped harmonic oscillator at an instant t, as we know, is given by $E = E_0e^{-2kt} = E_0e^{-t/\tau}$, where E_0 is its energy in the absence of damping.

Since here $E = \frac{E_0}{e} = E_0e^{-1}$,

we have $E_0e^{-1} = E_0e^{-t/\tau}$,

whence, $1 = t/\tau$ or, $t = \tau$.

And since $Q = \omega\eta\tau$, we have

$$\tau = \frac{Q}{\omega_0} = \frac{Q}{2\pi\eta_0} = \frac{8\times10^4}{2\pi\times512} = \frac{10^4}{2\pi\times64} = \frac{10^4}{128\pi}.$$

And therefore, the time in which the energy of the oscillator falls to 1/eth of its energy in the absence of damping, *i.e.*,

$$t = \tau\ 10^4/128\pi = 24.87 \text{ sec.}$$

And, clearly, *number of oscillations made by the oscillator in this time*

$$= nt = \left(\frac{\omega_0}{2\pi}\right)\tau = \frac{Q}{2\pi} = 8\times\frac{10^4}{2\pi} = 12740.$$

Now, in the case of an underdamped harmonic oscillator, we have

$$\omega = \sqrt{\omega_0^2 - k^2},$$

Or, $\omega^2 = \omega_0^2 - k^2 = \omega_0^2 - \left(\frac{1}{2\tau}\right)^2 = \omega_0^2 - \frac{1}{4}\tau^2.$

So that, dividing by ω_0 throughout, we have

$$\frac{\omega^2}{\omega_0^2} = 1 - \frac{1}{4\omega_0^2\tau^2} = 1 - \frac{1}{4Q^2} \qquad [\because \omega_0\tau = Q.]$$

And, therefore, $\frac{\omega}{\omega_0} = \left(1 - \frac{1}{4Q^2}\right)^{1/2} - \left(1 - \frac{1}{2}\times\frac{1}{4Q^2} + \ldots\right) = 1 - \frac{1}{8Q^2}.$

Or, $\frac{2\pi n}{2\pi n_0} = \frac{n}{n_0} = 1 - \frac{1}{8Q^2},$

where n_0 is the frequency of the undamped oscillator.

$\therefore$ *percentage reduction in the frequency of the oscillator*

$$= \frac{n_0 - n}{n_0}\times100$$

$$= \left(1 - \frac{n}{n_0}\right)\times100\left[1 - \left(1 - \frac{1}{8Q^2}\right)\right]\times100 = \frac{100}{8Q^2} = \frac{12.5}{Q^2}$$

$$= \frac{12.5}{(8\times10^4)^2} = 1.953\times10^{-9}.$$

The obvious *inference* from the above result is that *for oscillators with large values of the quality factor Q (which is usually the case), the reduction in frequency due to damping is negligibly small.*

Example 35(b):

If in problem 14 above, the rate of dissipation of energy is desired to be halved, what should be (i) the amplitude of the applied periodic force, its frequency remaining unchanged, (ii) the frequency of the applied periodic force, its amplitude remaining unchanged?

Solution:

(i) We know that the *rate of dissipation of energy* α (*amplitude of the force*)^2.

Thus, if R_1 and R_2 be the rates of dissipation of energy and f_{01} and f_{02} the amplitudes of the force in the two cases respectively, *the frequency of the force remaining the same in either case,* we have

$$\frac{R_1}{R_2} = \left(\frac{f_{01}}{f_{02}}\right)^2.$$

We know that the *effective current in an LCR circuit*

$$= \frac{\text{effective emf}}{\text{impedance of}} = \frac{E}{Z}.$$

Here, *effective* emf E = 500 *volts* and *impedance*

$$Z = \sqrt{R^2 + L^2 p^2}$$

$$= \sqrt{R^2 + L^2 (2\pi n)^2},$$

Or, $Z = \sqrt{(4)^2 + (0.68)^2 (2\pi \times 120)^2} = \sqrt{16 + 262800} = 512.6$ ohms.

∴ *effective current in the circuit,* $I = \frac{500}{512.6} = 0.9754$ amp.

And, *power factor of the circuit* $= \frac{R}{Z} = \frac{4}{512.6} = 0.0078.$

Example 35(c):

An electric lamp which runs at 40 volts D.C. and consumes 10 amperes current is connected to A.C. mains at 100 volts, 50 cycles. Calculate the inductance of the choke.

Solution:

Clearly, *resistance of the lamp,* $R = \frac{\text{voltage}}{\text{current}} = \frac{40}{10} = 4$ ohms.

And, *impedance of A.C. circuit,* $Z = \sqrt{R^2 + (Lp)^2}$, where L is the *inductance of the choke.*

Here, R = 4 ohms, π = 2πn × 50 = 100π and i = 10.

$$\therefore \quad I = \frac{E}{Z} = \frac{E}{\sqrt{R^2 + (Lp)^2}} = \frac{100}{\sqrt{4^2 + (L + 100\pi)^2}} = 10$$

So that, $\dfrac{10^4}{16 + 10^4 \pi^2 L^2} = 100.$

Or, $10^6\pi^2L^2 = 10^4 - 1600 = 8400.$

So that, $L^2 = \dfrac{8400}{10^6 \pi^2}.$

Or, $L = \sqrt{\dfrac{84}{10^4 \pi^2}} = \dfrac{\sqrt{84}}{100\pi} = 0.02917.$

Thus, the *inductance of the choke* = 0.02917 henry.